Fourth Edition

Using Information Technology

A Practical Introduction to Computers & Communications

Introductory Version

Stacey C. Sawyer

Brian K. Williams

McGraw-Hill Irwin

Boston   Burr Ridge, IL   Dubuque, IA   Madison, WI   New York   San Francisco   St. Louis
Bangkok   Bogotá   Caracas   Lisbon   London   Madrid
Mexico City   Milan   New Delhi   Seoul   Singapore   Sydney   Taipei   Toronto

McGraw-Hill Higher Education

A Division of The McGraw-Hill Companies

USING INFORMATION TECHNOLOGY

Published by McGraw-Hill/Irwin, an imprint of the McGraw-Hill Companies, Inc. 1221 Avenue of the Americas, New York, NY, 10020. Copyright © 2001, 1999, 1997, 1995, by The McGraw-Hill Companies, Inc. All rights reserved. No part of this publication may be reproduced or distributed in any form or by any means, or stored in a data base or retrieval system, without the prior written consent of the McGraw-Hill Companies, Inc., including, but not limited to, in any network or other electronic storage or transmission, or broadcast for distance learning. Some ancillaries, including electronic and print components, may not be available to customers outside the United States.

This book is printed on acid-free paper.

international	1 2 3 4 5 6 7 8 9 0	QPD/QPD	0 9 8 7 6 5 4 3 2 1 0
domestic	1 2 3 4 5 6 7 8 9 0	QPD/QPD	0 9 8 7 6 5 4 3 2 1 0

ISBN 0-07-239875-2

Vice president/Editor-in-chief: *Robin J. Zwettler*
Executive editor: *Linda S. Schreiber*
Development editors: *Burrston House, Ltd./Craig S. Leonard/Melissa Forte*
Senior marketing manager: *Jeff Parr*
Project manager: *Christine A. Vaughan*
Lead production supervisor: *Heather D. Burbridge*
Senior designer: *Laurie J. Entringer*
Production and Quark makeup: *Stacey C. Sawyer*
Cover image: *© Masterfile*
Cover design: *Asylum Studios*
Senior photo research coordinator: *Keri Johnson*
Photo research: *Judy Mason*
Senior supplement producer: *Marc Mattson*
Media technology producer: *David Barrick*
Compositor: *GTS Graphics, Inc.*
Typeface: *10/12 Trump Mediaeval*
Printer: *Quebecor World Dubuque, Inc.*

Library of Congress Cataloging-in-Publication Data

Sawyer, Stacey C.
    Using information technology introduction / Stacey C. Sawyer, Brian K. Williams,—
4th ed.
      p. cm.
    Revised ed. of: Using information technology / Brian K. Williams, Stacey C. Sawyer, c1999.
    Includes bibliographical references and index.
    ISBN 0-07-239875-2 (alk. paper)
    1. Computers. 2. Telecommunication systems. 3. Information technology.   I. Williams,
Brian K.   II. Williams, Brian K. Using information technology.   III. Title.

QA76.5 .S2193 2001
004—dc21   00-046298

INTERNATIONAL EDITION

Copyright © 2001. Exclusive rights by The McGraw-Hill Companies, Inc., for manufacture and export. This book cannot be re-exported from the country to which it is sold by McGraw-Hill. The International Edition is not available in North America.

When ordering the title, use ISBN 0-07-118067-2

www.mhhe.com

Photo and other credits are listed in the back of the book.

Brief Contents

About the Authors

Who are **Stacey Sawyer and Brian Williams**? We are a married couple living near Lake Tahoe, Nevada, with an avid interest in seeing students become well educated—especially in information technology.

What best describes what we do? We consider ourselves **watchers and listeners.** We spend our time *watching* what's happening in business and society and on college campuses and *listening* to the views expressed by instructors, students, and other participants in the computer revolution. We then try to translate these observations into meaningful language that can be best understood by students.

Over the past two decades, we have individually or together **authored more than 20 books** (and 30 revisions), most of them on computers and information technology. Both of us have **a strong commitment to helping students succeed in college.** Brian, for instance, has also co-authored four books in the college success field: *Learning Success, The Commuter Student, The Urban Student,* and *The Practical Student.* Stacey has an interest in language education and has worked on several college textbooks in English as a Second Language (ESL) and in Spanish, German, French, and Italian. We thus bring to our information-technology books an awareness of the needs of the increasingly diverse student bodies now in our colleges.

Stacey has a B.A. from Ohio Wesleyan and the University of Freiburg, Germany, and an M.A. from Middlebury College and the University of Mainz, Germany. She has taught at Ohio State University and managed and consulted for a number of for-profit and nonprofit health, educational, and publishing organizations. Brian has a B.A. and M.A. from Stanford University and has held managerial jobs in education, communications, and publishing.

In our spare time, we enjoy cooking, travel, music, and exploring the wilds of the American West.

To the Instructor

Introduction

Information technology—"now as vital as the air we breathe," as one observer says—is the major revolution of our time. Rolling inexorably forward, it is redefining entire industries, changing the nature of work and leisure, altering conventional meanings of time and space.

The Fourth Edition of *USING INFORMATION TECHNOLOGY* puts the reader in the front row of this revolution. UIT is a concepts textbook to accompany a one-semester or one-quarter introductory course on computers or microcomputers. This eight-chapter *Introductory Version*, which replaces the previous edition's *Brief Version*, though it is about the same length, is intended for people who will use computers as everyday tools, not those who will write programs or design computer systems. (A longer edition—11 chapters—is available as the *Complete Version*, which corresponds to the full-size version of the previous edition of our text.)

About Our Book: A New Identity—"A Book Reborn, Not Revised"

We consider this edition of *USING INFORMATION TECHNOLOGY* to be **a book reborn,** not merely revised. The book's **new identity** is embodied in several new features—many of which were **requested by instructors**—as explained below under "What's New in This Edition?"

What Users Liked #1: Emphasis on Role of Communications in Computing

Earlier editions of this book broke new ground by emphasizing the role of the Internet and of digital convergence, the technological fusion of computers and telecommunications. This theme was enthusiastically received by both students and instructors.

But **telecommunications is now taking new forms,** which we describe in this edition.

- **Moving beyond the PC:** We appear to be moving beyond the personal computer. Companies are launching a slew of simplified electronic devices that do only one or two tasks, such as checking e-mail or surfing the Web—devices uncomplicated enough to attract online users in the more than 60% of U.S. households without Internet access. Cellphones, palm computers, and information appliances are now being produced that we can use not only for communicating but also for surfing, shopping, and banking.

  [See examples pp. 5, 73.]

- **The Internet becomes personal:** The World Wide Web is more and more becoming a personal resource, from getting help with advancing careers and finances to finding relationships and spirituality.

  [See examples pp. 72, 296.]

- **The coming "Omninet":** Because of wireless communication, we are fast approaching the day when the Internet becomes such an all-pervasive presence that, like telephones and television, we will almost forget that it is there.

  [See examples pp. 4, 214.]

What Users Liked #2: Emphasis on Practicality

A feature that has been well received by instructors and students using past editions is that we not only cover fundamental concepts but also offer a great deal of **practical advice.** This advice, of the sort found in computer magazines and general-interest computer books, is not only expressed in the text but also in the box **Bookmark It! Practical Action Box.** This box consists of optional material on practical matters. *Examples:* "Tips for Managing Your E-Mail." "Choosing an Internet Service Provider." "Being Successful at Distance Learning."

[See examples pp. 42, 272.]

Ethics

What Users Liked #3: Emphasis throughout on Ethics

Many texts discuss ethics in isolation, usually in one of the final chapters. We believe this topic is too important to be treated last or lightly, and users have agreed. Thus, **we cover ethical matters throughout the book,** as indicated by the special logo shown here in the margin. *Examples:* We discuss such all-important questions as copying of Internet files, online plagiarism, privacy, computer crime, and netiquette. (A list of pages with ethics coverage appears on the inside front cover.)

[See examples pp. 23, 48, 255.]

What Users Liked #4: Emphasis on Reinforcement for Learning

Prior editions of our book offered the following features for reinforcing student learning:

- **Interesting writing:** Studies have found that textbooks **written in an imaginative style** significantly improve students' ability to retain information. Both instructors and students have commented on the distinctiveness of the writing in this book. We employ a number of journalistic devices—colorful anecdotes, short biographical sketches, interesting observations in direct quotes—to make the material as interesting as possible. We also use real anecdotes and examples rather than fictionalized ones.

[See examples pp. 127, 211.]

- **Key terms AND definitions emphasized:** To help readers avoid any confusion about which terms are important and what they actually mean, we print each key term **_bold italic underscore_** and its definition in **boldface.** *Example* (from Chapter 1): "**_Data_ consists of raw facts and figures that are processed into information.**"

[See example p. 106.]

- **Material in bite-size portions:** Major ideas are presented in **bite-size form,** with generous use of advance organizers, bulleted lists, and new paragraphing when a new idea is introduced. Most **sentences have been kept short,** the majority not exceeding 22–25 words in length.

[See example p. 107.]

What's New in This Edition?

A great number of brand-new features are to be found in this edition—most of which were suggested by instructors in conversations, reviews, and roundtable discussions. As stated, we think they make this edition "a book reborn, not revised."

What Users Requested #1: New Organization & New Emphasis on "E-Concepts"

In accordance with instructor suggestions, **we reorganized the table of contents to permit greater treatment of "e-concepts."**

Because the Internet and the World Wide Web are now so widespread—all students have seen the terms "www" and "dot-com," and most students are already online—many instructors have told us they would like to see **"e-concepts" treated earlier and more extensively** in the book. Accordingly, we have revised the table of contents as follows:

How the Table of Contents Has Changed

Chapters in the Present Edition	Chapters in the Previous Edition
1. Introduction to Information Technology	1. The Digital Age
2. The Internet & the World Wide Web	2. Applications Software
3. Software	3. System Software
4. Hardware	4. Processors
5. Telecommunications	5. Input & Output
6. E-Commerce, Files, & Databases	6. Storage & Databases
7. The Challenges of the Digital Age	7. Telecommunications
8. The Promises of the Digital Age	8. Communications Technology
Appendix. Systems & Programming	9. Systems
	10. Society & the Digital Age

Instructors will note the following changes:

- The Internet and World Wide Web are now discussed in Chapter 2 instead of Chapter 7, reflecting their importance in students' daily lives.
- Communications technology is now discussed earlier—in Chapter 5 instead of Chapter 8.
- Software is now discussed in one chapter instead of two.
- Hardware is now discussed in one chapter instead of three.
- Files and databases are discussed within the framework of e-commerce. (Storage is discussed as part of hardware.)
- The challenges and promises of the digital age, especially the "e-concepts," are discussed in two chapters instead of one.
- Information systems, programming, and languages are discussed in an appendix. (Some instructors don't cover these concepts in their courses.)

What Users Requested #2: A Much More Visual Book—Many More Illustrations Designed to Aid Student Learning

In this edition we have **increased the number of visuals—particularly artwork designed to aid student learning,** so that there is a significantly increased ratio of illustrations to text.

Specifically, we offer artwork—both line art and photographs—to serve the following purposes:

- **Artwork to illustrate how to use software:** In this edition, we don't just describe Web browser or word processing commands, for example. We show visually how menus, toolbars, icons, and similar features work.

[See example p. 118.]

- **Artwork to illustrate how hardware works:** In Chapter 1, we ask readers to **pretend to build their own PC**—and we provide illustrations to show how the pieces go together. In Chapter 4, we present a more complex example. We show **an advertisement for a PC**—and then explain the ad by repeating its hardware features along with specific illustrations describing what they mean and how they work.

[See examples pp. 13, 18.]

[See examples pp. 147, 150.]

- **Photos to show what's exciting and unique about the Digital Age:** Many textbooks have photos showing how computers are used in ordinary ways, and we do also. But to whet student interest, we also show **how computers are used in interesting, uncommon ways.**

What Users Requested #3: Help Students Think Critically about Information Technology & Take Ownership over the Material

More and more instructors are becoming familiar with the taxonomy of educational objectives encompassing critical-thinking skills. These skills are organized in the following hierarchy: (a) *memorization,* (b) *comprehension,* and (c) *application, analysis, synthesis,* and *evaluation.*

Drawing on our experience in writing books to guide students to college success, we have implemented a **novel pedagogy to reinforce learning and help students take ownership over the material.** We use the following hierarchical approach in both text and supplements:

> **First level—memorization.** Tests how well reader recalls basic terms and concepts.
>
> **Second level—comprehension.** Tests how well reader understands concepts and integrates ideas.
>
> **Third level—application, analysis, synthesis, evaluation.** Tests higher-order critical-thinking skills, including the ability to solve problems and make decisions.

The following are some pedagogical features that incorporate this three-tier hierarchy of critical thinking skills.

[See examples pp. 31, 34.]

- **Key Questions—to help students read with purpose:** We have **crafted the learning objectives as Key Questions** to help readers focus on essentials. Each Key Question appears in two places: on the first page of the chapter, and beneath the section head. Key Questions are also tied to the end-of-chapter summary, as we will explain.

[See examples pp. 32–33.]

- **Chapter visual overview—"Graphical Interface":** Each chapter opens with a two-page **visual overview of chapter concepts,** so that students can have a "pictorial road map" of the contents of the chapter before they begin reading. This "Graphical Interface: A Visual Overview of This Chapter" may also be used as a study aid to review concepts.

[See example p. 51.]

- **Concept Checks:** Appearing periodically throughout the text, **Concept Checks** spur students to recall facts and concepts they have just read.

- **Visual Summary:** Each chapter ends with an innovative **Visual Summary** of important terms, with an explanation of what they are and why they are important. The terms are accompanied, when appropriate, by a picture. Each concept or term is also given a cross-reference page number that refers the reader to the main discussion within the text. In addition, the term or concept is given a Key Question number corresponding to the appropriate Key Question (learning objective).

[See example p. 25.]

[See examples pp. 81–82.]

- **Chapter Review:** The end-of-chapter **Chapter Review** is constructed according to the three levels of educational objectives.

> **First level—memorization.** Self-Test (fill-in) Questions, Multiple-Choice Questions, and True/False Questions. Tests how well the reader recalls basic terms and concepts.
>
> **Second level—comprehension.** Short-answer questions and concept maps. Tests how well the reader understands concepts and integrates ideas.

Third level—application, analysis, synthesis, evaluation. Knowledge in Action questions. Tests higher-order critical-thinking skills, including the ability to solve problems and make decisions.

The Student Online Learning Center

For each chapter, the student **Online Learning Center** offers both a review of the text material, in the form of an **e-learning session,** and additional exercises organized around the following themes:

- Group/team projects
- Internet/Web-related content
- Mini-case studies of actual companies
- Profiles of careers that are influenced by information technology

The content and activities further establish the three-level approach.

Resources for Instructors

We understand that, in today's teaching environment, offering a textbook alone is not sufficient to meet the needs of the many different instructors who use our books. To teach effectively, instructors must have a full complement of supplemental resources to assist them in every facet of teaching from preparing for class to conducting and lecture to assessing students' comprehension. *Using Information Technology* offers a complete, fully integrated supplements package, as described below:

Instructor's Resource Kit

The Instructor's Resource Kit contains a printed Instructor's Manual and a CD-ROM containing the Instructor's Manual in both MS Word and .pdf formats, PowerPoint slides, Brownstone's Diploma test generation software, and accompanying test item files for each chapter. The distinctive features of each component of the Instructor's Resource Kit are described below.

- **Instructor's Manual** Prepared by Kerry Thompson of Digital Light Studios, the Instructor's Manual contains learning objectives, a Chapter Outline with lecture notes, a list of the chapter competencies, tips for covering difficult material, and answers to the Concept Checks. Also included are references to corresponding topics on the Interactive Companion CD-ROM, answers to all the exercises in the Chapter Review section, and answers to the On the Web Exercises. The manual also includes a helpful introduction that explains the features, benefits, and suggested uses of the IM and an index of concepts and corresponding competencies.
- **PowerPoint Presentation** Prepared by Linda Mehlinger of Morgan State University, the PowerPoint presentation is designed to provide instructors with a comprehensive teaching resource and includes chapter learning objectives, concepts overviews, figures from the text, additional examples/illustrations, anticipated student questions with answers, discussion topics, and Concept Checks recalling key points throughout each chapter. Each chapter ends with key terms illustrating those key points.

- **Testbank** The *Using Information Technology* edition testbank contains over 3000 questions categorized by level of learning (definition, concept, and application). This is the same learning scheme that is introduced in the text to provide a valuable testing and reinforcement tool. The test questions are identified by text page number to assist you in planning your exams, and rationales for each answer are also included. Additional test questions, which can be used as pretests and posttests in class, can be found on the Online Learning Center, accessible through our supersite (www.mhhe.com/it).

Business Week Edition of Using Information Technology

An exciting new supplement with this edition of *Using Information Technology* is our *Business Week Edition*. With the purchase of a *Business Week Edition* of a McGraw-Hill/Irwin textbook, students will receive a 15-week subscription to *Business Week* for only $8.25 more than the price of the book alone. Professors who adopt the *Business Week Edition* will enjoy a complimentary subscription for a full year to *Business Week* magazine and complimentary access to the Business Week Resource Center Web site (www.resourcecenter.businessweek.com) as well as Business Week Online through the duration of their subsription.

Students will also enjoy free access to the Business Week Resource Center Web site for the duration of their magazine subscription. The Business Week Resource Center Web site contains a wealth of supplemental materials, including the Business Week Online Archives. Students will have instant access to any business topic from the past ten years of *Business Week*—from 1991 to 2000. From the Resource Center, students may also access Business Week Online (www.businessweek.com) for current issues, online-only features, and career tips. Access to these sites provides a marvelous opportunity to increase students' Internet literacy as instructors explore new ways to integrate the Web into a wide array of student exercises and research projects.

Interactive Companion CD-ROM

This free student CD-ROM, designed for use in class, in the lab, or at home by students and professors alike includes a collection of interactive tutorial labs on some of the most popular topics in information technology. By combining video, interactive exercises, animation, additional content, and actual "lab" tutorials, we expand the reach and scope of the textbook.

Digital Solutions to Help You Manage Your Course

- **PageOut**—PageOut is our Course Web Site Development Center that offers a syllabus page, URL, McGraw-Hill Online Learning Center content, online exercises and quizzes, gradebook, discussion board, and an area for student Web pages. For more information, visit the PageOut Web site (www.mhla.net/pageout).
- **Online Learning Centers**—The Online Learning Center that accompanies *Using Information Technology* is accessible through our information Technology Supersite (www.mhhe.com/it). This site provides additional learning and instructional tools developed using the same three-level approach found in the text and supplements. This offers a consistent method for students to enhance their comprehension of the concepts presented in the text.

- **Online Courses Available**—OLCs are your perfect solutions for Internet-based content. Simply put, these Centers are "digital cartridges" that contain a book's pedagogy and supplements. As students read the book, they can go online and take self-grading quizzes or work through interactive exercises. These also provide students appropriate access to lecture materials and other key supplements.
- **UIT Website**—The new Website that accompanies the fourth edition of *Using Information Technology* offers additional, text-specific resources to include: a sample chapter, an overview, Meet the Authors section, the preface, a What's New section, a Feature Summary section, and links to professional resources. In addition, the Instructor's Manual and PowerPoint Presentation slides can be downloaded from the instructor section of the site, located at www.mhhe.com/cit/uit4e.

Online Learning Centers can be delivered through any of these platforms:

McGraw-Hill Learning Architecture (TopClass)
Blackboard.com
ECollege.com (formerly Real Education)
WebCT (a product of Universal Learning Technology)

Office 2000 Application Series

Available for discount packaging, our **Advantage Series** leads students through the features of Microsoft Office 2000 with a critical-thinking, "what, why, and how" orientation. McGraw-Hill/Irwin also offers additional MS Office 2000 series to suit your teaching preference. Each MS Office series is Microsoft Office User Specialist (MOUS) certified.

Skills Assessment

McGraw-Hill/Irwin offers two innovative systems to meet your skills assessment needs. These two products are available for use with any of our applications manual series.

ATLAS (Active Testing and Learning Assessment Software)—Atlas is one option to consider for an application skills assessment tool from McGraw-Hill. Atlas allows students to perform tasks while working live within the Microsoft applications environment. ATLAS is Web-enabled, customizable, and is available for Microsoft Office 2000. **SimNet** (Simulated Network Assessment Product) is another option for a skills assessment tool that permits you to test students' software skills in a simulated environment. SimNet is available for Microsoft Office 97 (deliverable via a network) and Microsoft Office 2000 (deliverable via a network and the Web). Both ATLAS and SimNet provide flexibility for you in your course by offering:

- Pre-testing options
- Post-testing options
- Course placement testing
- Diagnostic capabilities to reinforce skills
- Proficiency testing to measure skills

Acknowledgments

Two names are on the front of this book, but a great many others are important contributors to its development. First, we wish to think our publisher, David Brake, for his support and his encouragement during a difficult revision and production schedule. Thanks go also to Jodi McPherson, our sponsoring editor, for her enthusiasm and new ideas. Everyone in production provided excellent support and direction: Laurie Entringer, Heather Burbridge, Christine Vaughan, Keri Johnson, Marc Mattson, and David Barrick in production and manufacturing. Gladys True, who worked with us on many of our editions over the past ten years, started the production process with us on this edition, before she left for her well-deserved but unwelcome-to-authors retirement. Outside of McGraw-Hill we were fortunate to have the invaluable services of Burrston house, who provided developmental direction and reviews; Judy Mason, photo researcher; and Bernard Gilbert, Martha Ghent, and James Minkin, who provided some of the best copyediting, proofreading, and indexing in the business. Thanks also to all the extremely knowledgeable and hard-working professionals at GTS Graphics, who provided all the prepress services.

Finally, we are grateful to the following people for their participation in manuscript reviews. Their contributions have helped us to make this the most market-driven book possible.

Ray Cone
Odessa College

Robert Fritz
American River College

Julie Giles
DeVry Institute

Jan Harris
Lewis & Clark College

Lister W. Horn
Pensacola Junior College

Mary Horstman
Western Illinois University

Jim Lindsey
Ferris State University

Pat Ormond
Utah Valley State College

John Sharlow
Eastern Connecticut State University

DeLyse Totten
Portland Community College–Sylvania

Hyacinth Burton
Oakwood College

Warren Jones
University of Alabama

William A. Jones
Modest Junior College

Dee Joseph
St. Philips College

Cynthia Kirby
Onondaga Community College

Kurt Kominek
Northeast State Tech Community College

Terry Lundgren
Eastern Illinois University

Barbara Maccarone
North Shore Community College

Edmond Mannion
California State University–Chico

Daniela Marghitu
Auburn University

Ron Miller
Western Michigan University

Melissa Wertz
Pittsburgh Technical Institute

Scott Williams
Hampton University

Student's Guide

A One-Minute Course on How to Succeed in This Class

Got one minute to read this section? It could mean the difference between getting an A instead of a B, or a B instead of a C. Or even passing instead of failing.

Here Are the Rules

There are only four rules, and they aren't difficult.

Rule 1. You have to attend every class. (But that alone won't get you an A, as some students think.)

Rule 2. You can't put off studying, then cram the night before a test. This may work in high school, but college isn't high school.

Rule 3. You have to read or repeat material more than once. The important thing isn't reading. It's *re*reading.

Rule 4. You have to learn the secrets to using your textbook. It would be nice if all textbooks were organized the same way, but they aren't. Different texts have different features.

Getting the Most Information in the Least Time from This Book

Let's consider how you can best read *Using Information Technology*.

- Get an overview of the chapter first
- Check the key questions in each section before you read it
- Read the section, trying to answer the key question(s)
- Do the Concept Checks
- Read the Visual Summary
- Answer the questions in the Chapter Review

A look through the next seven pages will show you what the features we discussed look like.

Get an Overview of the Chapter First

Before you set out on a trip to a place you've never been to before, you would probably look at a map so you would get a "big picture" view of the route. Reading is the same way.

Scan the first page of the chapter and look at the **Chapter Outline** and the **Key Questions.**

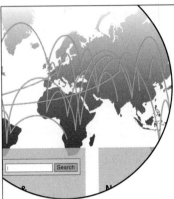

The Internet & the World Wide Web

Exploring Cyberspace

Key Questions
You should be able to answer the following questions.

2.1 **Choosing Your Internet Access Device & Physical Connection: The Quest for Broadband** What are the means of connecting to the Internet, and how fast are they?

2.2 **Choosing Your Internet Service Provider (ISP)** What are the three types of Internet service provider, and what kinds of services do they provide?

2.3 **Sending & Receiving E-Mail** How can one obtain e-mail software, what are the components of an e-mail address, and what are netiquette and spam?

2.4 **The World Wide Web** What are Web sites, Web pages, browsers, URLs, and search engines?

2.5 **The Online Gold Mine: More Internet Resources, Your Personal Cyberspace, E-Commerce, & the E-conomy** What are FTP, Telnet, newsgroups, real-time chat, and e-commerce?

Chapter Outline

Each chapter begins with an outline of the section headings in the chapter.

Key Questions

Use these Key Questions to help you read with purpose. Key Questions are repeated throughout the text.

Then turn the page and read through the **Graphical Interface: A Visual Overview.** These will give you a "big picture" of the chapter material.

Graphical Interface

A Visual Overview of This Chapter

1 **Choosing Your Internet Access Device & Physical Connection.** Some Internet **physical connections,** either wired or wireless, have more **bandwidth**—are able to transmit more data—than others. Data transmission is expressed in **bps** (bits per second), **Kbps** (kilobits—thousands of bits per second), **Mbps** (megabits—millions), and **Gbps** (gigabits—billions). Data is **downloaded** from a remote computer to a local computer or **uploaded,** the reverse.

There are four principal types of Internet physical connections: (1) Telephone (dial-up) modem connection is low-speed but inexpensive (up to 56 Kbps). (2) High-speed phone connections are **ISDN** (up to 128 Kbps), which transmits over traditional phone lines; **DSL** (up to 8.4 Mbps), also using traditional phone lines; and **T1** (1.5 Kbps), a special trunk line. (3) **Cable modems** (10 Mbps) connect to cable TV systems. (4) Wireless systems include microwave systems, such as **communications satellites** or space stations (up to 56 Kbps).

2 **Choosing Your Internet Service Provider (ISP).** With a physical connection installed, you then need an **Internet service provider (ISP),** a company to help you connect or **log on** to the Internet. Three types of ISPs are free-service providers, which make money presenting ads; basic-service providers, or small, local companies; and full-service providers, which offer more technical support and other services—examples are AOL and MSN—and which offer **menus,** or lists of commands or services.

3 **Sending & Receiving E-Mail.** Four alternatives for getting and sending e-mail are to buy e-mail software, get the software as part of a browser or other software, get it from your ISP, and get it free (for example, from CNN.com or NetZero). People will send e-mail to you at your **domain,** a location on the Internet consisting of your user name and domain name, such as *user@domain.*

E-mail allows users to send attachments, or separate long documents, with their e-mail messages. It also allows **instant messaging (IM),** which advises other users the moment a message arrives by displaying it in a **window,** a rectangular area on screen. You can exchange e-mail from people worldwide with similar interests through **listserves,** or e-mail mailing lists.

The two basic rules of online behavior, or **netiquette,** are don't waste people's time and don't say anything online you wouldn't say to someone's face. In particular, you should always first consult **FAQs,** or Frequently Asked Questions; avoid **flaming,** such as insults or obscenities; and smooth communication using **emoticons,** or friendly graphic symbols.

To manage your e-mail, filters or instant organizers are recommended. In addition, one needs to know how to manage **spam,** or unsolicited e-mail. Finally, assume e-mail messages are not private: anyone could read them.

Graphical Interface: A Visual Overview

Reading experts suggest you "preview" the material you are about to read. You can do this by reading the pages immediately following the chapter opening, which present the chapter's key terms and concepts. Graphics provide visual reinforcement.

Key Terms

Important terms are presented in bold type.

Check the Key Questions in Each Section before You Read It

Look at the **Key Question** near the section heading. Read this aloud (or beneath your breath) or write it down.

Bookmark This! Practical Action Box

These boxes present material on practical matters that students find useful.

BOOKMARK IT!

PRACTICAL ACTION BOX
Choosing an Internet Service Provider

If you belong to a college or company, you may get an ISP free. Some public libraries also offer free Net connections. Here are some important questions to ask when you're making those phone calls to locate an Internet service provider:[a]

Costs

- Is there a setup fee? (Most ISPs no longer charge this, though some "free" ISPs will.)
- How much is unlimited access per month? (Most charge about $20 for unlimited usage. But inquire if there are free or low-cost trial memberships or discounts for long-term commitments.)
- If access is supposedly free, what are the trade-offs besides putting up with heavy advertising? (For instance, if the ISP closely monitors your activity in order to accurately target ads, what guarantees do you have that information about you will be kept private? What charges will you face if you try to scrap the advertising window or drop the service?)
- Is there a contract, and for what length of time? That is, are you obligated to stick with the ISP for a while even if you're unhappy with it?

Access

- Is the access number a local phone call? (If not, your monthly long-distance phone tolls could exceed the ISP ...)
- ... l-up number if the main num-

- Is access available when you're traveling? Your provider should offer either a wide range of local access numbers in the cities you tend to visit or toll-free 800 numbers.

Support

- What kind of help does the ISP give in setting up your connection?
- Is there free, 24-hour technical support? Is it reachable through a toll-free number?
- How difficult is it to reach tech support? (Try calling the number before you sign up for the ISP and see how long it takes to get a response. Many ISPs keep customers on hold for a long time.)

Reliability

- What is the average connection success rate for users trying to connect on the first try? (The industry average call-success rate is 93.1%. You can try dialing the number during peak hours, to see if you get a modem screech, which is good, rather than a busy signal, which is bad. You can also check Visual Networks, *www.inversenet.com,* for call-failure/call-success rates of various ISPs.)
- Will the ISP keep up with technology? (Are they planning to offer broadband technology such as DSL for speedier access?)
- Will the ISP sell your name to marketers or bombard you with junk messages (spam)?

Key Questions

Use these Key Questions presented at the start of each section to help you read with purpose.

CONCEPT CHECK

What are the measures of data transmission speed?

Explain the differences among the methods of going online.

Describe the different types of Internet service providers.

2.3 Sending & Receiving E-Mail

KEY QUESTIONS
How can one obtain e-mail software, what are the components of an e-mail address, and what are netiquette and spam?

Once connected with an ISP, one of the first things most people want to do is join the millions of users who send and receive electronic mail. E-mail can be sent at any time and to several people simultaneously. You can receive e-mail wherever you are, using your user name and password to connect to the Internet. In addition, you can attach long (or short) documents or other materials to your e-mail message.

Concept Checks

These questions, which appear throughout the text, encourage you to take a moment to see how well you understand the concepts you have read in the preceding material.

Read the Section, Trying to Answer the Key Question(s)

Read the section, trying to answer the Key Question or Key Questions as you go. Make marks in the book if this helps you answer the question. In particular, look at the **key terms and definitions,** which appear in boldface. Look at the **graphics** (artwork and photos), which help to clarify the discussion.

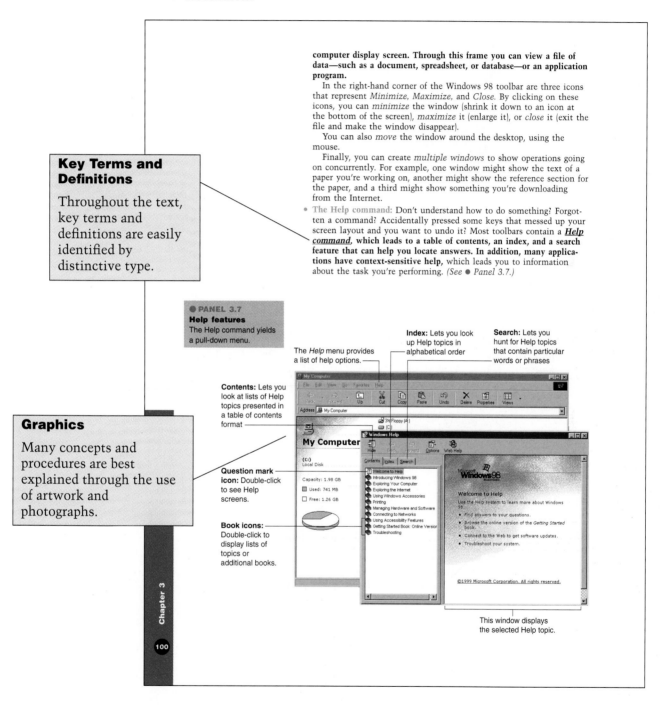

Key Terms and Definitions

Throughout the text, key terms and definitions are easily identified by distinctive type.

computer display screen. Through this frame you can view a file of data—such as a document, spreadsheet, or database—or an application program.

In the right-hand corner of the Windows 98 toolbar are three icons that represent *Minimize, Maximize,* and *Close.* By clicking on these icons, you can *minimize* the window (shrink it down to an icon at the bottom of the screen), *maximize* it (enlarge it), or *close* it (exit the file and make the window disappear).

You can also *move* the window around the desktop, using the mouse.

Finally, you can create *multiple windows* to show operations going on concurrently. For example, one window might show the text of a paper you're working on, another might show the reference section for the paper, and a third might show something you're downloading from the Internet.

• The Help command: Don't understand how to do something? Forgotten a command? Accidentally pressed some keys that messed up your screen layout and you want to undo it? Most toolbars contain a *__Help command__*, **which leads to a table of contents, an index, and a search feature that can help you locate answers. In addition, many applications have context-sensitive help,** which leads you to information about the task you're performing. *(See ● Panel 3.7.)*

● **PANEL 3.7**
Help features
The Help command yields a pull-down menu.

The *Help* menu provides a list of help options.

Index: Lets you look up Help topics in alphabetical order

Search: Lets you hunt for Help topics that contain particular words or phrases

Contents: Lets you look at lists of Help topics presented in a table of contents format

Graphics

Many concepts and procedures are best explained through the use of artwork and photographs.

Question mark icon: Double-click to see Help screens.

Book icons: Double-click to display lists of topics or additional books.

My Computer

(C:)
Local Disk

Capacity: 1.98 GB
■ Used: 741 MB
□ Free: 1.26 GB

This window displays the selected Help topic.

Chapter 3
100

Do the Concept Checks

The **Concept Checks** are questions that appear throughout the text that encourage you to see how well you understand the concepts you have read. Take a break from your reading and try to answer as many of these as possible. If you have trouble with the questions, you probably should go back and review the section before proceeding.

Read the Visual Summary

After you read the whole chapter, go through the **Visual Summary,** which gives the important concepts and terms of the chapter in alphabetical order and tells you why they are important.

Visual Summary

active-matrix display (p. 181, KQ 4.4) Also known as *TFT (thin-film transistor) display;* each pixel on the screen is controlled by its own transistor. Why it's important: *Active-matrix screens are much brighter and sharper than passive-matrix screens, but they are more complicated and thus more expensive. They also require more power, affecting the battery life in laptop computers.*

AGP (accelerated graphics port) bus (p. 160, KQ 4.1) Bus that transmits data at high speeds; designed to support video and three-dimensional (3D) graphics. Why it's important: *An AGP bus is twice as fast as a PCI bus.*

arithmetic/logic unit (ALU) (p. 153, KQ 4.1) Part of the CPU that performs arithmetic operations and logical operations and controls the speed of those operations. Why it's important: *Arithmetic operations are the fundamental math operations: addition, subtraction, multiplication, and division. Logical operations are comparisons such as is "equal to," "greater than," or "less than."*

ASCII (American Standard Code for Information Interchange) (p. 148, KQ 4.1) Binary code used with microcomputers. Besides more conventional characters, ASCII includes such characters as math symbols and Greek letters. Why it's important: *ASCII is the binary code most widely used in microcomputers.*

audio-input device (p. 177, KQ 4.3) Hardware that records analog sound and translates it for digital storage and processing. Why it's important: *Analog sound signals are continuously variable waves within a certain frequency range. For the computer to process them, these variable waves must be converted to digital 0s and 1s. The principal use of audio-input devices is to produce digital input for multimedia computers. An audio signal can be digitized in two ways—by an audio board or a MIDI board.*

bar-code reader (p. 176, KQ 4.3) Photoelectric scanner that translates bar codes into digital code. Why it's important: *With bar-code readers and the appropriate software system, store clerks can total purchases and produce invoices with increased speed and accuracy; and stores and other businesses can monitor inventory and services with increased efficiency.*

bar codes (p. 175, KQ 4.3) Vertical zebra-striped marks imprinted on most manufactured retail products. Why it's important: *In North America, supermarkets, food manufacturers, and others have agreed to use a bar-code system called the Universal Product Code. Other kinds of bar-code systems are used on everything from FedEx packages, to railroad cars, to the jerseys of long-distance runners.*

bay (p. 150, KQ 4.1) Shelf or opening in the computer case used for the installation of electronic equipment, generally storage devices such as a hard drive or DVD drive. Why it's important: *A computer may come equipped with four or seven bays, which permit the expansion of system capabilities.*

binary system (p. 147, KQ 4.1) A two-state system used for data representation in computers; has only two digits— 0 and 1. Why it's important: *In the computer, 0 can be represented by electrical current being off and 1 by the current being on. All data and program instructions that go into the computer are represented in terms of these binary numbers.*

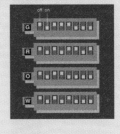

Answer the Questions in the Chapter Review

The **Chapter Review** at the end of the chapter offers a three-level process that helps you truly understand and "take ownership" of the material. Here's how it works:

- **First-Level Review—Memorization:** Level 1 questions test how well you recall basic terms and concepts. They include **Self-Test, Multiple-Choice,** and **True/False** questions.

Chapter Review

The Chapter Review follows a three-level process that helps students truly understand and "take ownership" of the chapter material they have just completed reading.

First-Level Review: Memorization

Level 1 questions test how well you recall basic terms and concepts. They include **Self-Test, Multiple-Choice, and True/False** questions.

Chapter Review

Self-Test Questions

1. Today's data-transmission speeds are measured in _____, _____, _____, and _____ per second.
2. A(n) _____ connects a personal computer to a cable-TV system that offers an Internet connection.
3. A space station that transmits data as microwaves is a _____.
4. A company that connects you through your communications line to its server, which connects you to the Internet is a(n) _____.
5. A small rectangular area on the computer screen that contains a document or displays an activity is called a(n) _____.
6. _____ is writing an online message that uses derogatory, obscene, or inappropriate language.
7. A(n) _____ is software that enables users to view Web pages and to jump from one page to another.
8. A computer with a domain name is called a(n) _____.
9. _____ comprises the communications rules that allow browsers to connect with Web servers.
10. A(n) _____ is a program that adds a specific feature to a browser, allowing it to play or view certain files.

Multiple-Choice Questions

1. Kbps means _____ bits per second.
   a. 1 billion
   b. 1 thousand
   c. 1 million
   d. 1 hundred
   e. 1 trillion
2. A location on the Internet is called a(n) _____.
   a. network
   b. user ID
   c. domain
   d. browser
   e. web
3. In the e-mail address *Kim_Lee@ earthlink.net.us*, *Kim_Lee* is the _____.
   a. domain
   b. URL
   c. site
   d. user ID
   e. location

4. Which of the following is not one of the four components of a URL?
   a. Web protocol
   b. name of the Web server
   c. name of the browser
   d. name of the directory on the Web server
   e. name of the file within the directory
5. Which of the following is the fastest method of data transmission?
   a. ISDN
   b. DSL
   c. modem
   d. T1 line
   e. cable modem
6. Which of the following is not a netiquette rule?
   a. consult FAQs
   b. flame only when necessary
   c. don't shout
   d. avoid huge file attachments
   e. avoid sloppiness and errors

True/False Questions

T F 1. AOL, EarthLink, and Prodigy are full-service ISPs.

T F 2. Users can do nothing to stop spam.

T F 3. A Web page is a document on the World Wide Web.

T F 4. Bookmarks let you store the URLs of Web pages you frequently visit.

T F 5. Search engines do not allow you to use keywords to initiate information searches.

Short-Answer Questions

1. Briefly define *bandwidth*.
2. What are three methods of data transmission that are faster than via a regular modem connection?
3. What does *log on* mean?
4. What is netiquette, and why is it important?
5. On the Web, many documents are "linked." What does that mean?

The Internet & the World Wide Web

Student's Guide

Second-Level Review: Comprehension

Level 2 questions test how well you understand concepts and integrate ideas. They include **Short-Answer Questions** and **Concept Mapping** assignments.

- **Second-Level Review—Comprehension:** Level 2 questions test how well you understand concepts and integrate ideas. They include **Short-Answer Questions** and **Concept Mapping** assignments.
- **Third-Level Review—Application, Analysis, Synthesis, Evaluation:** Level 3 questions and assignments appear under **Knowledge in Action.** They show your mastery of the material, including your ability to use it to solve problems and make decisions.

Concept Mapping

On a separate sheet of paper, draw a concept map, or visual diagram, linking concepts. Show how the following terms are related.

Cable modem	Physical connection
Communications satellite	Protocol
Directory	Real-time chat (RTC)
Domain	Site
Download	Search engine
DSL	T1 line
Home page	Upload
Hypertext	URL
Instant Messaging	Usenet
Internet service provider	Web browser
ISDN	Webcasting
List-serves	Web page
Newsgroup	Web portal
Newsreader	Web site

Knowledge in Action

1. Identify the four parts of the following URL:
   http://www.nps.gov/yell/swimming.htm
2. Some Web sites go overboard with multimedia effects, while others don't include enough. Locate a Web site that you think makes effective use of multimedia. What is the purpose of the site? Why is the site's use of multimedia effective? Take notes, and repeat the exercise for a site with too much multimedia and one with too little.
3. Visit the following job-hunting Web sites:
   www.monster.com
   www.occ.com
   www.careerpath.com
   www.cweb.com
   Investigate job offerings in a field you are interested in. Which is the easiest site to use? Why? What recommendations would you make for improving the site?
4. Distance learning uses electronic links to extend college campuses to people who otherwise would not be able to take college courses. Is your school or someone you know involved in distance learning? If so, research the system's components and uses. What hardware and software do students need in order to communicate with the instructor and classmates? What courses are offered? Prepare a short report on this topic.
5. It's difficult to conceive how much information is available on the Internet and the Web. One method you can use to find information among the millions of documents is to use a search engine, which helps you find Web pages based on typed keywords or phrases. Use your browser to visit the following search sites: *www.yahoo.com* and *www.goto.com*. Click in the Search box and then type the phrase "personal computers," then hit "Go," "Find it!", or hit the Enter key. When you locate a topic that interests you, print it out by choosing File, Print from the menu bar or by clicking the Print button on the toolbar. Report on your findings.

Third-Level Review: Application, Analysis, Synthesis, Evaluation

Level 3 questions and assignments appear under **Knowledge in Action.** They test higher-order critical-thinking skills, including the ability to solve problems and make decisions.

Investigating the World Wide Web

Activities for exploring the Web often appear under **Knowledge in Action** in the **Chapter Review.**

Clearly, this method takes longer than simply reading the material once and rapidly underlining it. But because it requires your involvement and understanding, it is a better way to learn.

Contents

Chapter 3

SOFTWARE: THE POWER BEHIND THE
POWER   83

Chapter 4

HARDWARE: HOW TO BUY A
MULTIMEDIA COMPUTER
SYSTEM   141

Contents

Chapter 5

TELECOMMUNICATIONS: NETWORKS & COMMUNICATIONS—THE "NEW STORY" IN COMPUTING   203

Chapter 6

E-COMMERCE, FILES, & DATABASES: DIGITAL ENGINES FOR THE NEW ECONOMY   235

Contents

xxiii

Chapter 7

THE CHALLENGES OF THE DIGITAL AGE: SOCIETY & INFORMATION TECHNOLOGY TODAY 263

Chapter 8

THE PROMISES OF THE DIGITAL AGE: SOCIETY & INFORMATION TECHNOLOGY TOMORROW 281

Appendix

SYSTEMS & PROGRAMMING: DEVELOPMENT, PROGRAMMING, & LANGUAGES 301

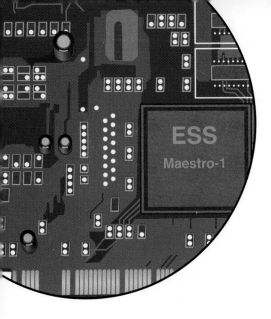

Introduction to Information Technology

Mind Tools for Your Future

Key Questions

You should be able to answer the following questions.

1.1 **Infotech Becomes Commonplace: Cellphones, E-Mail, the Internet, & the E-World** How does information technology facilitate e-mail, networks, and the use of the Internet and the Web, and what is the meaning of the term *cyberspace*?

1.2 **The "All-Purpose Machine": The Varieties of Computers** What are the five sizes of computers, and what are clients and servers?

1.3 **Understanding Your Computer: Pretending to Build Your Own PC** What four basic operations do all computers follow, and what are some of the devices associated with each operation? How does communications affect these operations?

1.4 **Where Is Information Technology Headed?** What are three directions of computer development and three directions of communications development?

Graphical Interface

1 **Infotech Becomes Commonplace. Information technology** merges computers and high-speed communications; new infotech phones are an example. **Computer**—a programmable, multiuse machine that processes data into information. **Communications**—electromagnetic devices and systems for communicating over distances. **Online**—using a computer to access information and services via a network.

Infotech has given us some now-commonplace technologies: **E-mail**—messages transmitted over a computer **network,** a communications system connecting computers. A big part of **cyberspace**—the wired and wireless world of communications—are the **Internet,** a network of 400,000 smaller networks, and its graphical subsection, the **World Wide Web,** which stores information in **multimedia** form—text, graphics, sound, video.

2 **The "All-Purpose Machine."** Computers are of five types. (1) **Supercomputers**—perform 1 trillion calculations per second. (2) **Mainframes**—for processing millions of calculations; often accessed by a **terminal,** a display screen and keyboard that can't do its own processing. (3) **Workstations**—used for scientific, mathematical, engineering, and certain manufacturing applications. (4) **Microcomputers**—personal computers, which may be **desktop PCs, tower PCs, laptops,** or **personal digital assistants (PDAs);** frequently used to connect to a **local area network (LAN),** which links equipment in an office or building. (5) **Microcontrollers**—tiny specialized computers installed in cars and appliances. **Servers** are central computers holding data and programs for many other computers.

3 **Understanding the Computer.** You should know three things: (1) The purpose of a computer is to process **data,** raw facts, into **information,** summarized data. (2) **Hardware** consists of machinery and equipment; **software** instructs the hardware what to do. (3) Computers perform four basic operations: **input,** putting data into the system; **processing,** manipulation of data into information; **storage,** of two types—**memory,** which temporarily holds data to be processed, and **secondary storage,** which holds data or information permanently; **output,** putting out the results of processing. In addition, communication could be considered a fifth operation.

Input hardware includes the **keyboard,** which converts characters into signals readable by the processor, and the **mouse,** used to manipulate objects on the display screen.

Processing and memory hardware includes the **case (system cabinet),** which contains: a processor **chip** with miniature electronic circuits that process information in megahertz, millions of processing cycles per second; and **memory chips (RAM chips),** which hold data prior to processing. Both chips are connected to the **motherboard,** which contains **expansion slots** for plugging in additional circuit boards.

Storage hardware includes: a **floppy-disk drive,** which stores data on removable 3.5-inch disks; a **hard-disk drive,** which stores data on a nonremovable disk platter; and a **CD-ROM drive,** which reads data from optical disks.

Output hardware includes: a **video card,** which converts processor information so it can be displayed as text and images on the **monitor;** a **sound card,** which outputs sound to **speakers;** and a **printer,** which produces images and text on paper.

Communications hardware includes a **modem,** which sends and receives data over phone lines to and from computers.

Software is of two types. **System software** helps the computer perform essential operating tasks. **Application software** performs specific tasks.

4 **Where Is Infotech Headed?** Computer developments have focused on three areas: miniaturization, speed, and affordability. Three developments in communications may be noted: **connectivity,** the ability to connect with computers via communications lines; **interactivity,** enabling a user to have a two-way dialogue; and multimedia. The melding of computers and communications has produced three developments: **convergence,** the combining of several industries through the language of computers; portability; and personalization. One result of these developments is information overload. Three ethical concerns raised by information technology are associated with speed and scale, unpredictability, and complexity.

Technology will provide no miracles that feel like miracles for long," observes editor and historian Frederick Allen.[1]

Adds science-fiction writer Bruce Sterling, "We should never again feel all mind-boggled at anything that human beings create. No matter how amazing some machine may seem, the odds are very high that we'll outlive it."[2]

The personal computer is over two decades old. The Internet has been familiar to the public for over 10 years. It has been more than five years since the now-commonplace "www" for World Wide Web began appearing in company ads. And, like cars, elevators, air-conditioning, and television—all of which have wrought tremendous changes on society and the landscape—they are rapidly achieving what technology is supposed to do: *become ordinary.* They are becoming part of the wallpaper of our lives, almost invisible.

"The fact that technology tends to disappear ensures that we will take much of it, when it works right, for granted," says Allen. "And because of that, we stay most aware of technologies that are changing rapidly or are grossly immature. Right now we're especially aware of the Internet precisely because of its youth."

But the notion that the Internet is still a brash upstart has already pretty much faded. For instance, its two most popular aspects—e-mail and the World Wide Web—are rapidly becoming as ordinary as the telephone and the freeway. Beyond them, however, lie some even more astonishing developments, as we discuss in this book.

1.1 Infotech Becomes Commonplace: Cellphones, E-Mail, the Internet, & the E-World

KEY QUESTIONS

How does information technology facilitate e-mail, networks, and the use of the Internet and the Web, and what is the meaning of the term cyberspace?

This book is about computers, of course. But not *just* about computers. It is also about the way computers communicate with each other. When computer and communications technologies are combined, the result is **_information technology_—"infotech"—technology that merges computing with high-speed communications links carrying data, sound, and video.** Examples of information technology include personal computers, of course, but also new forms of telephones, televisions, and various handheld devices such as personal digital assistants.

Note there are two parts to this definition—*computers* and *communications*:

- Computer technology: You have certainly seen, and possibly used, a computer. Nevertheless, let's define what it is. **A _computer_ is a programmable, multiuse machine that accepts data—raw facts and figures—and processes, or manipulates, it into information we can use,** such as summaries, totals, or reports. Its purpose is to speed up problem solving and increase productivity.

- Communications technology: Unquestionably you've been using communications technology for years. **_Communications technology_, also called *telecommunications technology*, consists of electromagnetic devices and systems for communicating over long distances.** The principal examples are telephone, radio, broadcast television, and cable TV. More recently there has been the addition of communications among computers, as when people tell you they "went online" with the Internet. **_Online_ means using a computer or other information device, connected through a voice or data network, to access information and services from another computer or information device.**

As our first example, let's consider something that seems to be everywhere these days—the cellphone.

The Telephone Grows Up

Seventy-something Louis DeMartino hadn't even had a chance to unwrap the package when a ringing sound came from inside it—a call from his granddaughter as a way of making the family's gift presentation of a cellphone all the more dramatic.[3] DeMartino may have just been getting started on using this apparatus, but Amit Sinai, 18, like many other teenagers, cannot imagine life without it. Flopping down on her towel after a quick dip in the ocean at a New York beach, for instance, the first thing she does, even before her face starts to dry, is reach for her cellphone and start dialing. "Hi, it's me," she says. "I'm on the beach. Where are you?"[4]

Cellphone mania has swept the world. All across the globe, people have acquired the portable gift of gab; some make 45 calls or more a day. It has taken more than 100 years for the telephone to get to this point—getting smaller, acquiring push buttons, losing its cord connection. In 1964, the ★ and # were added to the keypad to allow users to interact with data services and equipment. In 1973, the first cellphone call was processed. In its standard form, the phone is still so simply designed that even a young child can use it. However, it is now becoming more versatile—a way of connecting to the Internet and the World Wide Web.

Why introduce a book that is about computers with a discussion of telephones? Because Internet phones—such as AT&T's PocketNet Phone or Motorola's i1000 plus—represent another giant step for information technology. *(See ● Panel 1.1.)* Now you no longer need a personal computer to get on the Internet. These infotech phones, with their small display screens, provide a direct, wireless connection that enables you not only to make voice calls and check your daily "to-do" list but also to browse the World Wide Web and receive all kinds of information: news, sports scores, stock prices, term-paper research. And you can also send and receive e-mail.

See-through cover, so you can access without opening up phone

Online menus, for accessing Internet services

Speaker phone, for hands-free use

Connector to your laptop or personal digital assistant for wireless access

● PANEL 1.1
The Internet phone
With an Internet-accessible phone, you have e-mail and specific Web services such as news and stock quotes.

"You've Got Mail!": E-Mail's Mass Impact

It took the telephone 40 years to reach 10 million customers, and fax machines 20 years. Personal computers made it into that many American homes five years after they were introduced. E-mail, which appeared in 1981, became popular far more quickly, reaching 10 million users in little more than a year.[5] No technology has ever become so universal so fast. Not surprisingly, then, one of the first things new computer users learn is how to send and receive **e-mail (electronic mail), messages transmitted over a computer network,** most often the Internet. **A _network_ is a communications system connecting two or more computers;** the Internet is the largest such network.

In 1998, the volume of e-mail in the United States surpassed the volume of hand-delivered mail. By 2002, an average of 8 billion messages a day are expected to zip back and forth across the U.S. (which has some 60% of the world's e-mail accounts), up from 3.5 billion three years earlier.[6] Because of this explosion in usage, suggests a *Business Week* report, "E-mail ranks with such pivotal advances as the printing press, the telephone, and television in mass impact."[7]

How is electronic mail different from calling on a telephone or writing a conventional letter? E-mail "occupies a psychological space all its own," says one journalist. "It's almost as immediate as a phone call, but if you need to, you can think about what you're going to say for days and reply when it's

E-mailer

A college student sends e-mail. College students are among the biggest users of e-mail.

convenient."[8] E-mail has blossomed, points out another writer, not because it gives us more immediacy but because it gives us *less.* "The new appeal of e-mail is the old appeal of print," he says. "It isn't instant; it isn't immediate; it isn't in your face." E-mail has succeeded for the same reason that the videophone—which allows callers to see each other while talking—has been so slow to catch on: because "what we actually want from our exchanges is the minimum human contact commensurate with the need to connect with other people."[9]

From this it is easy to conclude, as *New York Times* computer writer Peter Lewis did, that e-mail "is so clearly superior to paper mail for so many purposes that most people who try it cannot imagine going back to working without it."[10] What is interesting, though, is that in these times when images often seem to overwhelm words, e-mail is actually *reactionary.* "The Internet is the first new medium to move decisively backward," points out one writer, "for it is, essentially, written. . . . Ten years ago, even the most literate of us wrote maybe a half a dozen letters a year; the rest of our lives took place on the telephone."[11] E-mail has changed all that.

The Internet, the World Wide Web, & the "Plumbing of Cyberspace"

Communications is extending into every nook and cranny of civilization. It has been called the "plumbing of cyberspace." The term *cyberspace* was coined by William Gibson, in his novel *Neuromancer,* to describe a futuristic computer network into which users plug their brains. In everyday use, it has a rather different meaning.

Today many people equate cyberspace with the Internet. But it is much more than that, says David Whittle in his book *Cyberspace: The Human Dimension.*[12] Cyberspace includes not only the World Wide Web, chat rooms, online bulletin boards, and member-based services like America Online—all features we explain in this book—"but also such things as conference calls and automatic teller machines," he says.[13] We may say that **cyberspace encompasses not only the online world and the Internet in particular but also the whole wired and wireless world of communications in general.**

The two most important aspects of cyberspace are the Internet and that part of it known as the World Wide Web.

- **The Internet—"the mother of all networks":** The Internet is at the heart of the "Information Age." Called "the mother of all networks," **the *Internet* is a worldwide network that connects up to 400,000**

smaller networks in more than 200 countries. These networks are formed by educational, commercial, nonprofit, and military entities.

- **The World Wide Web—the multimedia part of the Net:** The Internet has actually been around for more than 30 years. But what made it popular, apart from e-mail, was the development of the _**World Wide Web**_—**an interconnected system of computers all over the world that store information in multimedia form. The word _multimedia_, from "multiple media," refers to technology that presents information in more than one medium, such as text, still images, moving images, and sound.**

There is no doubt the influence of the Net and the Web is tremendous. At present, one-third to one-half of the U.S. population is online, and if the Internet industry were a nation, it would be the 18th-largest economy in the world.[14] But just how revolutionary is it? Is it equivalent to the invention of television, as some technologists say? Or is it even more important—equivalent to the invention of the printing press? "Television turned out to be a powerful force that changed a lot about society," says *USA Today* technology reporter Kevin Maney. "But the printing press changed everything—religion, government, science, global distribution of wealth, and much more. If the Internet equals the printing press, no amount of hype could possibly overdo it."[15]

Perhaps in a few years we'll begin to know the answer. No massive study was ever done of the influence of the last great electronic revolution to touch us, namely, television. But the Center for Communication Policy at the University of California, Los Angeles, in conjunction with other international universities, has begun to take a look at the effects of information technology—and at how people's behavior and attitudes toward it will change over a span of years.[16]

The E-World & Welcome to It

One thing we know already is that cyberspace is saturating our lives. More than 52% of American adults are Internet users, according to a recent survey.[17] While the average age of users is rising, to age 40 by late 1999, there's no doubt that young people love the Net. For instance, an amazingly high percentage of American teenagers—*81%*—use the Internet, according to Teenage Research Unlimited (TRU), a Chicago market research firm.[18] Why is that? E-mail is certainly one important reason. "It's all about staying connected," says New Yorker Kimy Tolentino, 19. His friends agree. "I'm an AOL junkie," says Molly Lowe, 15. "When I was on vacation, I asked people if I could use their laptops, just to check my e-mail." "I do that, too," says Raysa Rodriguez, 17.[19]

But it's more than just about e-mail. Teenagers are also big participants in online commerce. "When teens want information on a brand," says TRU's director of research, "they turn to the Internet first." Indeed, teens are voracious consumers of all things electronic. Unlike many adults, they are adept at "multitasking" (a term borrowed from computer jargon, as we explain later)—doing more than one task at once. Thus, they may be surfing the Web, but they're also watching TV and listening to music and maybe chatting on the phone at the same time.

The fact is that, not just for teens but for most Americans, the use of the Internet's favorite letter, "e"—as in e-business, e-commerce, e-shopping—is rapidly becoming outmoded. "E" is now part of nearly everything we do. As an executive for a marketing research firm says, "E-business is just business."[20] The electronic world is everywhere. The Net and the Web are everywhere. Cyberspace permeates everything. What editor Frederick Allen was talking about has already happened: infotech has become . . . ordinary.

CONCEPT CHECK

What are the two parts of information technology?

How does e-mail differ from other means of communication?

What is cyberspace, and what are its two most important aspects?

1.2 The "All-Purpose Machine": The Varieties of Computers

When the alarm clock★ blasts you awake, you leap out of bed and head for the kitchen, where you plug in the coffee maker★. After using your electric toothbrush★ and showering and dressing, you stick a bagel in the microwave★, then pick up the TV remote★ and click on the TV★ to catch the weather forecast. Later, after putting dishes in the dishwasher★, you go out and start up the car★ and head toward campus or work. Pausing en route at a traffic light★, you turn on your portable CD player★ to listen to some music.

You haven't yet touched a PC, a personal computer, but you've already dealt with at least 10 computers—as you probably guessed from the ★s. All these familiar appliances rely on tiny "computers on chips" called microprocessors. "These marvels of engineering have been infiltrating our everyday lives for more than a quarter of a century," says technology writer Dan Gillmor, but "in some ways the revolution has only begun."[21]

Maybe, then, the name *computer* is inadequate. As computer pioneer John Von Neumann has said, the device should not be called a computer but rather the "all-purpose machine." It is not, after all, just a machine for doing calculations. The most striking thing about it is that it can be put to *any number of uses*.

What are the various types of computers? Let's take a look.

All Computers, Great & Small: The Categories of Machines

At one time, the idea of having your own computer was almost like having your own personal nuclear reactor. In those days, in the 1950s and '60s, computers were enormous machines affordable only by large institutions. Now they come in a variety of shapes and sizes, which could be roughly classified according to their processing power into five sizes: *superlarge, large, medium, small,* and *tiny.*

A supercomputer
Blue Mountain

- Supercomputers—superlarge computers: **Typically priced from $500,000 to over $35 million, _supercomputers_ are high-capacity machines with hundreds of thousands of processors that can perform more than 1 trillion calculations per second.** These are the most expensive but fastest computers available. "Supers," as they are called, have been used for tasks requiring the processing of enormous volumes of data, such as doing the U.S. census count, forecasting weather, designing aircraft, modeling molecules, breaking codes, and simulating explosion of nuclear bombs. More recently they have been employed for business purposes—for instance, sifting demographic marketing information—and for creating film animation. The fastest computer in the world, which cost $121.5 million and looks like 48 refrigerator-size blue boxes, is Blue Mountain, located at Los Alamos Laboratory in New Mexico.

Mainframe computer

- Mainframe computers—large computers: The only type of computer available until the late 1960s, **_mainframes_ are water- or air-cooled computers that cost $5000– $5 million and vary in size from small, to medium, to large, depending on their use.** Small mainframes ($5000–$200,000) are often called *midsize computers*; they used to be called *minicomputers*. Mainframes are used by large organizations— such as banks, airlines, insurance companies, and colleges—for processing millions of transactions. Often users access a mainframe using a **_terminal_, which has a display screen and a keyboard and can input and output data but cannot by itself process data.**

Workstation

- Workstations—medium computers: Introduced in the early 1980s, _workstations_ are expensive, powerful computers usually used for complex scientific, mathematical, and engineering calculations and for computer-aided design and computer-aided manufacturing. Providing many capabilities comparable to midsize mainframes, workstations are used for such tasks as designing airplane fuselages, prescription drugs, and movie special effects. Workstations have caught the eye of the public mainly for their graphics capabilities, which are used to breathe three-dimensional life into movies such as _Jurassic Park_ and _Titanic_. The capabilities of low-end workstations overlap those of high-end desktop microcomputers.

- Microcomputers—small computers: _Microcomputers_, also called _personal computers_, which cost $500–$5000, can fit next to a desk or on a desktop, or can be carried around. They are either stand-alone machines or are connected to a computer network, such as a local area network. A _local area network (LAN)_ connects, usually by special cable, a group of desktop PCs and other devices, such as printers, in an office or a building.

Microcomputers are of several types: _desktop PCs, tower PCs, laptops_ (or _notebooks_), and _personal digital assistants_—handheld computers or palmtops.

Microcomputer

Desktop PCs are those in which the case or main housing sits on a desk, with keyboard in front and monitor (screen) often on top. _Tower PCs_ are those in which the case sits as a "tower," often on the floor beside a desk, thus freeing up desk surface space.

Laptop computers, also called _notebook computers_, are lightweight portable computers with built-in monitor, keyboard, hard-disk drive, battery, and AC adapter that can be plugged into an electrical outlet; they weigh anywhere from 1.8 to 9 pounds.

PDAs

Personal digital assistants (PDAs), also called _handheld computers_ or _palmtops_, combine personal organization tools—schedule planners, address books, to-do lists. Some are able to send e-mail and faxes. Some PDAs have touch-sensitive screens. Some also connect to desktop computers for sending or receiving information. (For now, we are using the word _digital_ to mean "computer based.")

- Microcontrollers—tiny computers: _Microcontrollers_, also called _embedded computers_, are the tiny, specialized microprocessors installed in "smart" appliances and automobiles. These microcontrollers enable microwave ovens, for example, to store data about how long to cook your potatoes and at what temperature. Recently

Microcontroller

Laptop

PDA

microcontrollers have been used to develop a new universe of experimental electronic appliances—e-pliances—for example, as tiny Web servers embedded in clothing, jewelry, and household appliances such as refrigerators.[22]

Servers

The word *server* does not describe a size of computer but rather a particular way in which a computer is used. Nevertheless, because servers have become so important to telecommunications, especially with the rise of the Internet and World Wide Web, they deserve mention here. (Servers are discussed in detail in Chapter 5.)

A *server*, **or network server, is a central computer that holds collections of data (databases) and programs for connecting PCs, workstations, and other devices, which are called** *clients***. These clients are linked by a wired or wireless network.** The entire network is called a *client/server network.* In small organizations, servers can store files and transmit e-mail. In large organizations, servers can house enormous libraries of financial, sales, and product information.

You may never see a supercomputer or mainframe or server or even a tiny microcontroller. But you will most certainly get to know the personal computer, if you haven't already. We consider this machine next.

CONCEPT CHECK

Describe the five sizes of computers.

What are the different types of microcomputers?

Define servers and clients.

1.3 Understanding Your Computer: Pretending to Build Your Own PC

KEY QUESTIONS
What four basic operations do all computers follow, and what are some of the devices associated with each operation? How does communications affect these operations?

Could you build your own personal computer? Some people do, putting together bare-bones systems for just a few hundred dollars. "If you have a logical mind, are fairly good with your hands, and possess the patience of Job, there's no reason you can't . . . build a PC," says one science writer. And, if you do it right, "it will probably take only a couple of hours," because industry-standard connections allow components to go together fairly easily.[23]

How Computers Work: Three Key Concepts

We're not actually going to ask you to build a PC—just to *pretend* to do so. The purpose of this exercise is to help you understand how a computer works. That information will help you when you go shopping for a new system or, especially, if you order up a custom-built system (as from Dell, Gateway, or Micron.)

Note: Complete advice about how to buy a computer system is available on our publisher's Web site (*www.mhhe.com/cit/uit4e*).

Before you begin, you will need to understand three key concepts.

FIRST: The purpose of a computer is to process data into information. *Data* **consists of the raw facts and figures that are processed into information**—for example, the votes for different candidates being elected to student-government office. *Information* **is data that has been summarized or otherwise manipulated for use in decision making**—for example, the total votes for each candidate, which are used to decide who won.

SECOND: You should know the difference between hardware and software. *Hardware* **consists of all the machinery and equipment** in a computer system. The hardware includes, among other devices, the keyboard, the screen, the printer, and the "box"—the computer or processing device itself. **_Software_, or programs, consists of all the instructions that tell the computer how to perform a task.** These instructions come from a software developer in a form (such as a CD-ROM disk) that will be accepted by the computer. Examples you may have heard of are Microsoft Windows 98 or Office 2000.

THIRD: Regardless of type and size, all computers follow the same four basic operations: (1) input, (2) processing, (3) storage, and (4) output. To this we will add (5) communications.

1. Input operation: **_Input_ is whatever is put in ("input") to a computer system.** Input can be nearly any kind of data—letters, numbers, symbols, shapes, colors, temperatures, sounds, or whatever raw material needs processing. When you type some words or numbers on a keyboard, those words are considered *input data.*

2. Processing operation: **_Processing_ is the manipulation a computer does to transform data into information.** When the computer adds 2 + 2 to get 4, that is the act of processing. The processing is done by the **_central processing unit_—frequently called just the _CPU_—a device consisting of electronic circuitry that executes instructions to process data.**

3. Storage operation: Storage is of two types—temporary storage and permanent storage, or primary storage and secondary storage. **_Primary storage_, or memory, is the computer circuitry that temporarily holds data waiting to be processed.** This circuitry is inside the computer. **_Secondary storage_, simply called *storage*, is the area in the computer where data or information is held permanently.** A floppy disk is an example of this kind of storage. (Storage also holds the *software*—the computer programs.)

4. Output operation: **_Output_ is whatever is output from ("put out of") the computer system, the results of processing,** usually information. Examples of output are numbers or pictures displayed on a screen, words printed out on paper in a printer, or music piped over some speakers.

5. Communications operation: These days, most (though not all) computers have communications ability, which offers an *extension* capability—in other words, it extends the power of the computer. With wired or wireless communications connections, data may be input from afar, processed in a remote area, stored in several different locations, and output in yet other places. However, you don't need communications ability to write term papers, do calculations, or perform many other computer tasks.

A simple diagram of these five operations is shown in the accompanying illustration. *(See ● Panel 1.2.)*

Pretending to Build a Desktop Computer

Now let's see how you would build a desktop PC. Remember, the purpose of this is to help you understand the internal workings of a computer so that you'll be better equipped when you go to buy one. Pretend, then, that someone has acquired the PC components for you from a catalog company (such as Tiger Direct or MicroWarehouse), and you're now sitting at a workbench or kitchen table about to begin assembling them. (All you would need is a screwdriver, perhaps a small wrench, and a static-electricity-free work area. You would also need the manuals that come with some of the components.)

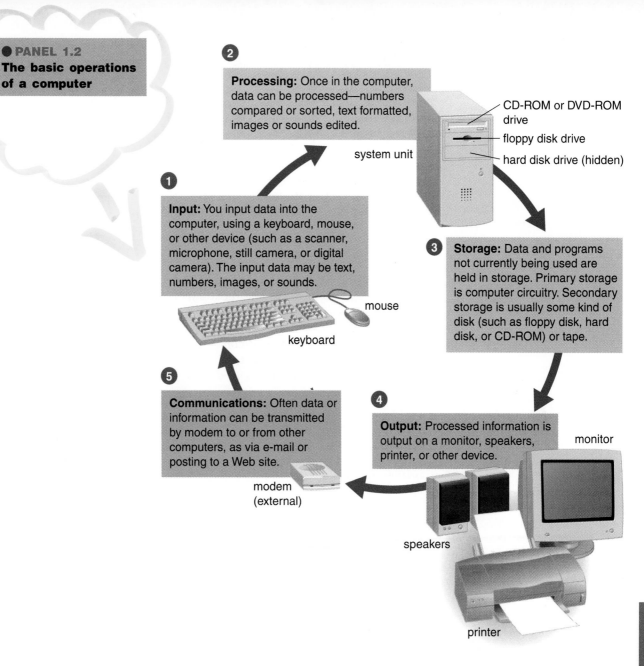

2 Processing: Once in the computer, data can be processed—numbers compared or sorted, text formatted, images or sounds edited.

CD-ROM or DVD-ROM drive

floppy disk drive

hard disk drive (hidden)

system unit

1 Input: You input data into the computer, using a keyboard, mouse, or other device (such as a scanner, microphone, still camera, or digital camera). The input data may be text, numbers, images, or sounds.

mouse

keyboard

3 Storage: Data and programs not currently being used are held in storage. Primary storage is computer circuitry. Secondary storage is usually some kind of disk (such as floppy disk, hard disk, or CD-ROM) or tape.

5 Communications: Often data or information can be transmitted by modem to or from other computers, as via e-mail or posting to a Web site.

4 Output: Processed information is output on a monitor, speakers, printer, or other device.

monitor

modem (external)

speakers

printer

Input Hardware: Keyboard & Mouse

Input hardware consists of devices that allow people to put data into the computer in a form that the computer can use. At minimum, you will need two things: a keyboard and a mouse.

Keyboard

Mouse

- Keyboard: On a PC, a keyboard is the primary input device. **A _keyboard_ is an input device that converts letters, numbers, and other characters into electrical signals readable by the processor.** A PC keyboard looks like a typewriter keyboard, but besides keys for letters and numbers it has several keys (such as F keys and Ctrl, Alt, and Del keys) intended for computer-specific tasks. After you assemble other components, you will plug the keyboard into the back of the computer in a socket intended for that purpose.

- Mouse: **A _mouse_ is an input device that is used to manipulate objects viewed on the computer display screen.** You will plug the mouse into the back of the computer in a socket after you assemble the other components.

Processing & Memory Hardware: Inside the System Cabinet

This is the part where you'll have to do most of your assembly work, so be sure to have available the manuals that come with the parts. The brains of the computer are the processing and memory devices, which are housed in the case or system cabinet.

Case or system cabinet

- **Case or system cabinet:** Also known as the **system unit, the _case_ or _system cabinet_ is the box that houses the processor chip (CPU), the memory chips, and the motherboard with power supply, as well as storage devices**—floppy-disk drive, hard-disk drive, and CD-ROM or DVD drive, as we will explain. The case comes in desktop or tower models. It includes a power supply unit and a fan to keep the circuitry you are about to install from overheating.

- **Processor chip:** It may be small and not look like much. But at a cost of $100–$750, it could be the most expensive hardware component of a build-it-yourself PC—and doubtless the most important. **A processor _chip_ is a tiny piece of silicon that contains millions of miniature electronic circuits.** The speed at which a chip processes information is expressed in _megahertz (MHz)_, millions of processing cycles per second, or _gigahertz,_ billions of processing cycles per second. For $100, you might get a

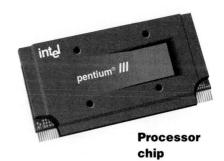

Processor chip

333-MHz chip; which is adequate for most student purposes, such as writing term papers. For top dollar, you might get a 1-gigahertz chip, which you would want if you're running software with gee-whiz graphics, such as those with some video games.

Memory chips

memory chip (RAM chip)

memory chips mounted on module

- **Memory chips:** These are also small. **_Memory chips_, also known as _RAM (random access memory) chips_, represent primary storage or temporary storage; they hold data prior to processing and information after processing,** before it is sent along to an output or storage device. You'll want enough memory chips to hold at least 64 megabytes, or roughly 64 million characters, of data, which is adequate for most student purposes. If you work with large graphic files, you'll need more memory capacity, 128 megabytes. (We explain the numbers used to measure storage capacities in a moment.)

- **Motherboard:** Also called the **system board, the _motherboard_ is the main circuit board in the computer.** This is the big green circuit board to which everything else—such as the keyboard, mouse, and printer—attaches through connections (called _ports_) in the back of the computer. The processor chip and memory chips are also installed on the motherboard.

    Note that the motherboard has **_expansion slots_—for expanding the PC's capabilities—which give you places to plug in additional circuit boards,** such as those for video, sound, and communications (modem).

Now you're ready to start putting components together. As the illustration opposite shows, you first ① plug your memory chips into the motherboard. Then you ② plug your processor chip into the motherboard. Now you ③ attach the motherboard to the system cabinet. Then you ④ connect the power supply unit to the system cabinet. (But don't plug your system into a wall plug just yet—you don't want to electrocute yourself.) Don't forget also ⑤ to connect the wire for the power switch, which turns the computer on and off, to the motherboard.

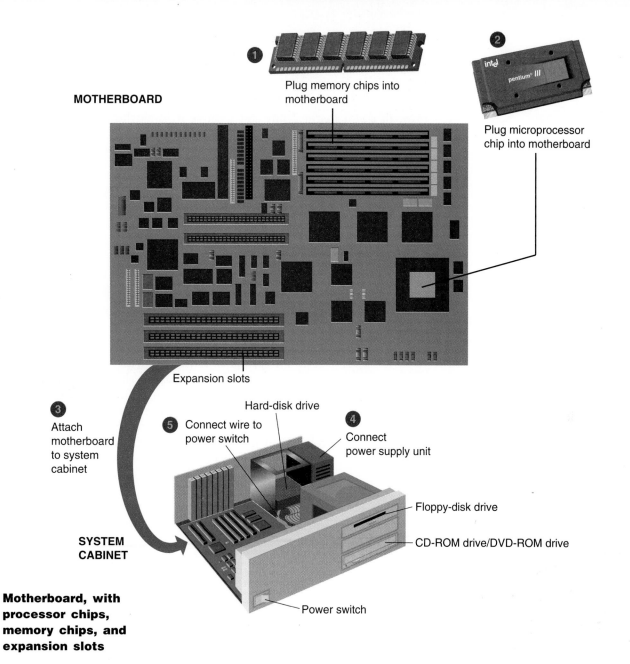

MOTHERBOARD

Plug memory chips into motherboard

Plug microprocessor chip into motherboard

Expansion slots

❸ Attach motherboard to system cabinet

❺ Connect wire to power switch

Hard-disk drive

❹ Connect power supply unit

SYSTEM CABINET

Floppy-disk drive

CD-ROM drive/DVD-ROM drive

Power switch

Motherboard, with processor chips, memory chips, and expansion slots

Storage Hardware: Floppy Drive, Hard Drive, & CD-ROM Drive

With the motherboard in the system cabinet, you can now proceed to install the storage hardware. Whereas memory chips deal with temporary storage, now we're concerned with secondary storage, or permanent storage, in which you'll be able to store your data for the long haul.

For today's student purposes, you'll need at minimum a floppy-disk drive, a hard-disk drive, and a CD-ROM drive. (If you work with large graphic files, you'll also want a Zip-disk drive, explained in Chapter 4.) These storage devices slide into the system cabinet from the front and are secured with screws. Each drive is attached to the motherboard by a flat cable (called a ribbon cable). Also each drive must be hooked up to a plug extending from the power supply.

A computer system's data/information storage capacity is represented by bytes, kilobytes, megabytes, gigabytes, and terabytes. Roughly speaking, a *byte = 1 character* of data, a *kilobyte = 1000 characters*, a *megabyte = 1 million characters*, a *gigabyte = 1 billion characters*, and a *terabyte = 1 trillion characters*. A character could be alphabetic (A, B, or C) or numeric (1, 2, or 3) or a special character (!, ?, *, $, %).

- **Floppy-disk drive:** A _**floppy-disk drive**_ **is a storage device that stores data on removable 3.5-inch-diameter diskettes.** These diskettes are made of "floppy" mylar, but they are encased in hard plastic. Each can store 1.44 million bytes (characters) or more of data. With the floppy-disk drive installed, you'll later be able to insert a diskette through a slot in the front and remove it by pushing the eject button.

Floppy-disk drive and diskette

- **Hard-disk drive:** A _**hard-disk drive**_ **is a storage device that stores billions of characters of data on a non-removable disk platter.** With 8 gigabytes of storage, you should be able to handle most student needs.

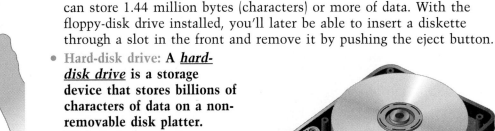

Hard-disk drive

- **CD-ROM drive:** A _**CD-ROM drive**_ **is a storage device that uses laser technology to read data from optical disks.** If you're trying to save money, you might conceivably skip buying this piece of storage equipment. However, these days new software is generally supplied on CD-ROM disks rather than floppy disks. And even if you can get a program on floppies, you'll find it easier to install a new program from one CD-ROM disk rather than repeatedly inserting and removing, say, 10 or 12 floppy disks.

CD-ROM or DVD-ROM drive

CD-ROM or DVD-ROM disk

CD-ROM / DVD-ROM drive

The system cabinet has lights on the front that indicate when these drives are in use. (You must not remove the diskette from the floppy-disk drive until the relevant light goes off, or else you risk damage to both disk and drive.) You'll need to attach wires for these lights to the motherboard.

Output Hardware: Video & Sound Cards, Monitor, Speakers, & Printer

Output hardware consists of devices that translate information processed by the computer into a form that humans can understand—print, sound, graphics, or video, for example. Now you need to install a video card and a sound card in the system cabinet. You then plug in the monitor, speakers, and a printer.

Incidentally, this is a good place to introduce the term _peripheral device._ A _**peripheral device**_ **is any component or piece of equipment that expands a computer's input, storage, and output capabilities.** Examples include printers and disk drives.

Video card

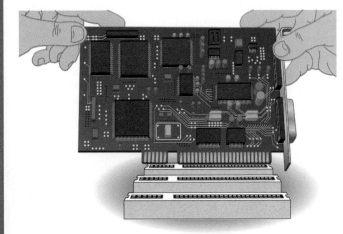

- **Video card:** You doubtless want your monitor to display color (rather than just black-and-white) images. Your system cabinet will therefore need to have a device to make this possible. **A _video card_ converts the processor's output information into a video signal that can be sent through a cable to the monitor.** Remember the expansion slots we mentioned? You plug your video card into one of these on the motherboard. (You can also buy a motherboard with built-in video.)

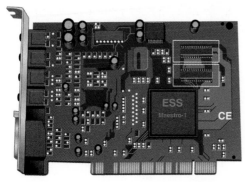

Sound card

- **Sound card:** You may wish to listen to music on your PC. If so, you'll need a **_sound card_, which enhances the computer's sound-generating capabilities by allowing sound to be output through speakers.** This, too, you would plug into an expansion slot on the motherboard. (Once again, you can buy a motherboard with built-in sound.) If you connect the CD-ROM drive to the card, you can listen to music CDs.

- **Monitor:** A 17-inch monitor should be adequate. As with television sets, the "inch" dimension on monitors is measured diagonally corner to corner. **The _monitor_ is the display device that takes the electrical signals from the video card and forms an image using points of colored light on the screen.** Later, after you've closed up the system cabinet (you haven't done so yet), you will connect the monitor by means of a cable to the back of the computer, using the clearly marked connector. You will also plug the power cord for the monitor into a wall plug.

Monitor

Speakers

- **Pair of speakers:** **_Speakers_—the devices that play sounds transmitted as electrical signals from the sound card**—may not be very sophisticated, but unless you're into high-fidelity recordings they're probably good enough. The two speakers are connected to a single wire that you plug into the back of the computer once you're finished with the installation.

- **Printer:** Especially for student work, you certainly need a **_printer_, an output device that produces text and graphics on paper.** There are various types of printers, as we discuss later (Chapter 3). The printer has two connections. One, which relays signals from the computer, goes to the back of the PC, where it connects with the motherboard. The other is a power cord that goes to a wall plug.

Printer

Communications Hardware: Modem

Computers can be stand-alone machines, unconnected to anything else. If all you're doing is word processing to write term papers, that may be fine. As we have seen, however, the communications component of the computer system vastly extends the range of a PC. Thus, while your system cabinet is still open there is one more piece of hardware to install.

Modem

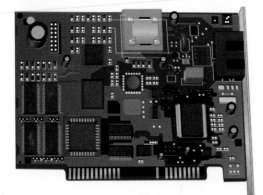

- **Modem:** **A _modem_ is a device that sends and receives data over telephone lines to and from computers.** The modem is mounted on an expansion card, which you fit into an expansion slot on the motherboard. Once you've finished putting the computer together, you can then run a telephone line from the telephone wall plug to the back of the PC, where it will connect to the modem.

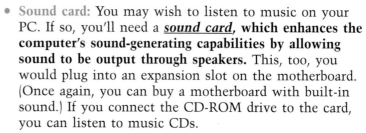

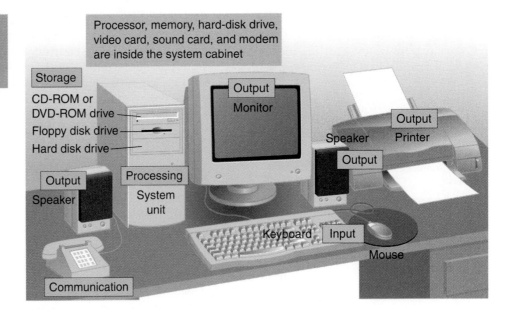

Processor, memory, hard-disk drive, video card, sound card, and modem are inside the system cabinet

Storage

CD-ROM or DVD-ROM drive

Floppy disk drive

Hard disk drive

Output
Monitor

Output

Speaker

Output
Printer

Output
Speaker

Processing
System unit

Keyboard

Input
Mouse

Communication

Now you can close up the system cabinet, plug in all the input and output devices, and turn on the power "on" button. Your PC system will look as shown in the illustration above. *(See ● Panel 1.3.)* Are you now ready to roll? Not quite.

Software

You have put all the pieces together, but now you need to check the motherboard manual for instructions on starting the system. One of the most important tasks is to install software, of which there are two types—system software and application software.

- **System software:** You're really interested in the type of software that allows you to write documents, do spreadsheets, and so on, which is called application software. Before you can get to that, however, you need to install the system software, such as Windows 98 or Windows Me, which most likely comes on CD-ROM disks. Insert these into your CD-ROM drive and follow the onscreen directions for installation.

  System software helps the computer perform essential operating tasks and enables the application software to run. System software consists of several programs. The most important is the *operating system,* the master control program that runs the computer. Examples of operating system software for the PC are various Microsoft programs (such as Windows 95, 98, and Me and NT/2000), Unix, and Linux. The Apple Macintosh computer is another matter altogether. As we explain elsewhere, it has its own hardware components and own software, which aren't directly generally transferable to the PC.

  After you install the system software, you will need to install setup software for the hard drive, the video and sound cards, and the modem. These setup programs (called *device drivers,* discussed in Chapter 3) will probably come on CD-ROMs (or maybe floppy disks). Once again, you insert them into the drive and then follow the instructions that appear on the monitor screen.

- **Application software:** Now you're finally getting somewhere! After you've installed the application software, you'll actually be able to start using your PC. **_Application software_ enables you to perform specific tasks—solve problems, perform work, or entertain yourself.** For example, when you prepare a term paper on your computer, you

Application software: Microsoft Office 2000

The program comes on a CD-ROM.

will use a word processing program (Microsoft Word and Corel Word-Perfect are two specific brands). Application software is specific to the system software you use. If you want to run Microsoft Word, for instance, you'll need to first have Microsoft Windows on your system, not Unix or Linux.

Application software comes on CD-ROMs or floppy disks. You insert them into your computer, and then follow the instructions on the screen for installation. Later on you may obtain entire application programs by getting them off the Internet, using your modem.

Although we have said a lot less about software than about hardware, they are equally important. We discuss software in much more detail in Chapter 3.

Is Building Your Own PC Worth the Effort?

If you add up the costs of all the components (not to mention the value of your time), and then start checking ads for PCs, you might wonder why anyone would bother going to the trouble of building one. And nowadays you would probably be right. "[I]f you think you'd save money by putting together a computer from scratch," says David Einstein, "think again. You'd be lucky to match the price PC-makers are charging these days in their zeal to undercut the competition."[24]

But if you had done this for real, it would not be a futile exercise: By knowing how to build a system yourself, you'd not only be able to impress your friends, but you'd also know how to upgrade any store-bought system to include components that are better than standard. For instance, as Einstein points out, if you're into video games, knowing how to construct your own PC would enable you to make a system that's right for you. You could include the latest three-dimensional graphics video card and a state-of-the-art sound card, for example. You wouldn't have to settle for the less-than-stellar technology supplied with some PCs. You'd also know how to order a custom-built system (as from Dell, Gateway, or Micron, the mail-order computer makers) that's right for you.

In Chapter 4, we'll expand on this discussion so that you can really know what you're doing when you go shopping for a PC system. However, if you're interested in tips on buying a system now, you can go to the Web site at *www.mhhe.com/cit/uit4e* for advice.

Before we end this introductory overview of information technology, let's wrap up the chapter with a look at the future.

CONCEPT CHECK

What are the three key concepts to understand about how computers work?

Name principal types of input, processing/memory, storage, output, and communications hardware.

Describe the two types of software.

1.4 Where Is Information Technology Headed?

KEY QUESTIONS

What are three directions of computer development and three directions of communications development?

Considering going into hotel work? Then you'll want to know that the hottest new job there is "computer concierge," or *compcierge* (pay: $40,000-plus a year), someone with a knowledge of computer systems who helps the 55% of guests who travel with laptop computers when they have PC or online problems.

Or what about accounting? There the newest specialty is *e-commerce accountant* (pay: up to $54,250), who offers advice on when it makes financial sense to sell goods online, how to find Web developers, and how to manage credit card and sales records.

Today, a knowledge of computers coupled with training in another field offers interesting career paths. Other examples include *bioinformaticist* (biology and computer background, $100,000-plus salary), who studies gene maps, and *virtual set designer* (training in architecture and 3-D computer modeling; pay: up to $150,000), who designs sets for TV shows by combining décors built on computers with physical spaces.[25]

Clearly, information technology is changing old jobs and inventing new ones. To prosper in this environment, you will need to combine a traditional education with training in computers and communications. And you will need to understand what the principal trends of the Information Age are. Let's consider these trends in the development of computers and communications and, most excitingly, the area where they intersect.

Three Directions of Computer Development: Miniaturization, Speed, & Affordability

One of the first computers, the outcome of military-related research, was delivered to the U.S. Army in 1946. ENIAC (short for Electronic Numerical Integrator And Calculator) weighed 30 tons and was 80 feet long and two stories high, but it could multiply a pair of numbers in the then-remarkable time of three-thousandths of a second. This was the first general-purpose, programmable electronic computer, the grandparent of today's lightweight handheld machines.

Since the days of ENIAC, computers have developed in three directions—and are continuing to do so.

- **Miniaturization:** Everything has become smaller. ENIAC's old-fashioned radio-style vacuum tubes gave way to the smaller, faster, more reliable transistor. A *transistor* is a small device used as a gateway to transfer electrical signals along predetermined paths (circuits).

    The next step was the development of tiny integrated circuits. *Integrated circuits* are entire collections of electrical circuits or

Grandpa
The ENIAC is the grandfather of today's lightweight machines.

Microprocessor

pathways that are now etched on tiny squares of silicon half the size of your thumbnail. *Silicon* is a natural element found in sand. In pure form, it is the base material for computer processing devices.

The miniaturized processor, or microprocessor, in a personal desktop computer today can perform calculations that once required a computer filling an entire room.

- Speed: Thanks to miniaturization, computer makers can cram more hardware components into their machines, providing faster processing speeds and more data storage capacity.

- Affordability: Processor costs today are only a fraction of what they were 15 years ago. A state-of-the-art processor costing less than $1000 provides the same processing power as a huge 1980s computer costing more than $1 million.

These are the three major trends in computers. What about communications?

Miniaturization
Users try out pen-based computers at a convention.

Three Directions of Communications Development: Connectivity, Interactivity, & Multimedia

Once upon a time, we had the voice telephone system—a one-to-one medium. You could talk to your Uncle Joe and he could talk to you, and with special arrangements (conference calls) more than two people could talk with one another. We also had radio and television systems—one-to-many media (or mass media). News announcers such as Dan Rather, Tom Brokaw, or Peter Jennings could talk to you on a single medium such as television, but you couldn't talk to them.

There have been three recent developments in communications:

- Connectivity: **_Connectivity_ is the ability to connect computers to one another by communications lines, so as to provide online information access.** The connectivity resulting from the expansion of computer networks has made possible e-mail and online shopping, for example.

- Interactivity: **_Interactivity_ is about two-way communication; a user can respond to information he or she receives and modify the process.** That is, there is an exchange or dialogue between the user and the computer or communications device. The ability to interact means users can be active rather than passive participants in the technological process. On the

Auto PC

The Clarion Auto PC is the first personal computer designed for a car.

television networks MSNBC or CNN, for example, you can immediately go on the Internet and respond to news from broadcast anchors. In the future, cars may respond to voice commands or feature computers built into the dashboard.

- **Multimedia:** Radio is a single-dimensional medium (sound), as is most e-mail (mainly text). As we mentioned earlier in this chapter, *multimedia* refers to technology that presents information in more than one medium—such as text, pictures, video, sound, and animation—in a single integrated communication. The development of the World Wide Web expanded the Internet to include pictures, sound, music, and so on, as well as text.

Exciting as these developments are, truly mind-boggling possibilities emerge as computers and communications cross-pollinate.

When Computers & Communications Combine: Convergence, Portability, Personalization— & Information Overload

Sometime in the 1990s, computers and communications started to fuse together, beginning a new era within the Digital Age. The result was three further developments, which haven't ended yet.

- **Convergence:** <u>*Convergence*</u> **describes the combining of several industries through various devices that exchange data in the format used by computers.** The industries are computers, communications, consumer electronics, entertainment, and mass media. Convergence has led to electronic products that perform multiple functions, such as TVs with Internet access or phones with screens displaying text and pictures.

Portability

This wearable computer, from Japan, has a mini-display screen in front.

- **Portability:** In the 1980s, portability, or mobility, meant trading off computing power and convenience in return for smaller size and weight. Today, however, we are close to the point where we don't have to give up anything. As a result, experts have predicted that small, powerful, wireless personal electronic devices will transform our lives far more than the personal computer has done so far. "[T]he new generation of machines will be truly personal computers, designed for our mobile lives," wrote one journalist back in 1992. "We will read office memos between strokes on the golf course and answer messages from our children in the middle of business meetings."[26] Today such activities are commonplace. The risk they bring is that, unless we're careful, work will invade our leisure time.

- **Personalization:** Personalization is the creation of information tailored to your preferences—for instance, programs that will automatically cull recent news and information from the Internet on just those topics you have designated. Companies involved in e-commerce can send you messages about forthcoming products based on your pattern of purchases, usage, and other criteria. Or they will build products (cars, computers, clothing) customized to your heart's desire.

We might also note a fourth, unintended, development:

- **Information overload:** At one time, most business transactions involved creating, using, sending, and storing paper. The arrival of the computer generated talk of the "paperless office," but there seems to be more paper than ever. Having paper records means having more information, which, however, does not mean that we have more knowledge. To avoid being buried in an avalanche of unnecessary data, we must learn to distinguish what we *really* need from what we *think* we need.

"E" Also Stands for Ethics

Ethics

Every computer user will have to wrestle with ethical issues related to the use of information technology. ***Ethics* is defined as a set of moral values or principles that govern the conduct of an individual or a group.** Because ethical questions arise so often in connection with information technology, we will note them, wherever they appear in this book, with the symbol shown in the margin.

Here, for example, are some important ethical concerns pointed out by Tom Forester and Perry Morrison in their book *Computer Ethics.*[27] These considerations are only a few of many; we'll discuss others throughout the book.

- **Speed and scale:** Great amounts of information can be stored, retrieved, and transmitted at a speed and on a scale not possible before. Despite the benefits, this has serious implications "for data security and personal privacy," as well as employment, they say, because information technology can never be considered totally secure against unauthorized access.
- **Unpredictability:** Computers and communications are pervasive, touching nearly every aspect of our lives. However, at this point, compared to other pervasive technologies—such as electricity, television, and automobiles—information technology seems a lot less predictable and reliable.
- **Complexity:** Computer systems are often incredibly complex—some so complex that they are not always understood even by their creators. "This," say Forester and Morrison, "often makes them completely unmanageable," producing massive foul-ups or spectacularly out-of-control costs.

CONCEPT CHECK

Describe the three directions in which computers have developed.

Describe the three recent directions in which communications have developed.

Discuss four effects of the fusion of computers and communications.

Name three ethical considerations that result from information technology.

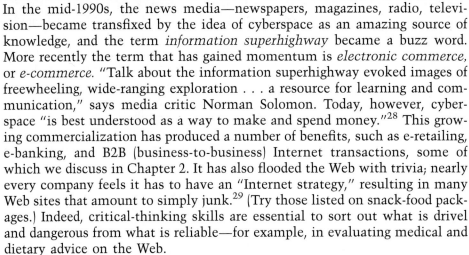

In the mid-1990s, the news media—newspapers, magazines, radio, television—became transfixed by the idea of cyberspace as an amazing source of knowledge, and the term *information superhighway* became a buzz word. More recently the term that has gained momentum is *electronic commerce,* or *e-commerce.* "Talk about the information superhighway evoked images of freewheeling, wide-ranging exploration . . . a resource for learning and communication," says media critic Norman Solomon. Today, however, cyberspace "is best understood as a way to make and spend money."[28] This growing commercialization has produced a number of benefits, such as e-retailing, e-banking, and B2B (business-to-business) Internet transactions, some of which we discuss in Chapter 2. It has also flooded the Web with trivia; nearly every company feels it has to have an "Internet strategy," resulting in many Web sites that amount to simply junk.[29] (Try those listed on snack-food packages.) Indeed, critical-thinking skills are essential to sort out what is drivel and dangerous from what is reliable—for example, in evaluating medical and dietary advice on the Web.

But the universe of information technology is much greater than all this. Earlier philosophers, such as Teilhard de Chardin (1881–1955) and H. G. Wells (1866–1946), foresaw the emergence of what might be called a "global brain," or planetary intelligence, as human knowledge and wisdom expanded exponentially. Today this explosion of information—advanced by supercomputers, fiber-optic cables, satellites, and the trends we have described—might be called the *Technosphere.*

Despite its possible downside (overconsumption, resource depletion, privacy loss, unequal information access), suggests one futurist, the Technosphere could also offer important benefits in the 21st Century: unlimited information anywhere anytime for low cost; high-speed "tele-health" and "tele-education" services delivered to remote areas; smart-energy and smart-technology transportation, houses, and offices; and "telecommuting" workers engaged in "electronic migration" to online jobs all over the planet.[30]

Let us begin to look at the mind tools for our future.

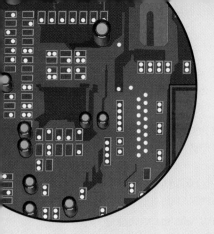

Visual Summary

Note to the reader: "KQ" refers to Key Questions on the first page of each chapter. The number ties the summary term to the corresponding section in the book.

application software (p. 18, KQ 1.3) Software that has been developed to solve a particular problem, perform useful work on general-purpose tasks, or provide entertainment. Why it's important: *Application software such as word processing, spreadsheet, database manager, graphics, and communications packages are commonly used tools for increasing people's productivity.*

case (p. 14, KQ 1.3) Also known as the system unit or system cabinet; the box that houses the processor chip (CPU), the memory chips, and the motherboard with power supply, as well as storage devices—floppy-disk drive, hard-disk drive, and CD-ROM or DVD drive. Why it's important: *The case protects many important processing and storage components.*

CD-ROM drive (p. 16, KQ 1.3) Storage device that uses laser technology to read data from optical disks. Why it's important: *New software is generally supplied on CD-ROM disks rather than diskettes. And even if you can get a program on floppies, you'll find it easier to install a new program from one CD-ROM disk rather than repeatedly inserting and removing many diskettes.*

central processing unit (CPU) (p. 12, KQ 1.3) Device consisting of electronic circuitry that executes instructions to process data. Why it's important: *The CPU is the "brain" of the computer.*

chip (p. 14, KQ 1.3) Tiny piece of silicon that contains millions of miniature electronic circuits used to process data. Why it's important: *Chips have made possible the development of small computers.*

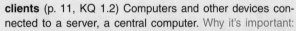

clients (p. 11, KQ 1.2) Computers and other devices connected to a server, a central computer. Why it's important: *Client/server networks are used in many organizations for sharing databases, devices, and programs.*

communications technology (p. 4, KQ 1.1) Also called *telcommunications technology;* consists of electromagnetic devices and systems for communicating over long distances. Why it's important: *Communications systems using electronic connections have helped to expand human communication beyond face-to-face meetings.*

computer (p. 4, KQ 1.1) Programmable, multiuse machine that accepts data—raw facts and figures—and processes (manipulates) it into useful information, such as summaries and totals. Why it's important: *Computers greatly speed up problem solving and other tasks, increasing users' productivity.*

connectivity (p. 21, KQ 1.4) Ability to connect computers to one another by communications lines, so as to provide online information access. Why it's important: *Connectivity is the foundation of the advances in the digital age. It provides online access to countless types of information and services. The connectivity resulting from the expansion of computer networks has made possible e-mail and online shopping, for example.*

convergence (p. 22, KQ 1.4) Term that describes the combining of several industries through various devices that exchange data in the format used by computers. The industries are computers, communications, consumer electronics, entertainment, and mass media. Why it's important: *Convergence has led to electronic products that perform multiple functions, such as TVs with Internet access or phones with screens displaying text and pictures.*

cyberspace (p. 6, KQ 1.1) Term used to refer to the online world and the Internet in particular but also the whole wired world of communications in general. Why it's important: *More and more human activities take place in cyberspace.*

data (p. 11, KQ 1.3) Raw facts and figures that are processed into information. Why it's important: *Users need data to create useful information.*

desktop PC (p. 10, KQ 1.2) Microcomputer unit that sits on a desk, with the keyboard in front and the monitor often on top. Why it's important: *Desktop PCs and tower PCs are the most commonly used types of microcomputer.*

e-mail (electronic mail) (p. 5, KQ 1.1) Messages transmitted over a computer network, most often the Internet. Why it's important: *E-mail has become universal; one of the first things new computer users learn is how to send and receive e-mail.*

Ethics

ethics (p. 23, KQ 1.4) Set of moral values or principles that govern the conduct of an individual or a group. Why it's important: *Ethical questions arise often in connection with information technology.*

expansion slots (p. 14, KQ 1.3) Internal "plugs" used for expanding the PC's capabilities. Why it's important: *Expansion slots give you places to plug in additional circuit boards, such as those for video, sound, and communications (modem).*

floppy-disk drive (p. 16, KQ 1.3) Storage device that stores data on removable 3½-inch-diameter flexible diskettes encased in hard plastic. Why it's important: *Floppy-disk drives are included on almost all microcomputers and enable the portability of many types of files.*

hard-disk drive (p. 16, KQ 1.3) Storage device that stores billions of characters of data on a nonremovable disk platter inside the computer case. Why it's important: *Hard disks hold much more data than diskettes do. Nearly all microcomputers use hard disks as their principal secondary-storage medium.*

hardware (p. 11, KQ 1.3) Refers to all machinery and equipment in a computer system. Why it's important: *Hardware runs under the control of software and is useless without it. However, hardware contains the circuitry that allows processing.*

information (p. 11, KQ 1.3) Data that has been summarized or otherwise manipulated for use in decision making. Why it's important: *The whole purpose of a computer (and communications) system is to produce (and transmit) usable information.*

information technology (p. 4, KQ 1.1) Technology that merges computing with high-speed communications links carrying data, sound, and video. Why it's important: *Information technology is bringing about the fusion of several important industries dealing with computers, telephones, televisions, and various handheld devices.*

input (p. 12, KQ 1.3) Whatever is put in ("input") to a computer system. Input devices include the keyboard and the mouse. Why it's important: *Useful information cannot be produced without input data.*

interactivity (p. 21, KQ 1.4) Two-way communication; a user can respond to information he or she receives and modify the process. Why it's important: *Interactive devices allow the user to actively participate in a technological process instead of just reacting to it.*

Internet (p. 6, KQ 1.1) Worldwide network that connects up to 400,000 smaller networks that link computers at academic, scientific, and commercial institutions in more then 200 countries. Why it's important: *The Internet makes possible the sharing of all types of information and services for millions of people all around the world.*

keyboard (p. 13, KQ 1.3) Input device that converts letters, numbers, and other characters into electrical signals readable by the processor. Why it's important: *Keyboards are the most common kind of input device.*

laptop computer (p. 10, KQ 1.2) Also called *notebook computer*, lightweight portable computer with a built-in monitor, keyboard, hard-disk drive, battery, and adapter; weighs from 1.8 to 9 pounds. Why it's important: *Laptop and other small computers have provided users with computing capabilities in the field and on the road.*

local area network (LAN) (p. 10, KQ 1.2) Network that connects, usually by special cable, a group of desktop PCs and other devices, such as printers, in an office or building. Why it's important: *LANs have replaced mainframes for many functions and are considerably less expensive.*

mainframe (p. 9, KQ 1.2) Second-largest computer available, after the supercomputer; capable of great processing speeds and data storage, and costs $5000–$5 million. Small mainframes are often called *midsize computers.* Why it's important: *Mainframes are used by large organizations (banks, airlines, insurance companies, universities) that need to process millions of transactions.*

memory chip (p. 14, KQ 1.3) Also known as *RAM* (for *random access memory*) chip; represents *primary storage* or *temporary storage.* Why it's important: *It holds data prior to processing and information after processing, before it is sent along to an output or storage device.*

microcomputer (p. 10, KQ 1.2) Also called *personal computer;* small computer that fits on or next to a desktop, or can be carried around. Costs $500–$5000. Why it's important: *The microcomputer has lessened the reliance on mainframes and has enabled more ordinary users to use computers. It can be used as a stand-alone machine or connected to a network.*

microcontroller (p. 10, KQ 1.2) Also called an *embedded computer;* the smallest category of computer. Why it's important: *Microcontrollers are built into "smart" electronic devices, such as appliances and automobiles.*

modem (p. 17, KQ 1.3) Device that sends and receives data over telephone lines to and from computers. Why it's important: *A modem enables users to transmit data from one computer to another by using standard telephone lines instead of special communications equipment.*

monitor (p. 17, KQ 1.3) Display device that takes the electrical signals from the video card and forms an image using points of colored light on the screen. Why it's important: *Monitors enable users to view output without printing it out.*

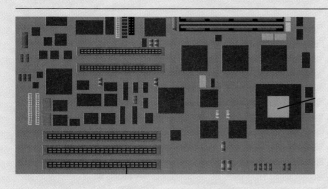

motherboard (p. 14, KQ 1.3) Main circuit board in the computer. Why it's important: *This is the big green circuit board to which everything else—such as the keyboard, mouse, and printer—attaches through connections (called ports) in the back of the computer. The processor chip and memory chips are also installed on the motherboard.*

mouse (p. 13, KQ 1.3) Input device used to manipulate objects viewed on the computer display screen. Why it's important: *For many purposes, a mouse is easier to use than a keyboard for inputting commands. Also, the mouse is used extensively in many graphics programs.*

multimedia (p. 7, KQ 1.1) From "multiple media"; refers to technology that presents information in more than one medium, including text graphics, animation, video, and sound. Why it's important: *Use of multimedia is becoming more common in business, the professions, and education as a means of improving the way information is communicated.*

network (p. 5, KQ 1.1) Communications system connecting two or more computers. Why it's important: *Networks allow users to share applications and data and to use e-mail. The Internet is the largest network.*

online (p. 4, KQ 1.1) Using a computer or other information device, connected through a voice or data network to access information and services from another computer or information device. Why it's important: *Online communications have become common and important to many businesses, services, and personal and educational settings.*

output (p. 12, KQ 1.3) Whatever is output from ("put out of") the computer system; the results of processing. Why it's important: *People use output to help them make decisions. Without output devices, computer users would not be able to view or use the results of processing.*

peripheral device (p. 16, KQ 1.3) Any component or piece of equipment that expands a computer's input, storage, and output capabilities. Examples include printers and disk drives. Why it's important: *Most of a computer system's input and output functions are performed by peripheral devices.*

personal digital assistant (PDA) (p. 10, KQ 1.2) Also known as *handheld computer* or *palmtop;* used as a schedule planner and address book and to prepare to-do lists and send e-mail and faxes. Why it's important: *PDAs make it easier for people to do business and communicate while traveling.*

primary storage (p. 12, KQ 1.3) Also called *memory;* the computer circuitry that temporarily holds data waiting to be processed. Why it's important: *By holding data, primary storage enables the processor to process.*

printer (p. 17, KQ 1.3) Output device that produces text and graphics on paper. Why it's important: *Printers provide one of the principal forms of computer output.*

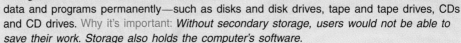

processing (p. 12, KQ 1.3) The manipulation the computer does to transform data into information. Why it's important: *Processing is the essence of the computer, and the processor is the computer's "brain."*

secondary storage (p. 12, KQ 1.3) Also called *storage;* refers to devices and media that store data and programs permanently—such as disks and disk drives, tape and tape drives, CDs and CD drives. Why it's important: *Without secondary storage, users would not be able to save their work. Storage also holds the computer's software.*

server (p. 11, KQ 1.2) Computer in a network that holds collections of data (databases) and programs for connecting PCs, workstations, and other devices, which are called *clients.* Why it's important: *Servers enable many users to share equipment, programs, and data.*

software (p. 12, KQ 1.3) Also called *programs;* step-by-step instructions that tell the computer hardware how to perform a task. Why it's important: *Without software, hardware is useless.*

sound card (p. 17, KQ 1.3) Special circuit board that enhances the computer's sound-generating capabilities by allowing sound to be output through speakers. Why it's important: *Sound is used in multimedia applications. Also, many users like to listen to music CDs on their computers.*

speakers (p. 17, KQ 1.3) Devices that play sounds transmitted as electrical signals from the sound card. Speakers are connected to a single wire that you plug into the back of the computer. Why it's important: *See* sound card.

supercomputer (p. 9, KQ 1.2) High-capacity computer with hundreds of thousands of processors that is the fastest calculating device ever invented. Costs $500,000–$35 million. Why it's important: *Supercomputers are used primarily for research purposes, airplane design, oil exploration, weather forecasting, and other activities that cannot be handled by mainframes and other less powerful machines.*

system software (p. 18, KQ 1.3) System software helps the computer perform essential operating tasks. Why it's important: *Application software cannot run without system software. System software consists of several programs. The most important is the operating system, the master control program that runs the computer. Examples of operating system software for the PC are various Microsoft programs (such as Windows 95, 98, and ME and NT/2000), Unix, Linux, and the Macintosh operating system.*

terminal (p. 9, KQ 1.2) Input and output device that uses a keyboard for input and a monitor for output; it cannot process data. Why it's important: *Terminals are generally used to input data to and receive data from a mainframe computer system.*

tower PC (p. 10, KQ 1.2) Microcomputer unit that sits as a "tower" often on the floor, freeing up desk space. Why it's important: *Tower PCs and desktop PCs are the most commonly used types of microcomputer.*

video card (p. 16, KQ 1.3) Circuit board that converts the processor's output information into a video signal that can be sent through a cable to the monitor. Why it's important: *Virtually all computer users need to be able to view video output on the monitor.*

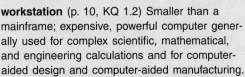

workstation (p. 10, KQ 1.2) Smaller than a mainframe; expensive, powerful computer generally used for complex scientific, mathematical, and engineering calculations and for computer-aided design and computer-aided manufacturing. Why it's important: *The power of workstations is needed for specialized applications too large and complex to be handed by PCs.*

World Wide Web (p. 7, KQ 1.1) The part of the Internet that stores information in multimedia form—sounds, photos, and video as well as text. Why it's important: *The Web is the most widely known part of the Internet.*

ESS
Maestro-1
CE

Chapter Review

Self-Test Questions

1. The _____ refers to the part of the Internet that stores information in multimedia form.

2. _____ and _____ refer to the two types of microcomputer. One sits on the desktop and the other usually is placed on the floor.

3. _____ technology merges computing with high-speed communications lines carrying data, sound, and video.

4. A _Computer_ is a programmable, multiuse machine that accepts data and processes it into information.

5. Messages transmitted over a computer network are called _____.

6. The _____ is a worldwide network that connects hundreds of thousands of smaller networks in more than 200 countries.

7. _____ refers to technology that presents information in more than one medium.

8. _____ are high-capacity machines with hundreds of thousands of processors.

9. Embedded computers, or _____, are installed in "smart" appliances and automobiles.

10. The kind of software that enables users to perform specific tasks is called _____ software.

Multiple-Choice Questions

1. Which of the following converts computer output into displayed images?
   a. printer
   b. monitor
   c. floppy-disk drive
   d. processor
   e. hard-disk drive

2. Which of the following computer types is the smallest?
   a. mainframe
   b. microcomputer
   c. microcontroller
   d. supercomputer
   e. workstation

3. Which of the following is a secondary storage device?
   a. processor
   b. memory chip
   c. floppy-disk drive
   d. printer
   e. monitor

4. Since the days when computers were first made available, computers have developed in what three directions?
   a. increased expense
   b. miniaturization
   c. increased size
   d. affordability
   e. increased speed

5. Which of the following operations constitute the four basic operations followed by all computers?
   a. input
   b. storage
   c. programming
   d. output
   e. processing

True/False Questions

T F   1. Main memory is a software component.

T F   2. The operating system is part of the system software.

T F   3. Processing is the manipulation a computer does to transform data into information.

T F   4. Primary storage is the area in the computer where data or information is held permanently.

Short-Answer Questions

1. What does _online_ mean?
2. What is the difference between system software and application software?
3. Briefly define _cyberspace_.
4. What is the difference between software and hardware?
5. Briefly describe what a local area network is.

Knowledge in Action

1. Determine what types of computers are being used where you work or go to school. In which departments are the different types of computers used? What are they used for? How are they connected to other computers?

2. Imagine a business you could start or run at home. What type of business is it? What type of computer(s) do you think you'll need? Describe the computer system in as much detail as possible, including hardware components in the areas we have discussed so far. Keep your notes and then refine your answers after you have completed the course.

3. What ethical concerns do you have now about how computers and communications systems are being used? List them, and then re-examine them at the end of the course. How have your concerns changed?

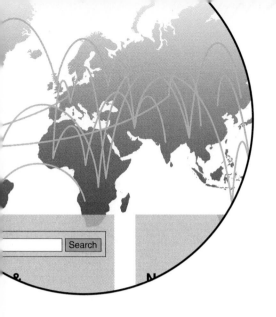

The Internet & the World Wide Web

Exploring Cyberspace

Key Questions

You should be able to answer the following questions.

2.1   **Choosing Your Internet Access Device & Physical Connection: The Quest for Broadband** What are the means of connecting to the Internet, and how fast are they?

2.2   **Choosing Your Internet Service Provider (ISP)** What are the three types of Internet service provider, and what kinds of services do they provide?

2.3   **Sending & Receiving E-Mail** How can one obtain e-mail software, what are the components of an e-mail address, and what are netiquette and spam?

2.4   **The World Wide Web** What are Web sites, Web pages, browsers, URLs, and search engines?

2.5   **The Online Gold Mine: More Internet Resources, Your Personal Cyberspace, E-Commerce, & the E-conomy** What are FTP, Telnet, newsgroups, real-time chat, and e-commerce?

A Visual Overview of this Chapter

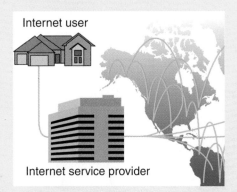

Internet user

Internet service provider

1 **Choosing Your Internet Access Device & Physical Connection.** Some Internet **physical connections,** either wired or wireless, have more **bandwidth**— are able to transmit more data—than others. Data transmission is expressed in **bps** (bits per second), **Kbps** (kilobits—thousands of bits per second), **Mbps** (megabits—millions), and **Gbps** (gigabits—billions). Data is **downloaded** from a remote computer to a local computer or **uploaded,** the reverse.

There are four principal types of Internet physical connections: (1) Telephone (dial-up) modem connection is low-speed but inexpensive (up to 56 Kbps). (2) High-speed phone connections are **ISDN** (up to 128 Kbps), which transmits over traditional phone lines; **DSL** (up to 8.4 Mbps), also using traditional phone lines; and **T1** (1.5 Kbps), a special trunk line. (3) **Cable modems** (10 Mbps) connect to cable TV systems. (4) Wireless systems include microwave systems, such as **communications satellites** or space stations (up to 56 Kbps).

2 **Choosing Your Internet Service Provider (ISP).** With a physical connection installed, you then need an **Internet service provider (ISP),** a company to help you connect or **log on** to the Internet. Three types of ISPs are free-service providers, which make money presenting ads; basic-service providers, or small, local companies; and full-service providers, which offer more technical support and other services—examples are AOL and MSN—and which offer **menus,** or lists of commands or services.

3 **Sending & Receiving E-Mail.** Four alternatives for getting and sending e-mail are to buy e-mail software, get the software as part of a browser or other software, get it from your ISP, and get it free (for example, from CNN.com or NetZero). People will send e-mail to you at your **domain,** a location on the Internet consisting of your user name and domain name, such as *user@domain.*

E-mail allows users to send attachments, or separate long documents, with their e-mail messages. It also allows **instant messaging (IM),** which advises other users the moment a message arrives by displaying it in a **window,** a rectangular area on screen. You can exchange e-mail from people worldwide with similar interests through **listserves,** or e-mail mailing lists.

The two basic rules of online behavior, or **netiquette,** are don't waste people's time and don't say anything online you wouldn't say to someone's face. In particular, you should always first consult **FAQs,** or Frequently Asked Questions; avoid **flaming,** such as insults or obscenities; and smooth communication using **emoticons,** or friendly graphic symbols.

To manage your e-mail, filters or instant organizers are recommended. In addition, one needs to know how to manage **spam,** or unsolicited e-mail. Finally, assume e-mail messages are not private: anyone could read them.

9 ... your browser connects you to a Web page located on a server in the United Kingdom (signified by "uk" in the URL), *www.ymca.org.uk/gallery/barbican*

4 **The World Wide Web.** What makes the Web so graphically inviting is that it is in multimedia form—graphics, video, and audio as well as text. What makes it easily navigable is that it uses **hypertext,** a system based on **hypertext markup language (HTML)** that uses "tags" or special instructions to provide links among words and phrases in many documents at many Internet sites.

A computer with a domain name (.com, .org, and the like) is called a **site,** and a **Web site** is the location of a Web domain name in a computer somewhere on the Internet. A **Web page** is a document (with text, pictures, sound) on the Web; the first page on the Web site is the **home page.** A **Web browser,** software for viewing and connecting to Web pages, is used to connect with the Web site's address, or **Universal Resource Locator (URL).** A URL, such as *http://www.nps.gov/yose/camping.htm,* consists of (1) the **protocol,** or communication rules—in particular, **HyperText Transfer Protocol (HTTP),** the protocol for connecting with Web servers; (2) the Web server name; (3) the directory; and (4) the file (perhaps with an extension, such as *htm*).

To get around the Web with a browser, you start out from the home page (which you can personalize or customize), then use directional features (Back, Forward, Home, Search), history lists (to keep track of where you've been), and bookmarks (to mark favorite URLs). To interact with a Web page, you use your mouse to click on hyperlinks, click on **radio buttons** (circles in front of options), and enter content in fill-in boxes. You can also click on **scroll arrows** to do **scrolling**—move up and down the Web page.

A starting point for obtaining information is a **Web portal,** a site (such as AOL or Yahoo!) that provides popular features such as search tools. You can check the portal's home page; use a **directory** or category of topics; or use a **keyword,** or subject word, to search for a topic. You can also use a **search engine** to find specific documents through keyword searches and menu choices. Search engines may be human-organized, computer-created, hybrid, or metacrawlers. Among the search strategies are the use of quotation marks around search terms and the use of operators (AND, OR, NOT, +, −).

Multimedia on the Web may require a **plug-in** (or **player** or **viewer**), a program on the browser that allows certain files to be played or viewed. **Helper** applications run multimedia elements separate from the browser. Web-site developers use **applets,** or small multimedia programs, written in **Java,** a programming language for creating animated, interactive Web pages. **Animation** is rapid sequencing of still images. **Streaming video** transfers data in a continuous flow. **Streaming audio** lets you listen to a file as it is being downloaded. **Push technology,** such as **webcasting,** automatically downloads data to your computer—customized text, video, and audio. **Internet telephony** allows you to make phone calls on the Net.

5 **The Online Gold Mine.** Four other Internet resources are (1) **FTP,** a method for copying files; (2) **Telnet,** a means for connecting to remote computers; (2) **newsgroups,** electronic bulletin boards that take place on a special network called **Usenet,** which requires a **newsreader** (part of most browsers) to access; and (3) **real-time chat (RTC),** typed online discussions, which require a **chat client** (also part of most browsers) to initiate.

The Internet offers personal resources—the ability to do online matchmaking, acquire an online education through **distance learning,** get health information, and amuse yourself. It also offers **e-commerce,** or online business activities, such as retail commerce online ("e-tailing"); online auctions; online finance; online job hunting; and **B2B commerce,** for business-to-business commerce, or the exchange of goods and services directly between companies.

t's the biggest technological generation gap in history."[1]

At least that's the opinion of an executive whose online movie company, Sightsound.com, markets almost exclusively to college students. He's talking about the technological gap—it's not strictly a generation gap—between on-campus students provided with high-speed Internet access by their colleges and most off-campus people, whose slow online connections squeeze through plain old copper-wire phone lines.

The difference can mean receiving information for a report in 3 seconds versus twiddling your thumbs for an entire minute. But academic pursuits account for only some of the activity on campus networks. A lot of the traffic consists of data containing instant messages, music files, toll-free phone calls, online games, e-commerce orders, movies, and much more. Students are passionate users of ICQ, shorthand for a service called "I Seek You," which allows people to exchange messages across the hall (or the world) as fast as they can type. As a result, it's not uncommon to see students emerge simultaneously from their separate dormitory rooms and head toward the dining hall together, having just messaged each other that it's time for dinner.[2] Thanks to this on-campus wiring, a high proportion of Internet users are college students.

Call it the Bandwidth Gap. In general terms, **_bandwidth_ is an expression of how much data—text, voice, video, and so on—can be sent through a communications channel in a given amount of time.** A college dormitory wired with coaxial or fiber-optic cable will have more bandwidth than will a house out in the country served by conventional copper-wire telephone lines; access to information will be hundreds of times faster in the dorm. Of course, many students are not privileged to have this kind of *broadband* or high-speed access. But so significant is the difference that students apparently are making decisions about college housing—even choice of college—based on the availability of high bandwidth. And those who are about to graduate wonder how they will ever survive in a narrow-bandwidth world.

Because of its standard interfaces and low rates, "the Internet has been the great leveler for communications—the way the PC was for computing," says Boston analyst Virginia Brooks.[3] Let's consider the Internet, a worldwide network connecting up to 400,000 smaller networks. *(See ● Panel 2.1.)* To gain access to the Internet, you need three things: (1) an *access device,* such as a personal computer with a modem; (2) a *physical connection,* such as a telephone line; and (3) an *Internet service provider (ISP).*

2.1 Choosing Your Internet Access Device & Physical Connection: The Quest for Broadband

KEY QUESTIONS

What are the means of connecting to the Internet, and how fast are they?

Can you watch movies or television on any Internet-connected PC? Perhaps you can if you're fortunate enough to be in a state-of-the-art college dorm with "24/7" (24-hours-a-day, seven-days-a-week) high-speed Internet access. But maybe you live off-campus, and while you may be able to connect to the campus network, it may mean a wait of several minutes. Or maybe you have no access to the campus network.

Let's assume you're in the last situation. What are your choices of a **_physical connection_—the wired or wireless means of connecting to the Internet?** A lot depends on where you live. As you might expect, urban and many suburban areas offer more broadband connections than rural areas do. Among the principal means of connection are: (1) telephone (dial-up) modem, (2) sev-

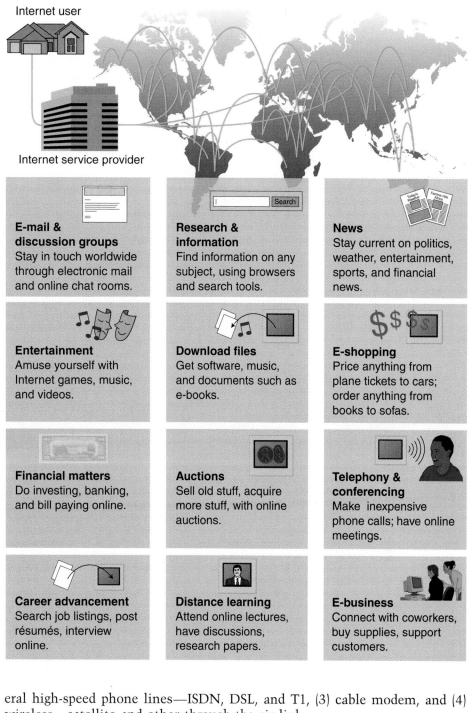

Internet user

Internet service provider

E-mail & discussion groups
Stay in touch worldwide through electronic mail and online chat rooms.

Research & information
Find information on any subject, using browsers and search tools.

News
Stay current on politics, weather, entertainment, sports, and financial news.

Entertainment
Amuse yourself with Internet games, music, and videos.

Download files
Get software, music, and documents such as e-books.

E-shopping
Price anything from plane tickets to cars; order anything from books to sofas.

Financial matters
Do investing, banking, and bill paying online.

Auctions
Sell old stuff, acquire more stuff, with online auctions.

Telephony & conferencing
Make inexpensive phone calls; have online meetings.

Career advancement
Search job listings, post résumés, interview online.

Distance learning
Attend online lectures, have discussions, research papers.

E-business
Connect with coworkers, buy supplies, support customers.

eral high-speed phone lines—ISDN, DSL, and T1, (3) cable modem, and (4) wireless—satellite and other through-the-air links.

Data is transmitted in characters, or collections of bits. Today's data-transmission speeds are measured in bits, kilobits, megabits, and gigabits per second.

- **bps:** A computer with an older modem might have a speed of 28,800 bps. The **_bps_ stands for bits per second.** (Eight bits equal one character.)
- **Kbps:** This is the most frequently used measure; **_kilobits per second,_ or _Kbps,_ are 1000 bits per second.** The speed of a modem that is 28,800 bps might be expressed as 28.8 Kbps.
- **Mbps:** Faster means of connection are measured in **_megabits per second,_ or _Mbps_—1 million bits per second.**
- **Gbps:** At the extreme are **_gigabits per second, Gbps_—1 billion bits per second.**

Why is it important to know these terms? Because the number of bits affects how fast you can upload and download information from a remote computer. <u>**Download**</u> **is the transmission of data from a remote computer to a local computer,** as, for example, from your college's mainframe to your own PC. <u>**Upload**</u> **is the transmission of data from a local computer to a remote computer.**

The table below shows the transmission rates for various connections, as well as the approximate costs (always subject to change, of course) and their pros and cons.[4] *(See ● Panel 2.2.)* Let's consider each option in turn.

Telephone (Dial-Up) Modem: Low Speed but Inexpensive & Widely Available

The telephone line that you use for voice calls is still the cheapest means of online connections and is available everywhere. As we discussed in Chapter 1, a *modem* is a device that sends and receives data over telephone lines to and from computers. This is known as a *dial-up* connection. These days, the modem is generally installed inside your computer, but there are also external modems. (We explain modems further in Chapter 5.) The modem is attached to the telephone wall outlet. *(See ● Panel 2.3, next page.)*

● PANEL 2.2
Methods of going online compared

Service	Cost per Month (Plus Installation, Equipment)	Maximum Speed (Download Only)	Pluses	Minuses
Telephone (dial-up) modem	$0–$30	56 Kbps	Inexpensive, available everywhere	Slow
ISDN	$40–$110	128 Kbps	Faster than dial-up, uses conventional phone lines	More expensive than dial-up
DSL	$30–$50	1.5–8.4 Mbps	Fast download, always on, higher security	Need to be close to phone company switching station, limited choice of service providers
T1	$1000	1.5 Mbps	24 separate circuits of 64 Kbps each, high speed both ways	Expensive, best for businesses
Cable modem	$30–$60	10 Mbps	Fast, always on, most popular broadband type	Slower service during peak times, vulnerability to hackers, limited choice of service providers
Satellite	$30–$130	400 Kbps	Wireless, fast, reliable	Slow (56 Kbps) uploads over phone lines
Radio waves	$159–$1400	155 Mbps	Wireless, fast, always on, reliable	Antenna must be line-of-sight with base station

● PANEL 2.3

The modem connection
You connect the modem inside your computer from a port (socket) in the back of your computer to a line that is then connected to a wall jack. Your telephone is also connected to your computer so that you can make voice calls.

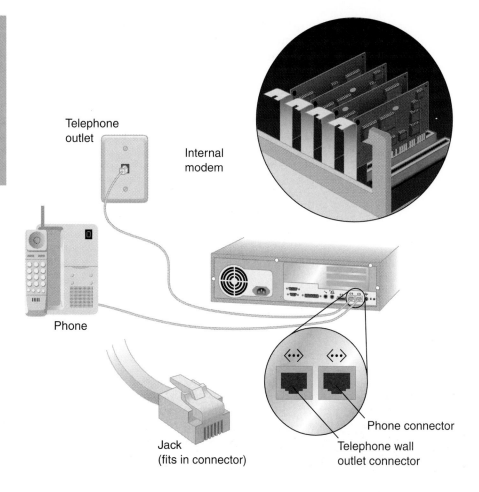

Telephone outlet

Internal modem

Phone

Jack (fits in connector)

Phone connector

Telephone wall outlet connector

Check online connections
You can check your online connection speed by going to the Windows taskbar and double-clicking the connection icon (bottom right of screen):

The result:

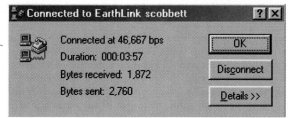

Most modems today have a maximum speed of 56 Kbps. That doesn't mean that you'll be sending and receiving data at that rate. The modem in your computer must negotiate with the modems used by your Internet service provider (ISP), the organization that actually connects you to the Internet, which may have modems operating at slower speeds, such as 28.8 Kbps. In addition, lower-quality phone lines or heavy traffic during peak hours—such as 5 p.m. to 11 p.m. in residential areas—can slow down your rate of transmission. (You can check your connection speed when you're online by using your mouse and clicking twice on the little symbol at the bottom right on your Windows screen—see illustration at left.)

One disadvantage of a telephone modem is that while you're online you can't use your phone to make voice calls. In addition, people who try to call you while you're using the modem will get a busy signal. (Call waiting will also interrupt an online connection, so you need to check with your phone company about how to disable it.) As we discuss in a few pages, you probably won't need to pay long-distance phone rates, since most ISPs offer local access numbers. The cost of a dial-up modem connection to the ISP is $10–$30 per month, plus a possible setup charge of $10–$25.

High-Speed Phone Lines: More Expensive but Available in Most Cities

Waiting while your computer's modem takes 25 minutes to transmit a 1-minute low-quality video from a Web site may have you pummeling the desk in frustration. To get some relief, you could enhance your POTS (for *plain old telephone system)* connection with a high-speed adaptation or get a new, dedicated line. Among the choices are ISDN, DSL, and T1, available in most major cities, though not in rural and many suburban areas.

- **ISDN line:** ___ISDN (Integrated Services Digital Network)___ **consists of hardware and software that allows voice, video, and data to be communicated over traditional copper-wire telephone lines.** Capable of transmitting up to 128 Kbps, ISDN is able to send signals over POTS lines. If you were trying to download an approximately 6-minute-long music video from the World Wide Web, it would take you about 4 hours and 45 minutes with a 28.8-Kbps modem. An ISDN connection would reduce this to an hour.

  ISDN costs $10–$40 a month to the phone company, plus perhaps another $30–$70 per month to the Internet service provider. In addition, you may need to pay your phone company $350–$700 or so to hook up an ISDN connector box, possibly run in a new phone line, and install the necessary software in your PC.

- **DSL line:** ___DSL (digital subscriber line)___ **also uses regular phone lines to transmit data in megabits per second.** (The most common type is asynchronous DSL, or ADSL, as we explain elsewhere.) Incoming data is significantly faster than outgoing data. That is, your computer can *receive* data at the rate of 1.5–8.4 Mbps, but it can *send* data at only 16–640 Kbps. This arrangement may be fine, however, if you're principally interested in obtaining very large amounts of data (video, music) rather than in sending them to others. With DSL, you can download that 6-minute music video in only 11 minutes (compared to an hour with ISDN). A big advantage of DSL is that it is always on and, unlike cable, its transmission rate is consistent. One-time installation cost is $100–$200 plus $100–$300 for a modem supplied by the phone company, and the monthly cost is $30–$50.

  There is one big drawback to DSL: You have to live within 3.3 miles of a phone company central switching office, because the access speed and reliability degrade with distance. However, phone companies are building thousands of remote switching facilities to enhance service throughout their regions. Another drawback is that you have to choose from a list of Internet service providers that are under contract to the phone company you use, although other DSL providers exist.

- **T1 line:** How important is high speed to you? Is it worth $1000 a month? Then consider getting a ___T1 line,___ **essentially a traditional trunk line that carries 24 normal telephone circuits and has a transmission rate of 1.5 Kbps.** Generally, T1 lines are used by corporate, government, and academic sites. (Another high-speed line, the T3 line, which transmits at 43 Mbps, costs $10,000 or more a month.)

Cable Modem: Close Competitor to DSL

If DSL's 11 minutes to move a 6-minute video sounds good, 2 minutes sounds even better. That's the rate of transmission for cable modems, which can transmit outgoing data at 500 Kbps and incoming data at 10 Mbps (and eventually, it's predicted, at 30 Mbps). **A** ___cable modem___ **connects a personal computer to a cable-TV system that offers an Internet connection.** With around 1 million subscribers, cable has been rapidly gaining market share in high-speed delivery services. Like a DSL connection, it is always on; unlike DSL, you don't need to live near a switching station. Costing $30–$60 a month plus installation and equipment ($300–$500), cable is available in most major cities.

A disadvantage, however, is that because you and your neighbors are sharing the system, during peak-load times, your service may be slowed to the speed of a regular dial-up modem. (You're also more vulnerable to attacks from hackers, although there are defensive or "firewall" programs, such as Winproxy, that reduce the risk.) Finally, cable companies may force you to use their own Internet service providers. For example, if you're with Comcast, you might have to use Excite@Home although you prefer the content offered by AOL, say.

Wireless Systems: Satellite & Other Through-the-Air Connections

Suppose you live out in the country and you're tired of the molasses-like speed of your cranky local phone system. You might consider taking to the air.

- **Satellite:** With a pizza-size satellite dish on your roof, you can receive data at the rate of 400 Kbps from a _**communications satellite**_, **a space station that transmits radio waves called microwaves from earth-based stations.** Unfortunately, your outgoing transmission will still be only 56 Kbps, because you'll have to use your phone line for that purpose, although genuine two-way satellite service is under development (as from DirectPC). Equipment available from InfoDish or PC Connection costs about $190; installation runs $100–$250; and monthly charges (including ISP charges) are $30–$130, depending on how much time you spend online.

- **Other wireless connections:** In urban areas, some businesses are using radio waves transmitted between towers that handle cellular phone calls, which can send data at up to 155 Mbps and are not only fast and dependable but also always on. The cost is $159–$1400 a month, depending on speed. However, because your antenna must be within line of sight of the service's base station, this system is not available outside cities. Perhaps more practical and affordable is wireless technology such as that to be offered by AT&T (though only in areas not served by its TCI cable service), which enables you to send and receive data (at 512 Kbps) from a box on the outside wall of your house; it does not require line-of-sight—that is, unimpeded—connection with the base station.

Universal Broadband Is Coming

Most PC users employ the physical connections we have described, but there are other possibilities. With WebTV, for instance, you use your television to access the Internet. If your wireless aims are modest—you want to access the Net from your back porch—Apple Computer's AirPort allows you to connect to other Macs and the Internet. The Palm VII, a personal digital assistant from 3Com, enables you to get on the Net from greater distances, but the small screen best accommodates specially prepared Web pages, of which there are few so far.[5]

As for PC users, nearly a third of the online households in the U.S. have DSL, cable, or wireless services. Most of them, however, are in and around major cities, and the companies providing connections have been slow to expand them to the rest of North America. But broadband is coming. When the telcos (telephone companies) finish scrambling to upgrade their phone lines to DSL, the cable companies make their "pipes" better handle two-way data, and the wireless companies refine their through-the-air links, ordinary Internet users will probably have connections _100 times faster_ than they are today.[6]

Suppose you have an access device such as a modem and you've signed up for a wired or wireless connection. Next, unless you're already on a college campus network, you'll need to arrange for an ***Internet service provider***, or ***ISP*, a company that connects you through your communications line to its server, or central computer, which connects you to the Internet.** Some well-known ISPs are America Online (AOL, which is also a portal or proprietary network, as we explain), EarthLink, Microsoft Network (MSN), AT&T World-Net, and Prodigy. There are also many local ISPs.

If you've decided simply to use the regular 56-Kbps dial-up modem in your PC connected to a plain old telephone line, you'll quickly notice the fierce competition between companies vying to become your ISP. For instance, perhaps your new PC comes with a keyboard button labeled "Internet," which, when pressed, begins the steps toward connecting you with a service provider—the provider that has come to a financial arrangement with the computer's manufacturer. Beside dealing with the blizzard of ads in magazines and on television, you may also receive promotional ISP start-up disks in the mail. In addition, your phone company probably offers an Internet service. (To do some comparison shopping, go online to *www.thelist.com*, which lists ISPs from all over the world and will guide you through the process of finding one that's best for you.)

The Three Types of ISP Services: Free, Basic, & Full

There are three types of ISPs, with varying levels of service: (1) *free service*, (2) *basic service*, and (3) *full service.*

- **Free-service providers:** One recent development in computers is the availability of things that are free, including ISPs. Among ISPs offering free services are NetZero, AltaVista (partnered with 1stUp.com), Blue-Light.com (from Yahoo!, Kmart, and Softbank), and Excite@Home (with 1stUp.com). The trade-off here is that in return for your free Internet service, you have to endure a barrage of ads in a banner-size box on your screen. In addition, you need to fill out questionnaires with personal data about yourself, opening yourself up to customized ads and e-mails. ISPs such as BlueLight.com have partnered with merchants in hopes of boosting e-commerce sales. Some brokerage firms (such as J.B. Oxford) use free access to retain customers.

  What's happening here is that, as high-speed Net access at $40 a month spreads, users are more reluctant to pay $20 or so a month for conventional dial-up access. Eventually, some experts think, the trend will halve the prices of conventional ISPs.[7] One new free-ISP company, Broadband Digital Group, hopes to go the competition one better by offering free DSL connections. It expects to make money by asking consumers to fill out surveys and then targeting them with ads for services such as video-on-demand, pay-per-view, online games, and Internet phone calling.[8]

- **Basic-service providers:** The trend toward free service can only be bad news for basic-service ISPs, which tend to be small, local companies offering dial-up access. In return for low fees, such companies typically have offered local access numbers, software to connect you to the Net, and technical support.

- **Full-service providers:** Full-service ISPs include AOL, EarthLink, MSN, Prodigy, and WorldNet. Compared to basic providers, full-service ISPs offer more technical support, backup access numbers in case the main number is out of service, and space for your own Web site on their server. They also offer national and even international access from local numbers so that, when you travel, you pay only the price of a local call for your connection.

The basis for AOL's early success was that it was an *online service.* As such, it was able to offer people most of the content and services they wanted within a single site: e-mail, news, chat, instant messaging, shopping, travel reservations, and financial services. On a lesser scale, MSN acts as an ISP and an entry point to mostly Microsoft-produced content.

AOL content is divided into channels such as Games, Entertainment, Computing, Health, Sports, Shopping, and International. Finding content is easy because each has its own **_menu_—a list of commands or options**—of services. "While sophisticates dismiss AOL as 'the Net on training wheels,'" says *Business Week* writer Stephen Wildstrom, "its simplicity and sheltered environment have helped the company become, by far, the largest Internet service provider—and have positioned it to become the globe's biggest media company."[9] That status comes thanks to AOL's merger with media giant Time Warner, announced in early 2000, which exploded the whole category of ISP into another realm. By the time you read this, other companies may have followed suit.

Some ISPs give you only one e-mail account, but AOL, EarthLink, AT&T WorldNet, and Prodigy provide five or more addresses, so that family members can share your account.[10] *(See ● Panel 2.4.)*

Once you have contacted an ISP and paid the required fee (charged to your credit card), the ISP will provide you with information about phone numbers for a local connection. This connection is called a *point of presence (POP)*— a server owned by the ISP or leased from a common carrier, such as AT&T. The ISP will also provide you with communications software for setting up your computer and modem to dial into their network of servers. For this you will be given (or give yourself) a *user name* ("user ID") and a *password*, a secret word or string of characters that enables you to **_log on_, or make a connection to the remote computer.** You will also need to get yourself an e-mail address, as we discuss next.

● PANEL 2.4
Full-service ISPs compared

Internet Service Provider	Monthly Cost for Unlimited Hours of Use	Number of E-Mail Users per Account	Number of Megabytes for a Web Site
America Online (AOL) 800-827-6364 *www.aol.com*	$21.95	1 master name, 4 additional names	2 per screen name
AT&T WorldNet 800-967-5363 *www.att.net*	$21.95, first month free	6	5 per user
Earthlink 800-395-8425 *www.earthlink.net*	$19.95	5	6
Microsoft Network (MSN) 800-373-3676 *www.msn.com*	$19.95, first month free; subaccounts	1 primary, up to 5	12
Prodigy 800-776-3439 *www.prodigy.com*	$19.95	5	6

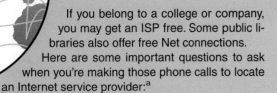

PRACTICAL ACTION BOX
Choosing an Internet Service Provider

If you belong to a college or company, you may get an ISP free. Some public libraries also offer free Net connections. Here are some important questions to ask when you're making those phone calls to locate an Internet service provider:[a]

Costs

- Is there a setup fee? (Most ISPs no longer charge this, though some "free" ISPs will.)

- How much is unlimited access per month? (Most charge about $20 for unlimited usage. But inquire if there are free or low-cost trial memberships or discounts for long-term commitments.)

- If access is supposedly free, what are the trade-offs besides putting up with heavy advertising? (For instance, if the ISP closely monitors your activity in order to accurately target ads, what guarantees do you have that information about you will be kept private? What charges will you face if you try to scrap the advertising window or drop the service?)

- Is there a contract, and for what length of time? That is, are you obligated to stick with the ISP for a while even if you're unhappy with it?

Access

- Is the access number a local phone call? (If not, your monthly long-distance phone tolls could exceed the ISP fee.)

- Is there an alternative dial-up number if the main number is out of service?

- Is access available when you're traveling? Your provider should offer either a wide range of local access numbers in the cities you tend to visit or toll-free 800 numbers.

Support

- What kind of help does the ISP give in setting up your connection?

- Is there free, 24-hour technical support? Is it reachable through a toll-free number?

- How difficult is it to reach tech support? (Try calling the number before you sign up for the ISP and see how long it takes to get a response. Many ISPs keep customers on hold for a long time.)

Reliability

- What is the average connection success rate for users trying to connect on the first try? (The industry average call-success rate is 93.1%. You can try dialing the number during peak hours, to see if you get a modem screech, which is good, rather than a busy signal, which is bad. You can also check Visual Networks, *www.inversenet.com,* for call-failure/call-success rates of various ISPs.)

- Will the ISP keep up with technology? (Are they planning to offer broadband technology such as DSL for speedier access?)

- Will the ISP sell your name to marketers or bombard you with junk messages (spam)?

CONCEPT CHECK

What are the measures of data transmission speed?

Explain the differences among the methods of going online.

Describe the different types of Internet service providers.

2.3 Sending & Receiving E-Mail

KEY QUESTIONS
How can one obtain e-mail software, what are the components of an e-mail address, and what are netiquette and spam?

Once connected with an ISP, one of the first things most people want to do is join the millions of users who send and receive electronic mail. E-mail can be sent at any time and to several people simultaneously. You can receive e-mail wherever you are, using your user name and password to connect to the Internet. In addition, you can attach long (or short) documents or other materials to your e-mail message.

It's not necessary, incidentally, to have a PC for e-mail. Ameritech's e-LISTEN allows you to listen to e-mail messages read to you over the telephone or printed out on a fax machine. A single-purpose e-mail device called Mail-Station—about the size of a hardcover book, with laptop-style keyboard and keys about 85% of normal size plus an adjustable-angle screen—allows you to send and receive messages of up to 1000 words. Using other devices (from Sharp and JVC), you can send and receive e-mail from just about any phone—something travelers toting laptop computers may envy.

Mailstation

E-Mail Software & Carriers

If you aren't on a campus network, there are four ways to go about getting and sending e-mail:

- **Buy e-mail software:** Popular e-mail software programs are Eudora, Outlook Express, or Lotus Notes. However, there is probably no need for you to spend money on these programs because of the following alternatives.

- **Get e-mail program as part of other computer software:** When you buy a new computer, the system will probably include e-mail software, perhaps as part of the software (called *browsers*) used to search the World Wide Web, such as Internet Explorer or Netscape Communicator. An example is Microsoft's Outlook Express, which is part of its Explorer.

Free e-mail from Yahoo!

- **Get e-mail software as part of your ISP package:** Internet service providers—AOL, Prodigy, EarthLink, AT&T WorldNet—provide e-mail software for their subscribers.

- **Get free e-mail services:** According to one study, 69% of e-mail users employ free e-mail services.[11] These are available from a variety of sources, ranging from so-called portals or Internet gateways such as Yahoo!, Excite, or Lycos to cable-TV channel CNN's Web site to Juno and Net-Zero.

E-Mail Addresses

You'll need an e-mail address, of course, a sort of electronic mailbox used to send and receive messages. All such addresses follow the same approach: *user@domain*. **A _domain_ is simply a location on the Internet.** Consider the following address:

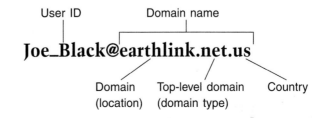

Let's look at the elements of this address.

Joe_Black The first section, the *user ID,* identifies who is at the address—in this case, *Joe_Black.* (There are many ways that Joe Black's user name might be designated, with and without capital letters: *Joe_Black, joe_black, joe.black, joeblack, jblack, joeb,* and so on.)

@earthlink The second section, the *domain name,* which is located after the @ (called "at") symbol, tells the location and type of address. Domain name components are separated by periods (called "dots"). The *domain* portion of the address (such as *EarthLink,* an Internet service provider) provides specific information about the *location*—where the message should be delivered.

.net The *top-level domain* is a three-letter extension that describes the *domain type:* .net, .com, .gov, .edu, .org, .mil, .int—network, commercial, government, educational, nonprofit, military, or international organization. *(See ● Panel 2.5.)*

.us Some domain names also include a two-letter extension for the country—for example, *.us* for United States, *.ca* for Canada, *.uk* for United Kingdom, *.jp* for Japan, *.tr* for Turkey.

Sometimes you'll see an address in which people have their own domains—for example, *Joe@Black.com.* However, you can't simply make up a domain name; it has to be registered. (Check on whether an address is available and register it by checking *www.register.com* or *www.internic registrations.com.*)

Incidentally, many people who are unhappy with their ISPs don't change because they don't want to have to notify their friends of a new e-mail address. However, you can switch ISPs by using an e-mail forwarding service (such as Pobox.com, or a college alumni group that offers lifetime e-mail addresses). That way, you can keep one e-mail address no matter how many times you change providers.[12]

Some tips about using e-mail addresses:

- **Type addresses carefully:** You need to type the address *exactly* as it appears, including all spaces, underscores, and periods. If you type an e-mail address incorrectly, your message will be returned to you with a header of (to most people) incomprehensible strings of characters.
- **Use the "reply" command:** When responding to an e-message someone has sent you, the easiest way to avoid making address mistakes is to use the "Reply" command, which will automatically fill in the correct address in the "To" line.

● **PANEL 2.5**
The meaning of Internet top-level domain abbreviations

Domain	Description	Example
.com	Commercial businesses	Editor@mcgraw-hill.com
.edu	Educational and research institutions	Professor@stanford.edu
.gov	U.S. government agencies and bureaus	President@whitehouse.gov
.int	International organizations	Secretary_general@unitednations.int
.mil	U.S. military organizations	Chief_of_staff@pentagon.mil
.net	Internet network resources	Contact@earthlink.net
.org	Nonprofit and professional organizations	Director@redcross.org

Sending e-mail

Save: Command for saving messages

Address Book: Lists e-mail addresses you use most; can be attached automatically to messages

Send: Command for sending messages

cc: For copying ("carbon copy") message to others

bcc: For copying others ("blind carbon copy") without the primary recipient knowing it

Message area

You can conclude every message with a custom "signature"

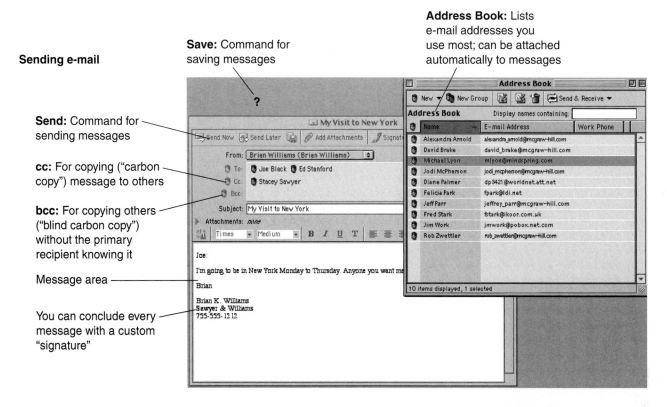

Receiving e-mail

Reply, Reply All, Forward, Delete: For helping you handle incoming e-mail

Inbox lists messages waiting in e-mailbox. (Unopened envelope icon shows unread mail.)

New message displayed here

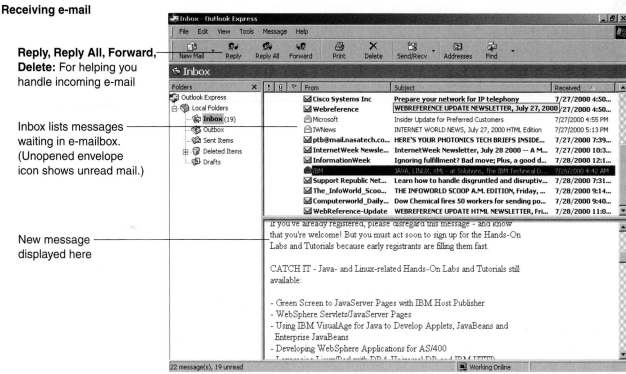

Replying to e-mail

Using the **Reply** command automatically fills in To, From, and Subject lines

Basic elements of e-mail

- **Use the "address book" feature:** You can store the e-mail addresses of people sending you messages in your program's "address book." This feature also allows you to organize your e-mail addresses according to a nickname or the person's real name so that, for instance, you can look up your friend Joe Black under his real name, instead of under his user name, *bugsme2,* which you might not remember. The address book also allows you to organize addresses into various groups—such as your friends, your parents, club members—so you can easily send all members of a group the same message with a single command.

Attachments

You have written a great research paper and you immediately want to show it off to someone. If you were sending it via the Postal Service, you would write a cover note—"Folks, look at this great paper I wrote about term-paper cheating! See attached"—then attach it to the paper, and stick it in an envelope. E-mail has its own version of this. If the file of your paper exists in the computer from which you are sending e-mail, you can write your e-mail message (your cover note) and then use the Attach File command to attach the document. (Note: It's important that the person receiving the e-mail attachment have the exact same software that created the attached file, such as Microsoft Word 2000, or have software that can read and convert the file.)

While you could also copy your document into the main message and send it that way, e-mail tends to lose formatting options such as **bold** or *italic* text or special symbols. And if you're sending song lyrics or poetry, the lines of text may break differently on someone else's display screen than they do on yours. Thus, the benefit of the attachment feature is that it preserves all such formatting, provided the recipient is using the same word processing software that you did. (If your e-mail is written in Hypertext Markup Language, or HTML, the code designed for making Web pages, as we shall discuss, you can add special fonts, images, and colors to your messages.)

You can also attach pictures, sounds, videos, and other files to your e-mail message.

Note: Many *viruses*—those rogue programs that can seriously damage your PC or programs—ride along with e-mail as attached files. Thus, you should never open an attached file from an unknown source. This was what made the so-called May 2000 Love Bug (ILOVEYOU virus) such a disaster, as we describe in Chapter 7.

Instant Messaging

Instant messages are like a cross between e-mail and phone, allowing for communication that is far speedier than conventional e-mail. With **_instant messaging (IM)_, any user on a given e-mail system can send a message and have it pop up instantly on the screen of anyone else logged onto that system.** Then, if both parties agree, they can initiate online typed conversations in real time. The messages appear on the display screen in a small **_window_—a rectangular area containing a document or activity**—so that users can exchange messages almost instantaneously while operating other programs. Eventually, there will probably be an "open standard" so that anyone can send instant messages to anyone else, no matter what system people are on.[13]

Examples of present instant-message systems are AOL Instant Messenger (AIM—AOL pioneered the idea by allowing members to add other members' names to a "Buddy List"), ICQ ("I Seek You," also from AOL), MSN Messenger, Prodigy Instant Messaging, Tribal Voice PowWow, and Yahoo! Messenger. Some of these, such as Yahoo!'s, allow voice chats among users, if their PCs are microphone-equipped.

Menus for e-mail attachment

Third, use your e-mail software's toolbar buttons or menus to attach the file that contains the attachment.

Sending an e-mail attachment

Fourth, click on *Send* to send the e-mail message and attachment.

First, address the person who will receive the attachment.

Second, write a "cover letter" e-mail advising the recipient of the attachment.

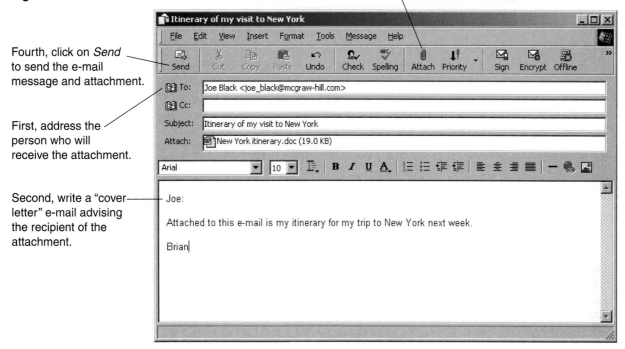

Receiving an e-mail attachment

When you receive a file containing an attachment, you'll see an icon indicating the message contains more than just text. You can click on the icon to see the attachment.

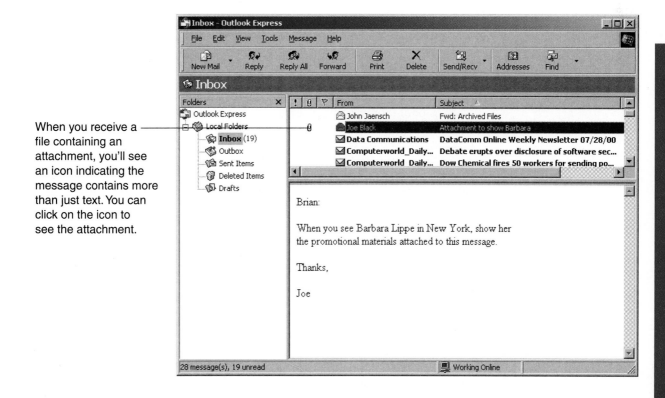

To get instant messaging, which is available for free, you download software and register with the service, providing it with a user name and password. You can then create a list of "buddies" with whom you want to communicate regularly. When your computer is connected to the Internet, the software checks in with a central server, which verifies your identity and looks to see if any of your "buddies" are also online. You can then start a conversation by sending a message to any buddy currently online.[14]

IM has become a hit with many users; indeed, users of AOL's product alone exchange 750 million messages a day. "Instant messaging has made me a happier, more adjusted person," says Randy Weston, 33, an editor who lost touch with many old friends after relocating from San Francisco to Boston. "People who slipped off my radar screen are now back on. With instant messaging, I can see which of my friends are online and I can talk with them in real-time as if we were on the phone."[15] Instant messaging is especially useful in the workplace as a way of reducing long-distance telephone bills when you have to communicate with colleagues who are geographically remote—in South America, say—but with whom you must work closely.

There are, however, a couple of possible drawbacks to be aware of:

- **Lack of common standards:** As of this writing, none of the existing IM products can communicate with each other. If you're using AOL's IM, not only can you not communicate with a buddy on Yahoo!—you can't even communicate with a buddy on AOL's ICQ. Perhaps this will have changed by the time you read this.

- **Time wasters when you have to get work done:** An instant message "is the equivalent of a ringing phone because it pops up on the recipient's screen right away," says one writer.[16] Because of its speed, intrusiveness, and ability to show who else is online, some analysts suggest that IM can destroy workers' concentration in some offices. You can put off acknowledging e-mail, voice mail, or faxes. But instant messaging is "the cyber-equivalent of someone walking into your office and starting up a conversation as if you had nothing better to do," says one critic. "It violates the basic courtesy of not shoving yourself into other people's faces."[17]

  You can turn off your instant messages, but that is like turning off the ringer on your phone; after a while people will wonder why you're never available. Buddy lists or other contact lists can also become ingroupish. When that happens, people are distracted from their work as they worry about staying current with their circle (or being shut out of one). Some companies have reportedly put an end to instant messaging, sending everyone back to the use of conventional e-mail.

Mailing Lists: E-Mail–Based Discussion Groups

Want to receive e-mail from people all over the world who share your interests? You can try finding a mailing list and then "subscribing"—signing up, just as you would for a free newsletter or magazine. **_List-serves_ are e-mail mailing lists of people who regularly participate in discussion topics.** To subscribe, you send an e-mail to the list-serve moderator and ask to become a member, after which you will automatically receive e-mail messages from anyone who responds to the server. A directory of mailing lists is available at Publicly Accessible Mailing Lists *(www.neosoft.com/internet/paml)* or OneList *(www.onelist.com)*.

Ethics

Netiquette: Appropriate Online Behavior

You may think etiquette is about knowing which fork to use at a formal dinner. Basically, though, etiquette has to do with politeness and civility—with rules for getting along so that people don't get upset or suffer hurt feelings.

New Internet users, known as *newbies*, may accidentally offend other people in a discussion group or in an e-mail simply because they are unaware of **_netiquette_, or "network etiquette"—guides to appropriate online behavior.** In general, netiquette has two basic rules: (a) don't waste people's time, and (b) don't say anything to a person online that you wouldn't say to his or her face.

Some more specific rules of netiquette are as follows:

- **Consult FAQs:** Most online groups post **_FAQs—Frequently Asked Questions—that explain expected norms of online behavior for a particular group.** Always read these first—before someone in the group tells you you've made a mistake.

- **Avoid flaming:** A form of speech unique to online communication, **_flaming_ is writing an online message that uses derogatory, obscene, or inappropriate language.** Flaming is a form of public humiliation inflicted on people who have failed to read FAQs or otherwise not observed netiquette (although it can happen just because the sender has poor impulse control and needs a course in anger management). Something that smooths communication online is the use of **_emoticons_, keyboard-produced pictorial representations of expressions.**

- **Don't SHOUT:** Use of all-capital letters is considered the equivalent of SHOUTING. Avoid, except when they are required for emphasis of a word or two (as when you can't use italics in your e-messages).

- **Avoid sloppiness, but avoid criticizing others' sloppiness:** Avoid spelling and grammatical errors. But don't criticize those same errors in others' messages. (After all, they may not be English native speakers.) Most e-mail software comes with spell-checking capability, which is easy to use.

- **Don't send huge file attachments, unless requested:** Your cousin living in the country may find it takes minutes rather than seconds for his or her computer to download a massive file (as of a video that you want to share). This may tie up the system at a time when your relative badly needs to use it. Better to query in advance before sending large files as attachments. Also, whenever you send an attachment, be sure the recipient has the appropriate software to open your attachment (you both are using Microsoft Word 2000, for example).

- **When replying, quote only the relevant portion:** If you're replying to just a couple of matters in a long e-mail posting, don't send back the entire message. This forces your recipient to wade through lots of text to find the reference. Instead, edit his or her original text down to the relevant paragraph and then put in your response immediately following.

Emoticons

:-)	Happy face
:-(	Sorrow or frown
:-O	Shock
:-/	Sarcasm
;-)	Wink
<g>	Grin
BTW	By the way
IMHO	In my humble opinion
FYI	For your information

Filtering e-mail
E-mail software lets you create folders for storing mail.

Help screen on filtering

Using Filters to Sort Your E-Mail

The average corporate employee gets and sends 201 messages a day, according to a 1999 study by Pitney Bowes.[18] Hopefully, your volume of e-correspondence won't be anywhere near this high. However, one way to stay organized is to use *filters* or instant organizers, using the name of the person or the mailing list to put that particular mail into one folder. Then you can read e-mails sent to this folder later when you have time, freeing up your inbox for mail that needs your more immediate attention. Instructions on how to set up filters are in your e-mail program's Help section.

Spam: Unwanted Junk E-Mail

Several years ago, Monty Python, the British comedy group, did a sketch in which restaurant customers were unable to converse because people in the background (a group of Vikings, actually) kept chanting "Spam, spam, eggs and spam . . ." The term *spam* was picked up by the computer world to describe another kind of "noise" that interferes with communication. Now **_spam_ refers to unsolicited e-mail in the form of advertising or chain letters.** Usually you won't recognize the sender on your list of incoming mail, and often the subject line will give no hint, stating something such as "The status of your application" or "It's up to you now." The solicitations can range from money-making schemes to online pornography.

Some ways to deal with this nuisance are as follows:[19]

- **Delete without opening the message:** Opening the spam message can actually send a signal to the spammer that someone has looked at the onscreen message and therefore that the e-mail address is valid—which means you'll probably get more spams in the future. If you don't recognize the name on your inbox directory or the topic on the inbox subject line, you can simply delete the message without reading it. Or you can use a preview feature in your e-mail program to look at the message without actually opening it, then delete it. (Hint: Be sure to get rid of all the deleted messages from time to time, which otherwise will build up in your "trash" area.)

- **Never reply to a spam message!** The following advice needs to be taken seriously: *Never reply in any way to a spam message!* Replying confirms to the spammer that yours is an active e-mail address. Some spam senders will tell you that if you want to be removed from their mailing list, you should type the word REMOVE or UNSUBSCRIBE in the subject line and use the reply command to send it back. Invariably, however, all this does is confirm to the spammer that your address is valid, setting you up to receive more unsolicited messages.

  Michael Ashley Lopez, an archaeology graduate student at the University of California at Berkeley, found he had been included on an e-mail list for fans of teen idol Britney Spears. He opened the first e-mail message, wasted 13 seconds reading it, then clicked on the link to unsubscribe from it. The result was he couldn't get off. Months later, despite a determined effort to shake this nuisance, he was still receiving invitations to check out Britney's latest single or preview her latest video.[20]

- **Enlist the help of your ISP or use spam filters:** Your ISP may offer a spam filter to stop the stuff before you even see it. If it doesn't, you can sign up for a filtering service, such as ImagiNet *(www.imagin.net)* for a small monthly charge. Or there are do-it-yourself spam-stopping programs. Examples: Brightmail *(www.brightmail.com)*, Novasoft SpamKiller *(www.spamkiller.com)*, High Mountain Software SpamEater Pro *(www.hms.com)*.

  Be warned, however: Even so-called spam killers don't always work. Certainly it didn't for Michael Lopez, victim of the repeated Britney Spears e-mail, even though he had signed up for an e-mail blocking service. "Nothing will work 100%, short of changing your e-mail address," says the operator of an online service called SpamCop. "No matter how well you try to filter a spammer, they're always working to defeat the filter."[21]

- **Fight back:** If you want to try to get back at spammers, check with abuse.net *(www.abuse.net)*, The Anti-Spam HOWTO *(zikzak. zikzak.net/~acb/features/anti-spam-howto.html)*, or Ed Falk's Spam Tracking Page *(www.rahul.net/falk)*. These will tell you where to report spammers, the appropriate people to complain to, and other spam-fighting tips.

Ethics

What About Keeping E-Mail Private?

The single best piece of advice that can be given about sending e-mail is this: *Pretend every electronic message is a postcard that can be read by anyone.* Because the chances are high that it could be. (And this includes e-mail on college campus systems as well.)

Think the boss can't snoop on your e-mail at work? The law allows employers to "intercept" employee communications if one of the parties involved agrees to the "interception." The party "involved" is the employer. And in the workplace, e-mail is typically saved on a server, at least for a while. Indeed, Federal laws require employers to keep some e-mail messages for years.

Think you can keep your e-mail address a secret among your friends? You have no control over whether they might send your e-messages on to someone else—who might in turn forward it again. (One thing you can do for them, however, is delete their names and addresses before sending one of their messages on to someone.)

Think your ISP will protect your privacy? Often service providers post your address publicly or even sell their customer lists.

Think spammers can't find you? They will if you post an e-mail to an Internet message or bulletin board, making yourself a target for pieces of software (known as "harvester bots") that scour such boards for active e-mail addresses.

And we have not even mentioned your e-mail being intercepted by those knowledgeable individuals known as hackers or crackers, which we discuss elsewhere.

If you're really concerned about preserving your privacy, you can try certain technical solutions—for instance, installing software that encodes and decodes messages (such as PGP, discussed in Chapter 7). But the simplest solution is the easiest: Don't put any sensitive or embarrassing information in your e-mail.

CONCEPT CHECK

What are the options for getting and sending e-mail?

What are some features available with e-mail?

Explain appropriate e-mail etiquette, use of filters, avoidance of spam, and why you should keep e-mail private.

2.4 The World Wide Web

KEY QUESTIONS
What are Web sites, Web pages, browsers, URLs, and search engines?

"I found my old first-grade teacher by surfing the Internet."

When people talk about the Internet in this way, they really mean the World Wide Web. After e-mail, visiting sites ("surfing") on the Web is the most popular use of the Internet. Among the forces driving its popularity are entertainment and e-commerce. *Entertainment* offerings range from listening to music to creating your own, from playing online games by yourself to playing with others, from checking out local restaurants to researching overseas travel. *E-commerce* offers online auctions, retail stores, and discount travel services as well as all kinds of "B2B," or business-to-business, connections, as when General Motors buys online from its steel suppliers.

What makes the World Wide Web so graphically inviting and easily navigable is that this international collection of servers (1) contains information in multimedia form and (2) is connected by hypertext links.

1. **Multimedia form—what makes the Web graphically inviting:** Whereas e-mail messages are generally text, the Web provides information in *multimedia* form—graphics, video, and audio as well as text. You can see color pictures, animation, and full-motion video. You can download music. You can listen to radio broadcasts. You can have telephone conversations with others.

2. **Use of hypertext—what makes the Web easily navigable:** Whereas with e-mail you can connect only with specific addresses you know about, with the Web you have hypertext. **_Hypertext_ is a system in which documents scattered across many Internet sites are directly linked, so that a word or phrase in one document becomes a connection to a document in a different place.** The format, or language, used on the Web is called hypertext markup language. (It is not, however, a programming language.) **_Hypertext markup language_, abbreviated HTML, is the set of special instructions (called "tags" or "markups") that are used to specify document structure, formatting, and links to other documents.**

For example, if you were reading this book onscreen, you could use your mouse to click on the word *multimedia*—which would be highlighted—in the top paragraph above, and that would lead you to another location, where perhaps "multimedia" is defined. Then you could click on a word in that definition, and that would lead you to some related words—or even some pictures.

The result is that one term or phrase will lead to another, and so you can access all kinds of databases and libraries all over the world. Among the droplets in what amounts to a Niagara Falls of information available: *Weather maps and forecasts. Guitar chords. Recipe archives. Sports schedules. Daily newspapers in all kinds of languages. Nielsen television ratings. A ZIP code guide. Works of literature. The Alcoholism Research Data Base. U.S. Government phone numbers. The Central Intelligence Agency world map. The daily White House press releases.* And on and on.

How hypertext markup language (HTML) works
The coding in the HTML files tells your Web browser, first, how to find the files of text, graphics, and multimedia files on the server and, second, how to display them on the Web page. The browser also interprets HTML tags, or instructions, as links to other Web sites or to other Web resources, such as files to download.

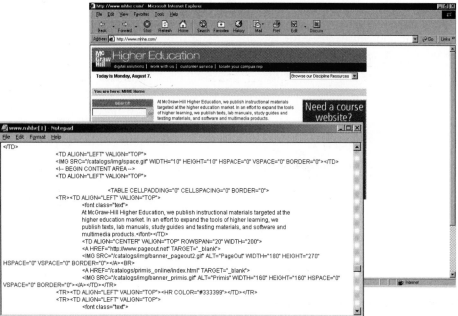

Meaning of tags: Every HTML tag is surrounded by a less-than and greater-than sign—for example, <TR>. Tags often appear as beginning and ending tags, which are identical except for a slash in the end tag—for example, <TR> Paragraph of text. </TR>.

How hypertext works

The Web & How It Works

If a Rip Van Winkle fell asleep in 1989 (the year that computer scientist Tim Berners-Lee developed the Web software) and awoke today, he would be completely baffled by the new vocabulary that we now encounter on an almost daily basis: *Web site, home page, www.* Let's see how we would explain to him what these and similar Web terms mean.

- **Web site—the domain on the computer:** You'll recall we described top-level domains, such as .com, .edu, .org, and .net, in our discussion of e-mail addresses. **A computer with a domain name is called a _site_.** When you decide to buy books at the online site of bookseller Barnes & Noble, you would visit its Web site *www.barnesandnoble.com*; the **_Web site_ (often spelled *website*) is the location of a Web domain name in a computer somewhere on the Internet.** That computer might be located in Barnes & Noble offices, but it might be located somewhere else entirely. (The Web site for New Mexico's Carlsbad Caverns is not located underground in the caverns, but the Web site for your college is probably on the campus.)

- **Web pages—the documents on a Web site:** A Web site is composed of a Web page or collection of related Web pages. **A _Web page_ is a document on the World Wide Web that can include text, pictures, sound, and video.** The first page you see at a Web site is like the title page of a book. This is the **_home page_, or welcome page, which identifies the Web site and contains links to other pages at the site.** If you have your own personal Web site, it might consist of just one page—the home page. Large Web sites have scores or even hundreds of pages. As of 1999, it was estimated there were around a billion pages on the World Wide Web.[22] (The contents of home pages often change. Or they may disappear, so that the connecting links to them in other Web pages become links to nowhere.)

- **Browsers—software for connecting with Web sites:** A **_Web browser_, or simply *browser*, is software that enables users to view Web pages and to jump from one page to another.** The two best known browsers are Microsoft's Internet Explorer, which most users prefer, and Netscape Communicator, once the leader but now used by only about a third of consumers.[23] When you connect to a particular Web site with your browser,

Web site home page

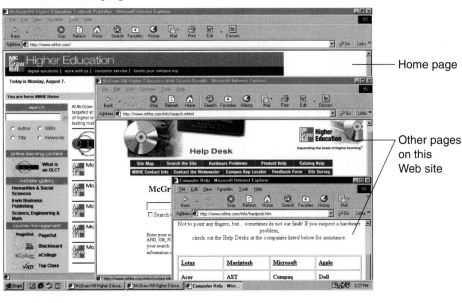

— Home page

— Other pages on this Web site

Microsoft Internet Explorer

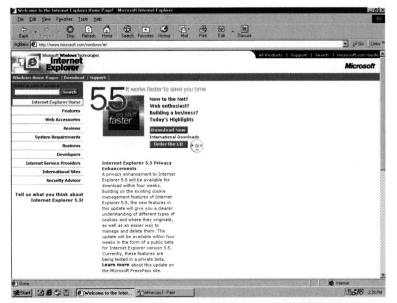

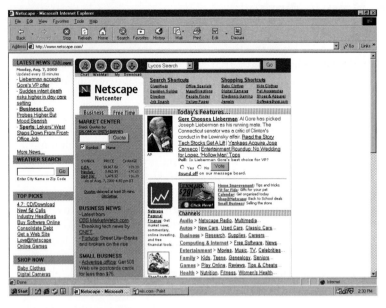

Netscape Navigator

the first thing you will see is the home page. Then, using your mouse, you can move from one page to another by clicking on hypertext links.

- **URLs—addresses for Web pages:** Before your browser can connect with a Web site, it needs to know the site's address, the URL. The **_URL (Universal Resource Locator)_ is a string of characters that points to a specific piece of information anywhere on the Web.** A URL consists of (1) the Web _protocol_, (2) the name of the Web _server_, (3) the _directory_ (or folder) on that server, and (4) the _file_ within that directory (perhaps with an _extension_ such as _html_ or _htm_). Usually you need to type a URL _exactly_ the way it appears—not type a capital letter, for instance, if a lowercase letter is indicated.

Consider the following example of a URL for a Web site offered by the National Park Service for Yosemite National Park:

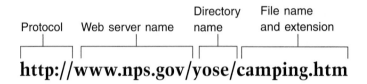

| Protocol | Web server name | Directory name | File name and extension |

http://www.nps.gov/yose/camping.htm

Let's look at these elements.

http://   A **_protocol_ is a set of communication rules for exchanging information.** When you see the _http://_ at the beginning of some Web addresses (as in _http://www.mcgraw-hill.com_), that stands for **_HTTP_, for _HyperText Transfer Protocol_, the communications rules that allow browsers to connect with Web servers.** Note: Most browsers assume that all Web addresses begin with _http://_ and so you don't need to type this part; just start with whatever follows, such as _www_.

www.nps.gov/   The _Web server_ is the particular computer on which this Web site is located. The _www_ stands for "World Wide Web," of course; the _.nps_ stands for "National Park Service," and the _.gov_ is the top-level domain name indicating that this is a government Web site. The server might be physically located in Yosemite National Park in California; in the Park Service's headquarters in Washington, D.C.; or somewhere else entirely.

yose/   The _directory_ name is the name on the server for the directory, or folder, from which you need to pull the file. Here it is _yose_ for "Yosemite." For Yellowstone National Park, it is _yell_.

camping.htm   The _file_ is the particular page or document that you are seeking. Here it is _camping.htm_, because you have gone to a Web page about Yosemite's camping facilities. The _.htm_ is an extension to the file name, and this extension informs the browser that the file is an HTML file.

A URL, you may have observed, is _not_ the same thing as an e-mail address. Some people might type in _president@whitehouse.gov.us_ and expect to get a Web site, but it won't happen. The Web site for the White House (which

Chapter 2

includes presidential information, history, a tour, and guide to federal services) is *http://www.whitehouse.gov.*

Using Your Browser to Get around the Web

As stated, the World Wide Web now consists of an estimated 1 billion Web pages. Moreover, the Web is constantly changing; more sites are created and old ones are retired. Without a browser and various kinds of search tools, there would be no way any of us could begin to make any kind of sense of this enormous amount of data.

As we mentioned, a Web page may consist of *hyperlinks*—words and phrases that appear as underlined or color text—that are references to other Web pages. On a home page, for instance, the hyperlinks serve to connect the top page with other pages throughout the Web site. Other hyperlinks will connect to other pages on other Web sites, whether located on a computer next door or one on the other side of the world.

If you buy a new computer, it will come with a browser already installed. Most browsers have a similar look and feel. Here we show one of the popular browsers in use—Microsoft Internet Explorer.

Notice that the Web browser screen has five basic elements: *menu bar, toolbar, URL bar, workspace,* and *status bar.* To execute menu bar and toolbar commands, you use the mouse to move the pointer over the word, known as a *menu selection,* and click the left button of the mouse. This will result in a *pull-down menu* of other commands for other options. *(See ● Panel 2.6.)*

After you've been using a mouse for a while, you may find moving the pointer around somewhat time-consuming. As a shortcut, if you click on the right mouse button, you can reach many of the commands on the toolbar (*Back, Forward,* and so on) via a pop-up menu.

- **Starting out from home:** The first page you see when you start up your browser is the *home page* or *start page.* (You can also start up from just a blank page, if you don't want to wait for the time it takes to connect with a home page.) You can choose any page on the Web you want as your start page, but a good start page offers links to sites you want to visit frequently. Often you may find that the ISP with which you arrange your Internet connection will provide its own start page. However, you'll no doubt be able to customize it to make it your own personal home page.

● PANEL 2.6
The commands on a browser screen

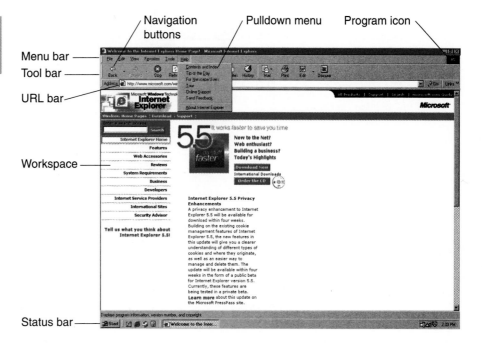

Navigation buttons

Pulldown menu

Program icon

Menu bar

Tool bar

URL bar

Workspace

Status bar

The Internet & the World Wide Web

Web page with hyperlinks

Suppose you live in North America and are planning a trip to Europe. While you're in England, you want to visit London, and you're interested in finding an inexpensive place to stay there. The World Wide Web and its hyperlinks can help you achieve this.

1 You might begin your search by going to the portal AltaVista at *www.altavista.com*. AltaVista's servers are located in Sunnyvale, California.

2 Scrolling down the AltaVista home page shows several underlined links, which contain the URLs (universal resource locators, or Web locations) for other Web pages, or documents on the Web.

When you click on the link *Europe*, . . .

3 . . . your Web browser takes you to the Web page on the AltaVista server containing lists of European countries at *http://dir.altavista.com/Top/Regional/Europe*

When you click on *United Kingdom*, . . .

4 . . . your browser takes you to another page on the server that offers features about the UK, *http://dir.altavista.com/Top/Regional/Europe/UK*

Clicking on *England* . . .

5 . . . takes you to a Web page of locations in England, *http://dir.altavista.com/Top/Regional/Europe/UK/England*

Clicking on *London* . . .

6 . . . takes you to a Web page of features about London, *http://dir.altavista.com/Top/Regional/Europe/UK/England/London*

Clicking on *Accommodation* . . .

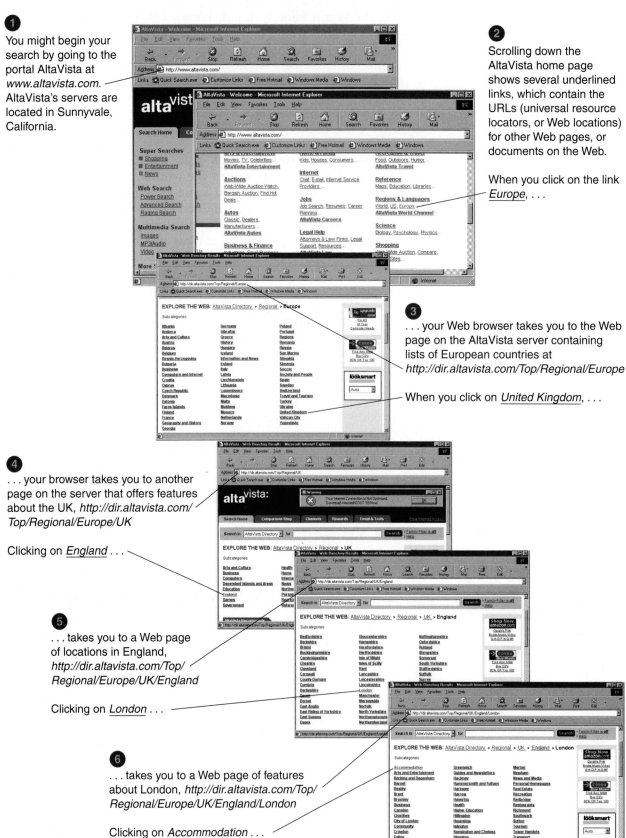

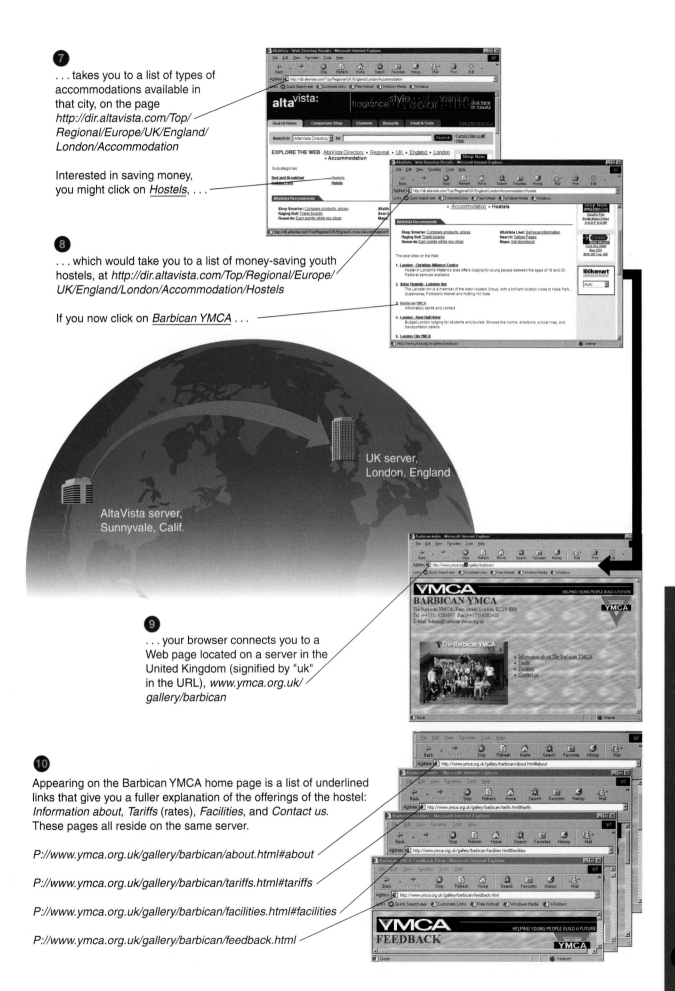

7

. . . takes you to a list of types of accommodations available in that city, on the page *http://dir.altavista.com/Top/ Regional/Europe/UK/England/ London/Accommodation*

Interested in saving money, you might click on *Hostels*, . . .

8

. . . which would take you to a list of money-saving youth hostels, at *http://dir.altavista.com/Top/Regional/Europe/ UK/England/London/Accommodation/Hostels*

If you now click on *Barbican YMCA* . . .

UK server,
London, England

AltaVista server,
Sunnyvale, Calif.

9

. . . your browser connects you to a Web page located on a server in the United Kingdom (signified by "uk" in the URL), *www.ymca.org.uk/ gallery/barbican*

10

Appearing on the Barbican YMCA home page is a list of underlined links that give you a fuller explanation of the offerings of the hostel: *Information about*, *Tariffs* (rates), *Facilities*, and *Contact us*. These pages all reside on the same server.

P://www.ymca.org.uk/gallery/barbican/about.html#about

P://www.ymca.org.uk/gallery/barbican/tariffs.html#tariffs

P://www.ymca.org.uk/gallery/barbican/facilities.html#facilities

P://www.ymca.org.uk/gallery/barbican/feedback.html

- **Personalizing your home page:** Want to see the weather forecast for your college and/or hometown areas when you first log on? Or your horoscope, "message of the day," or the day's news (general, sports, financial, health, or lottery results)? Or the Web sites you visit most frequently? Or a reminder page (as for deadlines or people's birthdays)? You can probably personalize your home page following the directions provided with the first start page you encounter. Or if you have an older Microsoft or Netscape browser you can get a customizing system from either company. A customized start page is also provided by Yahoo!, Excite, AltaVista, and similar services.

- **Getting around—Back, Forward, Home, and Search features:** Driving in a foreign city (or even Boston or San Francisco) can be an interesting experience in which street names change, turns lead into unknown neighborhoods, and signs aren't always evident, so that soon you have no idea where you are. That's what the Internet is like, although on a far more massive scale. Fortunately, unlike being lost in Rome, here your browser toolbar provides navigational aids. *Back* takes you back to the previous page. *Forward* lets you look again at a page you returned from. If you really get lost, you can start over by clicking on *Home,* which returns you to your home page. *Search* lists various other search tools, as we will describe. Other navigational aids are history lists and bookmarks.

- **History lists:** If you are browsing through many Web pages, it can be difficult to keep track of the locations of the pages you've already visited. The *history list* allows you to quickly return to the pages you have recently visited.

- **Bookmarks or favorites:** One great helper for finding your way is the *bookmark* or *favorites* system, which lets you store the URLs of Web pages you frequently visit so that you don't have to remember and retype your favorite addresses. Say you're visiting a site that you really like and that you know you'd like to come back to. You click on your *Bookmark* or *Favorites* feature, which displays the URL on your screen, then click on *Add,* which automatically stores the address. Later you can locate the site name on your bookmark menu,

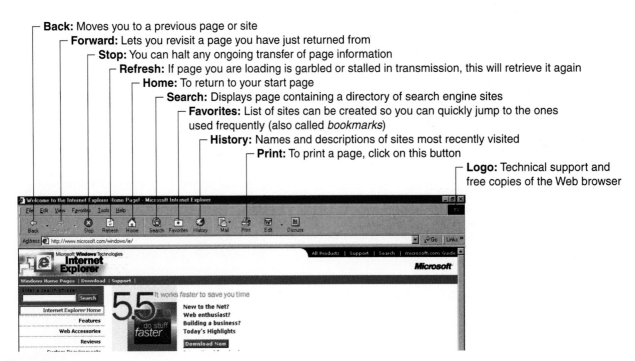

Back: Moves you to a previous page or site

Forward: Lets you revisit a page you have just returned from

Stop: You can halt any ongoing transfer of page information

Refresh: If page you are loading is garbled or stalled in transmission, this will retrieve it again

Home: To return to your start page

Search: Displays page containing a directory of search engine sites

Favorites: List of sites can be created so you can quickly jump to the ones used frequently (also called *bookmarks*)

History: Names and descriptions of sites most recently visited

Print: To print a page, click on this button

Logo: Technical support and free copies of the Web browser

Bookmarks and history lists

Adding Bookmarks (Favorites)
If you are at a Web site you may want to visit again, you click on your *Bookmarks* (in Netscape) or *Favorites* (in Internet Explorer) button and choose *Add Bookmark* or *Add to Favorites*. Later, to revisit the site, you can go to the bookmark menu, and the site will reappear.

Favorites

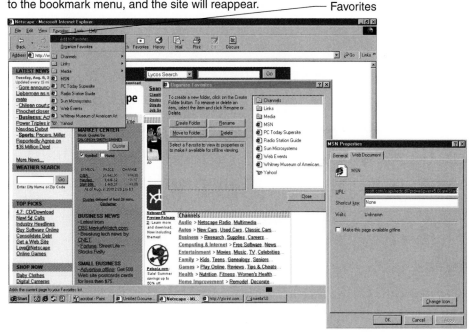

History List
If you want to return to a previously viewed site and are using Netscape, you click on *Communicator*, then choose *History* from the menu. If you're using Internet Explorer, click on *History*.

History

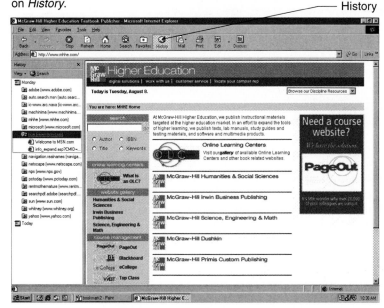

click on it, and the site will reappear. (When you want to delete it, you can use the right mouse button and select the *Delete* command.)

- **Interactivity—hyperlinks, radio buttons, and fill-in text boxes:** For any given Web page that you happen to find yourself on, there may be one of three possible ways to interact with it—or sometimes even all three on the same page.

(1) By using your mouse to click on the hyperlinks, which will transfer you to another Web page.

(2) By using your mouse to click on *radio buttons*— little circles located in front of the various options that, when selected, have a dot placed in the circle—and then clicking on a *Submit* command.

(3) By typing in text in a fill-in text box, then hitting the Enter key or clicking on a *Go* or *Continue* command, which will transfer you to another Web page.

Hyperlinks
Clicking on underlined or color
term transfers you to another Web page.

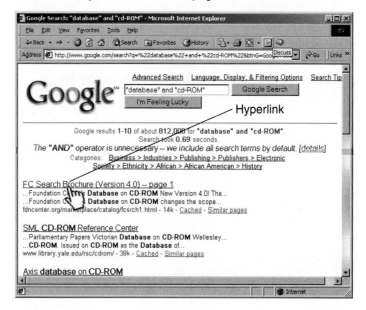

Radio buttons
Act like station selector buttons on a car radio

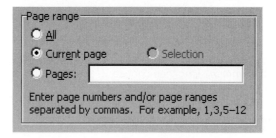

Text boxes
Require you to type in information

- **Scrolling and frames:** To the bottom and side of your screen display, **you will note small up/down and left/right arrows. These are *scroll arrows*. Clicking on scroll arrows with your mouse pointer moves the screen so that you can see the rest of the Web page, a movement known as *scrolling*.** You can also use the arrow keys on your keyboard for scrolling.

  Some Web pages are divided into different rectangles known as frames, each with its own scroll arrows. **A *frame* is an independently controllable section of a Web page.** A Web page designer can divide a page into separate frames, each with different features or options.
- **Turning off images:** If you're not on a high-speed access system, you may find that it takes a while for pictures to fade in on your screen. You can save time by turning off the display of images; in their place you will see icons.
- **Looking at two pages simultaneously:** If you want to look at more than one Web page at the same time, you can position them side by side on your display screen. Select *New* from your File menu to open more than one browser window.

Popular portals

America Online (AOL)	*www.aol.com*
Yahoo!	*www.yahoo.com*
Microsoft Network (MSN)	*www.msn.com*
Netscape	*www.netscape.com*
Lycos	*www.lycos.com*
Go Network	*www.go.com*
Excite Network	*www.excite.com*
AltaVista	*www.altavista.com*
WebCrawler	*www.webcrawler.com*

Web Portals: Starting Points for Finding Information

Using a browser is sort of like exploring an enormous cave with flashlight and string. You point your flashlight at something, go there, and at that location you can see another cave chamber to go to; meanwhile, you're unrolling the ball of string behind you, so that you can find your way back.

But what if you want to visit only the most spectacular rock formations in the cave and skip the rest? For that you need a guidebook. There are many such "guidebooks" for finding information on the Web, sort of Internet super-stations known as ***Web portals***—**Web sites that group together popular features such as search tools, e-mail, electronic commerce, and discussion groups.** The most popular portals are America Online, Yahoo!, Microsoft Network, Netscape, Lycos, Go Network, Excite Network, AltaVista, and WebCrawler.[24]

When you log on to a portal, you can do three things: (1) check the home page for general information, (2) use the directories to find a topic you want, and (3) use a keyword to search for a topic. *(See ● Panel 2.7.)*

● PANEL 2.7
A portal home page

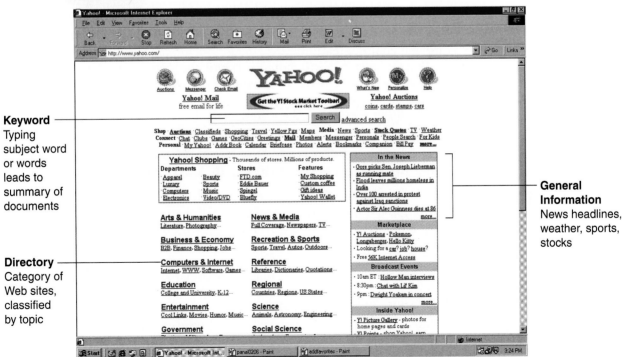

Keyword
Typing subject word or words leads to summary of documents

Directory
Category of Web sites, classified by topic

General Information
News headlines, weather, sports, stocks

- **Check the home page for general information:** You can treat a portal's home or start page as you would one of the mass media—something you tune in to in order to get news headlines, weather forecasts, sports scores, stock-price indexes, and today's horoscope. You might also proceed past the home page to check your e-mail, if you happen to be using the portal for this purpose.

- **Use the directories to find a topic:** Before they acquired their other features, many of these portals began as a type of search tool known as a <u>*directory*</u>, **providing lists of several categories of Web sites classified by topic,** such as *Business & Finance* or *Health & Fitness.* Such a category is also called a *hypertext index*, and its purpose is to allow you to access information in specific categories by clicking on a hypertext link.

  The initial general categories in Yahoo!, for instance, are *Arts & Humanities, Business & Economy, Computers & Internet, Education, Government, Health,* and so on. Using your mouse to click on one general category (such as *Recreation & Sports*) will lead you to another category (such as *Sports*), which in turn will lead you to another category (such as *College & University*), and on to another category (such as *Conferences*), and so on, down through the hierarchy. If you do this long enough, you will "drill down" through enough categories that you will find a document (Web site) on the topic you want.

  Unfortunately, not everything can be so easily classified in hierarchical form. A faster way may be to do a *keyword search.*

- **Use keyword to search for a topic:** At the top of each portal's home page is a blank space into which you can type a *keyword*, the subject word or words of the topic you wish to find. If you want a biography on former San Francisco football quarterback Joe Montana, then *Joe Montana* is the keyword. This way you don't have to plow through menu after menu of subject categories. The results of your keyword search will be displayed in a short summary of documents containing the keyword you typed.

Many users are increasingly bypassing the better known Web portals and going directly to specialty sites or small portals, such as those featuring education, finance, and sports.[25] Examples are Webstart Communications' computer and communications site *(www.cmpcmm.com/cc)*, Travel.com's travel site *(www.travel.com/sitemap.htm)*, and the *New York Times* home page used by the paper's own newsroom staff to find journalism-related sites *(www.nytimes.com/library/tech/reference/cynavi.html)*. Some colleges are also installing portals for their students.

Four Types of Search Engines: Human-Organized, Computer-Created, Hybrid, & Metacrawlers

At one time, as much as 40% of the Web was covered by a single search tool. By 1999, however, no one tool covered more than 16%. One study found that the top 11 search tools covered only 42% of the Web in 1999, whereas two years earlier six search tools covered 60%.[26] "As the Web billows in size," points out one writer, "search sites cover less of it, and what they cover is more likely to be popular commercial sites."[27]

When you use a keyword to search for a topic, you are using a piece of software known as a *search engine*. Whereas *directories* are lists of Web sites classified by topic (as offered by portals), <u>**search engines**</u> **allow you to find specific documents through keyword searches and menu choices.** The type of search engine you use depends on what you're looking for.

There are four types of such search tools: (1) human-organized, (2) computer-created, (3) hybrid, and (4) metacrawlers.[28]

- **Human-organized search sites:** If you're looking for a biography of Apple Computer founder Steve Jobs, a search engine based on human judgment is probably your best bet. Why? Because, unlike a computer-created search site, the search tool won't throw everything remotely associated with his name at you. More and more, the top five search sites on the Web (Yahoo!, AOL, MSN, Netscape, and Lycos) are going in the direction of human indexing. Unlike indexes created by computers, humans can judge data for relevance and categorize them in ways that are useful to you. Many of these sites hire people who are subject-area experts (with the idea that, for example, someone interested in gardening would be best able to organize gardening sites). Examples of human-organized search sites are Yahoo!, Open Directory, About.com, and LookSmart.

- **Computer-created search sites:** If you want to see what things show up next to Steve Jobs's name or every instance in which it appears, a computer-created search site may be best. These are assembled by software "spiders" that crawl all over the Web and send back reports to be collected and organized with little human intervention. The downside is that computer-created indexes deliver you more information than you want. Examples of this type are Northern Light, Excite, WebCrawler, FAST Search, and Inktomi.

- **Hybrid search sites:** Hybrid sites generally use humans supplemented by computer indexes. The idea is to see that nothing falls through the cracks. All the principal sites are now hybrid: AOL Search, AltaVista, Lycos, MSN Search, and Netscape Search. Others are Ask Jeeves, Direct Hit, Go, GoTo.com, Google, HotBot, and Snap. Ask Jeeves pioneered the use of natural-language queries (you ask a question as you would to a person: "Where can I find a biography of Steve Jobs?"). Google ranks listings by popularity as well as by how well they match the request. GoTo ranks by who paid the most money for top billing.

- **Metasearch sites:** Metasearch sites send your query to several other different search tools and compile the results so as to present the broadest view. Examples are Go2Net/MetaCrawler, SavvySearch, Dogpile, Inference Find, ProFusion, Mamma, The Big Hub, and C4 TotalSearch.

More information about these search tools is given in the box on the next page. (See ● *Panel 2.8.*)

Tips for Smart Searching

The phrase "trying to find a needle in a haystack" will come vividly to mind the first time you type a word into a search engine and back comes a response on the order of "63,173 listings found." Clearly, it becomes mandatory that you have a strategy for narrowing your search. The following are some tips.[29]

- **Start with general search tools:** Begin with general search tools such as those offered by AltaVista, Excite, GoTo.com, HotBot, Lycos, and Yahoo! (Later, if you haven't been able to narrow your search, you can go to specific search tools, as we'll describe.)

- **Choose your search terms well and watch your spelling:** Use the most precise words possible. If you're looking for information about novelist Thomas Wolfe (author of *Look Homeward Angel*, published 1929) rather than novelist/journalist Tom Wolfe (*A Man in Full*, 1998), details are important: *Thomas*, not *Tom*. Use *poodle* rather than *dog*, *Maui* rather than *Hawaii*, *Martin guitar* rather than *guitar*, or you'll get thousands of responses that have little or nothing to do with what

Human-Organized Search Sites

- **Yahoo!** *(www.yahoo.com)*. 1994. The most popular search site of all. Has the largest human-compiled Web directory. Supplemented by Inktomi. Users should narrow search results category before they begin searching.
- **LookSmart** *(www.looksmart.com)*. 1996. One of the easiest directories to use. Supplemented by AltaVista. Used by Excite and MSN Search.
- **About** *(www.about.com)*. 1997. Began as The Mining Company. Trained human "guides" cover 50,000 subjects.
- **Open Directory** *(dmoz.org)*. 1998. Uses 21,500 volunteer indexers. Owned by Netscape. Used by AltaVista, AOL Search, HotBot, Lycos, and Netscape.

Computer-Created Search Sites

- **Webcrawler** *(www.webcrawler.com)*. 1994. Began at University of Washington; now owned by Excite.
- **Excite** *(www.excite.com)*. 1995. One of the most popular search services. Owns Magellan and WebCrawler.
- **Inktomi** *(www.inktomi.com)*. 1996. Began at University of California, Berkeley; available only through partners such as AOL Search and Snap.
- **Northern Light** *(www.northernlight.com)*. 1997. One of the largest indexes. Also enables users for a fee to get additional documents from sources not easily accessible, such as magazines, journals, and news wires.
- **FAST Search** *(www.alltheweb.com)*. 1999. Also powers the Lycos MP3 search service. Has announced plans to index the entire Web.

Hybrid Search Sites

- **Lycos** *(www.lycos.com)*. 1995. Began as search engine, shifted to human directory in 1999. Main listings come from Open Directory, secondary results from Direct Hit or Lycos's own index. Good for finding graphics, music files, and other specialized content.
- **AltaVista** *(www.altavista.com)*. 1995. One of the largest search engines. Additional listings provided by Ask Jeeves and Open Directory. Multilingual searches, queries entered as simple questions.
- **Ask Jeeves** *(ask.com)*. 1996. Allows users to ask questions in natural language rather than keywords. Includes Urban Cool (www.urbancool.com), which allows users to ask questions posed in the popular language of the street.
- **HotBot** *(www.hotbot.com)*. 1996. One of the best-rounded search engines. Owned by Lycos. First page of results from Direct Hit, secondary results from Inktomi. Directory information from Open Directory.
- **GoTo** *(www.goto.com)*. 1997. Companies pay to be placed higher in search results.
- **Snap** *(www.snap.com)*. 1997. Human directory of Web sites, supplemented by Inktomi. Owned by CNet and NBC.
- **Direct Hit** *(www.directhit.com)*. 1998. Owned by Ask Jeeves. Refines results based on popularity. Highest-ranking results are those most frequently chosen. Used on Ask Jeeves, Lycos, and HotBot and an option on LookSmart and MSN Search.
- **Google** *(www.google.com)*. 1998 Links popularity to ranking. The more sites that link to a Web page, the higher that page will rank in searches.
- **Go** *(go.com)*. 1999. Owned by Disney; offers search service of former Infoseek (1995). Includes human directory. Focuses on entertainment and leisure.
- **AOL Search** *(search.aol.com)*. Covers both the Web and America Online's content. Directory listings mainly from the Open Directory, with backup from Inktomi.
- **MSN Search** *(search.msn.com)*. Results from LookSmart directory, secondary results from AltaVista and Direct Hit.
- **Netscape Search** *(search.netscape.com)*. Results from Open Directory and Netscape's own database, secondary results from Google.

Metasearch Sites

- **Go2Net/MetaCrawler** *(www.go2net.com)*. 1995. Started at University of Washington.
- **SavvySearch** *(savvysearch.com)*. 1995. Started at Colorado State University, Fort Collins.
- **Dogpile** *(www.dogpile.com)*. Searches a customizable list of search engines.
- **Inference Find** *(www.infind.com)*. Lists results grouped by subject, rather than by search engine or in one long list.
- **ProFusion** *(www.profusion.com)*.
- **The Big Hub** *(www.thebighub.com)*.
- **C4 TotalSearch Technology** *(www.c4.com)*.

Adapted from Elizabeth Weise, "Successful Net Search Starts with Need," *USA Today,* January 24, 2000, p. 3D.

you're looking for. You may need to use several similar words to explore the topic you're investigating: *car racing, auto racing, drag racing, drag-racing, dragracing,* and so on.

- **Use phrases with quotation marks rather than separate words:** If you type *ski resort,* you could get results of (1) everything to do with skis on the one hand and (2) everything to do with resorts—winter, summer, mountain, seaside—on the other. Better to put your phrase in quotation marks—*"ski resort"*—to narrow your search.

- **Put unique words first in a phrase:** Better to have *"Tom Wolfe novels"* rather than *"Novels Tom Wolfe."* Or if you're looking for the Hoagy Carmichael song rather than the southern state, indicate *"Georgia on My Mind."*

- **Use operators—AND, OR, NOT, and + and − signs:** Most search sites use symbols called *Boolean operators* to make searching more precise. To illustrate how they are used, suppose you're looking for the song "Strawberry Fields Forever."[30]

  AND connects two or more search words and means that all of them must appear in the search results. Example: *Strawberry AND Fields AND Forever.*

  OR connects two or more search words and indicates that any of the words may appear in the results. Example: *Strawberry Fields OR Strawberry fields.*

  NOT, when inserted before a word, excludes that word from the results. Example: *Strawberry Fields NOT W.C.* (to distinguish from the comedian W.C. Fields).

  + (plus sign), like *AND,* precedes a word that must appear: Example: *+Strawberry+Fields.*

  − (minus sign), like *NOT,* excludes the word that follows it. Example: *Strawberry Fields−New York.*

- **Read the Help or Search Tips section:** All search sites provide a Help section and tips. This could save you time later.

- **Try an alternate general search site or a specific search site:** If you're looking for very specific information, a general type of search site such as Yahoo! or AltaVista may not be the best way to go. Instead you should turn to a specific search site.[31] Examples: To explore public companies, try Company Sleuth *(www.companysleuth.com),* Hoover's Online *(www.hoovers.com),* or KnowX *(www.knowx.com).* For news stories, try Yahoo! News *(dailynews.yahoo.com)* or Total-News *(www.totalnews.com).* For pay-per-view information, try Dialog Web *(www.dialogweb.com),* Lexis-Nexis *(www.lexis-nexis.com),* and Dow Jones Interactive *(www.djnr.com).*

Multimedia on the Web

Many Web sites (especially those trying to sell you something) are multimedia, using a combination of text, images, sound, video, or animation. While you may be satisfied for now with accessing Web pages with just text and images, eventually you'll probably want more.

- **Plug-ins and helper applications:** In the 1990s, as the Web was evolving from text to multimedia, browsers were unable to handle many kinds of graphic, sound, and video files. To do so, external application files called plug-ins had to be loaded into the system. **A _plug-in_—also called a player or a viewer—is a program that adds a specific feature to a browser, allowing it to play or view certain files.** For example, to view certain documents, you may need to download Adobe Acrobat Reader, to listen to CD-quality audio you may need to download Liquid

Plug in
Adobe Acrobat Reader allows you
to view or print certain Web documents.

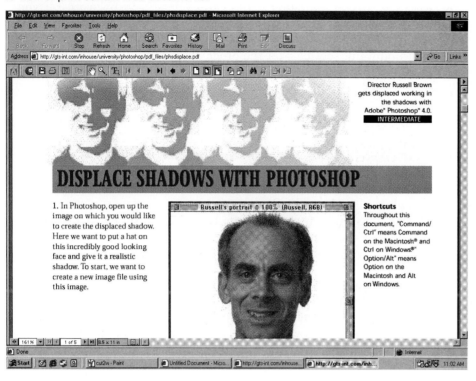

MusicPlayer. Plug-ins are required by many Web sites if you want to fully experience their content. A variant is the *helper application, or add-on,* which runs multimedia elements separate from the browser.

Recent versions of Microsoft Internet Explorer and Netscape Communicator can handle a lot of multi-media that previously these browsers could not. Now if you come across a file for which you need a plug-in or add-on, the browser will ask whether you want it, then tell you how to go about downloading it, usually at no charge.

- **Developing multimedia—applets, Java, JavaScript, and ActiveX:** How do Web-site developers get all those nifty special multimedia effects? Often Web pages contain links to **_applets_, small programs that can be quickly downloaded and run by most browsers.**

Web page combining text and images
The National Park Service's opening screen
about Mount Rushmore in South Dakota.

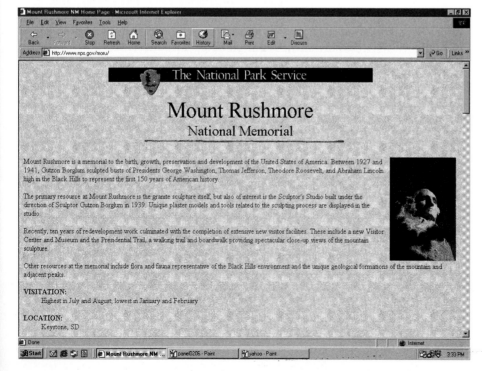

Applets are written in **_Java_, a complex programming language that enables programmers to create animated and interactive Web pages.** Java applets enhance Web pages by playing music, displaying graphics and animation, and providing interactive games.

If you are creating your own Web multimedia, you may want to learn techniques such as JavaScript and ActiveX, which may be used to create Web-page interest and activity—such as scrolling banners, pop-up menus, and the like—as we discuss further in the Appendix.

Animation
This Web page shows an example of animation in a virtual approach to Mars

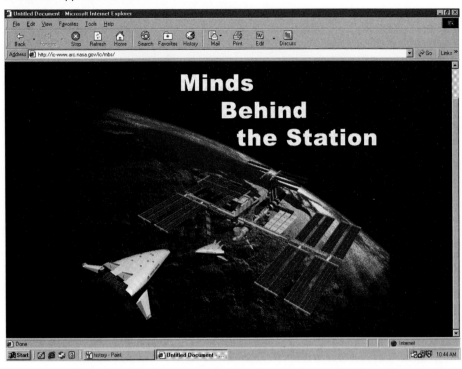

- **Text and images:** You can call up all kinds of text documents on the Web, such as newspapers, magazines, famous speeches, and works of literature. You can also view images, such as scenery, famous paintings, and photographs. Most Web pages combine both text and images.

- **Animation:** _**Animation**_ **is the rapid sequencing of still images to create the appearance of motion,** as in a Road Runner cartoon. Animation is used in online video games as well as in moving banners displaying sports scores or stock prices.

- **Video:** Video can be transmitted in two ways. (1) A file, such as a movie or video clip, may have to be completely downloaded before you can view it. This may take several minutes in some cases. (2) A file may be displayed as _streaming video_ and viewed while it is still being downloaded to your computer. _**Streaming video**_ **is the process of transferring data in a continuous flow so that you can begin viewing a file even before the end is sent.** For instance, RealPlayer offers live, television-style broadcasts over the Internet as streaming video for viewing on your PC screen. You download RealPlayer's software, install it, then point your browser to a site featuring RealVideo. That will produce a streaming-video television image in a window a few inches wide.

- **Audio:** Audio, such as sound or music files, may also be transmitted in two ways: (1) downloaded completely before they can be played or (2) downloaded as _**streaming audio**_, **allowing you to listen to the file while the data is still being downloaded to your computer.** A popular standard for transmitting audio is RealAudio, which can compress sound so it can be played in real time, even though sent over telephone lines. You can, for instance, listen to 24-hour-a-day net.radio, which features "vintage rock," or English-language services of 19 shortwave outlets from World Radio Network in London. Many large radio stations outside the U.S. have net radio, allowing people in foreign lands to listen to their home stations.

Push Technology & Webcasting

It used to be that you had to do the searching on the World Wide Web. Now, if you wish, the Web will come searching for you. The driving force behind this is _**push technology**_, **software that automatically downloads information to your computer** (as opposed to "pull" technology, in which you go to a Web site and pull down the information you want).

One result of push technology is _**webcasting**_, **in which customized text, video, and audio are sent to you automatically on a regular basis.** The idea here is that you choose the categories, or (in Microsoft Internet Explorer) the

Streaming audio
RealPlayer is used for transmitting streaming audio, which many radio stations now use for live broadcasts

Click on for live broadcast

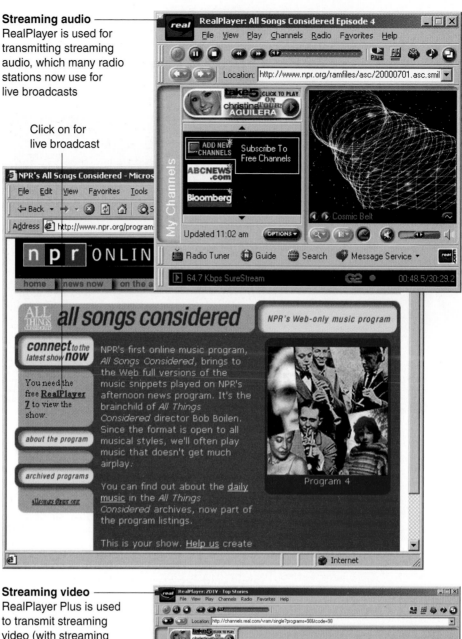

Streaming video
RealPlayer Plus is used to transmit streaming video (with streaming audio)

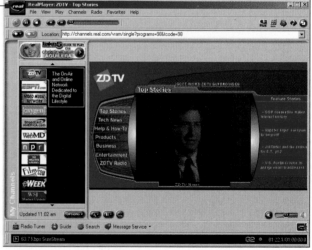

channels, of Web sites that will automatically send you updated information. Thus, it saves you time because you don't have to go out searching for the information. Several services offer personalized news and information, based on a profile that you define when you register with them. Entrypoint.com, for example, will send news on a particular topic, such as all financial news about sugar beets, the weather in Omaha, or the game results for the Tennessee Titans. You can view the news as it comes in or save it to read later.

Push technology

The push technology from Entrypoint.com periodically delivers news that you preset according to your specifications

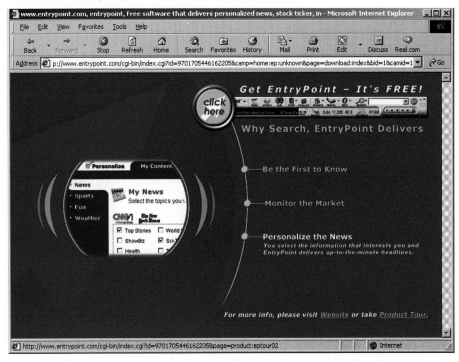

The Internet Telephone & Videophone

A few years ago, the idea of using the Internet to make phone calls was considered "a gimmicky technology for nerdy hobbyists," in the words of one writer.[32] Now Internet firms have taken aim at the traditional telephone companies.

The key element is that the Internet breaks up conversations (as it does any other transmitted data) into "information packets," which can be sent over separate lines, then regrouped at the destination, whereas conventional voice phone lines carry a conversation over a single path. Thus, the Internet can move a lot more traffic over a network than the traditional telephone link can. (We describe how telecommunications technology works in Chapter 5.)

With **_Internet telephony_—using the Net to make phone calls, either one-to-one or for audioconferencing**—you can make long-distance phone calls that are surprisingly inexpensive. Indeed, it's theoretically possible to do this without owning a computer, simply by picking up your standard telephone and dialing a number that will "packetize" your conversation. However, it's more common practice to use a PC with a sound card and a microphone, a modem linked to a standard Internet service provider, and Internet telephone software such as Netscape Conference (part of Netscape Communicator) or Microsoft NetMeeting (part of Microsoft Internet Explorer).

Besides carrying voice signals, Internet telephone software also allows videoconferencing, enabling you and others to be linked by a videophone that will transmit pictures of the people in the conversation, assuming everyone has a video camera attached to their PCs. It can also allow people to make sketches on "whiteboards" as they talk, as when three people meet online to discuss the floor plan for a new house.

Designing Web Pages

If you want to advertise a business online or just have your own personal Web site, you will need to design a Web page, determine any hyperlinks, and hire 24-hour-a-day space on a Web server or buy one of your own. Professional

Web page designers can produce a page for you, or you can do it yourself using a menu-driven program included with your Web browser or a Web-page design software package such as Microsoft FrontPage or Adobe PageMill. After you have designed your Web page, you can put it on your ISP's server. (We describe Web authoring software in the Appendix.)

CONCEPT CHECK

Describe how the World Wide Web works.

How do you use a browser to get around the Web?

How do you use a Web portal to find information?

Describe the types of search engines and some tips for searching.

Discuss Web multimedia, push technology, webcasting, and telephony.

What are the measures of data transmission speed?

Explain the differences among the methods of going online.

2.5 The Online Gold Mine: More Internet Resources, Your Personal Cyberspace, E-Commerce, & the E-conomy

KEY QUESTIONS
What are FTP, Telnet, newsgroups, real-time chat, and e-commerce?

Deborah Thede of San Francisco drove from store to store over a period of several months, testing out one couch after another. "I had pretty specific wants and just never saw what I wanted in a store," she said. Then someone told her about Furniture.com on the Internet, and she ended up ordering a custom-built couch from the site. It wasn't quite the size and color she had in mind, but there was no sales tax—a considerable savings on an $800 piece of furniture. "It's very comfortable, it's beautiful, and I'm not sorry I bought it," says Thede. And at less than $800, it seemed to be a bargain.[33]

Is this a glimpse of our cyberfuture? Will people be ordering pianos and stoves this way? Certainly the availability of things on the Net seems wide open. Let's consider some of them.

Other Internet Resources: FTP, Telnet, Newsgroups, & Real-Time Chat

E-mail and the World Wide Web seem to attract all the attention. But other cyber resources are also widely used: FTP, Telnet, newsgroups, and real-time chat.

- **FTP—for copying all the free files you want:** Many Net users enjoy "FTPing"—cruising the system and checking into some of the tens of thousands of FTP sites, which predate the Web, offering interesting free files to copy (download). ***FTP*, for *File Transfer Protocol*, is a method whereby you can connect to a remote computer called an FTP site and transfer publicly available files to your own microcomputer's hard disk.** The free files offered cover nearly anything that can be stored on a computer: software, games, photos, maps, art, music, books, statistics.

  Some FTP files are open to the public, some are not. For instance, a university might maintain an FTP site with private files (such as lecture transcripts) available only to professors and students with assigned user names and passwords. It might also have public FTP files open to anyone with an e-mail address. You can download FTP files using either your Web browser or special software (called an *FTP client program*), such as Fetch.

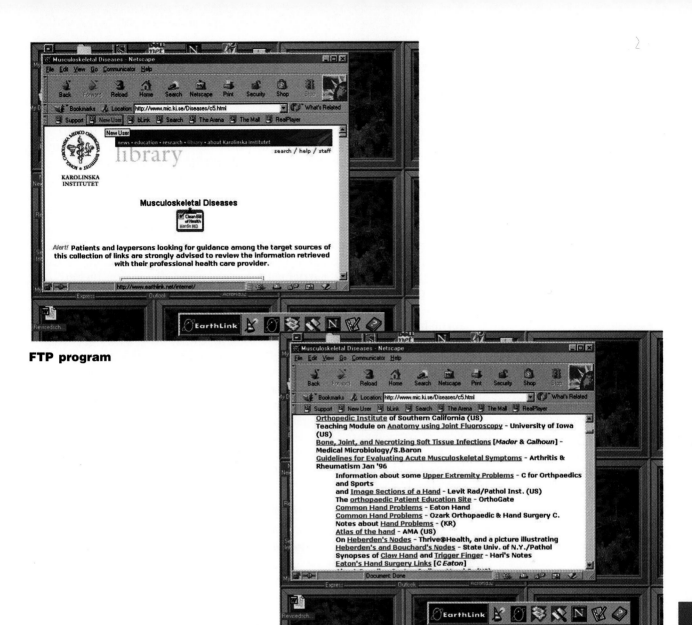

FTP program

- Telnet—to connect to remote computers: *Telnet* **is a program or command that allows you to connect to remote computers on the Internet.** This feature, which allows microcomputers to communicate successfully with mainframes, enables you to tap into Internet computers and access public files as though you were connected directly instead of, for example, through your ISP site.

    The Telnet feature is especially useful for perusing large databases at universities, government agencies, or libraries. As an electronic version of a library card catalog, Telnet can be used to search most major public and university library catalogs. (See, for example, Internet Public Library, *www.ipl.org*, and Library Spot, *www.libraryspot.com*.)

- Newsgroups—for online typed discussions on specific topics: **A *newsgroup* is a giant electronic bulletin board on which users conduct written discussions about a specific subject.** There are more than 30,000 newsgroup forums—which charge no fee—and they cover an amazing array of topics. Some examples are *rec.arts.startrek.info, soc.culture.african.american,* and *misc.jobs.offered.* Newsgroups take place on a special network of computers called *Usenet*, **a worldwide network of servers that can be accessed through the Internet.** To participate, you need a **_newsreader_, a program included with most browsers that allows you to access a newsgroup and read or type messages.** (Messages, incidentally, are known as *articles*.)

One way to find a newsgroup of interest to you is to use a portal such as Yahoo!, Excite, or Lycos to search Usenet for specific topics. Or you can use the search engine Dejanews *(www.dejanews.com)*, which will present the newsgroups that correspond to the topic you specify. About a dozen major topics, identified by abbreviations ranging from *alt* (alternative topics) to *talk* (opinion and discussion), are divided into hierarchies of subtopics.

- **Real-time chat—typed discussions while online at the same time:** With newsgroups (and mailing lists, which we described under e-mail), participants may contribute to a discussion, then go away and return hours or days later to catch up on others' typed contributions. With **_real-time chat (RTC)_, participants have a typed discussion ("chat") while online at the same time,** just like a telephone conversation except that messages are typed rather than spoken. Otherwise the format is much like a newsgroup, with a message board to which participants may send ("post") their contributions. To start a chat, you use what is known as a *chat client*, a program available on your browser that will connect you to a chat server. One of the most popular chat clients is Internet Relay Chat (IRC).

    Unlike instant messaging (discussed under e-mail), which tends to involve one-on-one conversation, real-time chat usually involves several participants. As a result, RTC "is often like being at a crowded party," says one writer. "There are any number of people present and many threads of conversation occurring all at once."[34]

Your Personal Cyberspace

As we mentioned in Chapter 1, information technology has become more personal as it has evolved. Unlike the generally impersonal mass media, the Internet allows you to pursue your personal interests in the areas of relationships, education, health, and entertainment, for example.

- **Relationships—online matchmaking:** It's like walking into "a football stadium full of single people of the gender of your choice," says Trish McDermott, an expert for Match.com, a San Francisco online-dating service. People who connect online before meeting in the real world, she points out, have the chance to base their relationship on personality, intelligence, and sense of humor rather than purely physical attributes. "Online dating allows people to take some risks in an anonymous capacity," she adds. "When older people look back at their lives, it's the risks that they didn't take that they most regret."[35] (Still, there *are* some risks in trying to establish intimacy through online means.)

    People can also use search sites such as Infospace.com, Switchboard.com, and GTE Superpages.com to try to track down old friends and relatives.[36] Others find common bonds by joining online communities such as The WELL *(www.well.com)*, the women's site Ivillage.com *(www.ivillage.com)*, the older people's site Third Age *(www.thirdage.com)*, and the gardening site GardenWeb *(www.gardenweb.com)*.[37] Finally, the Net is no longer dominated by English; 43% of the users are non-English speakers, and that number was expected to pass 50% in 2000. After English, the most common languages among Internet users are (in order) Japanese, Spanish, and German, with French and Chinese tied for fifth.[38]

- **Education—the rise of distance learning:** Sally Wells of Oregon has four children, a full-time job, and 14 cows to milk. She'd like to get a master's degree, but with no time to drive an hour to campus she takes four marketing courses online.[39] Adult learners—defined by educators as those over age 24— aren't the only ones involved in ***distance learning*, the name given online education programs.** Younger college students also like it because they don't have to spend time commuting, the scheduling is flexible, and they often have a greater selection of course offerings. Although for instructors an online class is more labor-intensive than a regular chalk-and-talk class, they often find there is better interaction with students.[40]

Distance learning
Different from a chalk-and-talk class.

- **Health—patient self-education:** Health is one of the most popular subject areas of research on the Web, although it can be difficult to get accurate information (many sites are trying to sell you something). Among the sites offering reputable advice are Intelihealth *(www.intelihealth.com)*, Mayo Clinic Health Oasis *(www.mayohealth.org)*, The Physician and Sportsmedicine Online *(www.physportsmed.com)*, Cyber Diet *(cyberdiet.com)*, Phys.com *(phys.com)*, the American College of Physicians *(acponline.org)*, and Medline *(www.nlm.nih.gov/medlineplus)*.[41]

- **Entertainment—amusing yourself:** Two-thirds of all Internet users in the U.S. seek out entertainment on the Web, according to a 1999 survey.[42] No wonder so many major media companies have created Web sites to try to help promote or sell movies, music, TV shows, and the like. Of course, there are many other types of entertainment sites, devoted to games, hobbies, jokes, and so on. If you want to see animated hamsters dancing, for instance, try visiting *www.hamsterdance.com*.

E-Commerce

"What's your opinion on the Internet vs. the real world?" a *USA Today* reader asked. "[T]he Internet *is* the real world," replied reporter Lorrie Grant, who had been assigned to spend a month using the Web for shopping, working, banking, and other activities. "I had everything at my fingertips: office supplies, groceries, stocks, banking and bill payment, apparel, flowers, music, gifts, greeting cards, and more. Just point and click, *voilà*."[43]

The explosion in ***electronic commerce*, or *e-commerce*—the conducting of business activities online**—is not only widening consumers' choice of products and services but is also creating new businesses and compelling established businesses to develop Internet strategies. Let's look at some of the developments.

- **E-tailing—retail commerce online:** Is the Internet spawning an entirely new way of doing business? Certainly many so-called *brick-and-mortar* retailers—those operating out of physical buildings—have been surprised at the success of such online companies as Amazon.com, seller of books, CDs, and other products. As a result, traditional retailers from giant Wal-Mart to funky little Buch Spieler Music in Montpelier, Vermont, have rushed to put their products online—and it has helped to revive some small-town main streets that had suffered from plant closings and competition from mega-malls.[44]

Retail goods can be classified into two categories—hard and soft. *Hard goods* are those that can be viewed and priced online, such as computers, clothes, furniture, and—yes—even groceries, but then sent to buyers by mail or truck. *Soft goods* are those that can be downloaded from the retailer's site, such as music, software, and greeting cards.

- **Auctions—linking individual buyers and sellers:** Today millions of buyers and sellers are linking up at online auctions, where everything is available from comic books to wines. More than 500 Web sites now wield an electronic gavel and within a few years are expected to lure about 10% of all online purchasers. The Internet is also changing the tradition-bound art and antiques business (dominated by such venerable names as Sotheby's, Christie's, and Butterfield & Butterfield). There are generally two types of auction sites: (1) person-to-person auctions, such as eBay *(www.ebay.com),* that connect buyers and sellers for a listing fee and a commission on sold items, and (2) vendor-based auctions, such as OnSale *(www.onsale.com),* that buy merchandise and sell it at discount. Some auctions are specialized, such as Priceline *(www.priceline.com),* an auction site for airline tickets.

- **Online finance—trading, banking, and e-money:** The Internet has changed the nature of stock trading. For the first time, says technology observer Denise Caruso, "anyone with a computer, a connection to the global network, and the requisite ironclad stomach for risk has the information, tools, and access to transaction systems required to play the stock market, a game that was once the purview of an elite few."[45] Companies such as E*Trade and Ameritrade are building one-stop financial supermarkets offering a variety of money-related services, including home mortgage loans and insurance. More than 1000 banks have Web sites, offering services that include account access, funds transfer, bill payment, loan and credit card applications, and investments. You can, for instance, apply for a Visa card called NextCard and get approved (or turned down) in about two minutes.

- **Online job hunting:** There are 2500 Web sites that promise to match job hunters with an employer. Some are specialty "boutique" sites looking for, say, scientists or executives. Some are general sites, the leaders being Monster.com, CareerPath.com, Headhunter.net, Career-Mosaic.com, www.usajobs.opm.gov, and CareerBuilder.com.[46] Job sites can help you keep track of job openings and applications by downloading them to your own computer.

- **B2B commerce:** Of course, every kind of commerce has taken to the Web, ranging from travel bookings to real estate. One of the most important variations is **_B2B_—for _business-to-business_—_commerce_, the electronic sales or exchange of goods and services directly between companies, cutting out traditional intermediaries.** Expected to grow even more rapidly than other forms of e-commerce, B2B commerce covers an extremely broad range of activities, such as supplier-to-buyer display of inventories, provision of wholesale price lists, and sales of closed-out items and used materials—usually without agents, brokers, or other middlemen. Companies say e-purchasing, for example, slashes up to 20% off what they buy from each other.[47]

CONCEPT CHECK

Describe FTP, Telnet, newsgroups, and real-time chat.

What are some ways the Internet can be of personal use and of e-commerce use?

At the beginning of this new century, the American economy was sizzling: the stock market at a record high, corporate profits soaring, unemployment at a 30-year low, the rate of inflation barely perceptible. How did we arrive at this wonderland? One possible answer: information technology. Suggests one analysis: "The spread of computers, telecommunications equipment, and the like—which control manufacturing, supply-chain management, and business-to-business e-commerce—is helping companies hold down inventories, thus eliminating a prime cause of past recessions, when businesses found themselves overstocked with products they had to sell at discounts or simply write off their books. Fewer recessions mean greater certainty of earnings growth."[48]

New kinds of companies now dominate the landscape. For example, at the end of 1999, the valuation of Yahoo!, only a few years old, was about $90 billion—twice that of venerable General Motors, although its earnings were about one-hundredth of GM's. America Online was worth more than GM, Ford, Sears, and Disney combined.[49]

Call it the "electronic economy"—or *e-conomy*. While the ups and downs of the business cycle have yet to be repealed, the Internet and computers have dramatically changed the nature of enterprise.

Visual Summary

animation (p. 67, KQ 2.4) The rapid sequencing of still images to create the appearance of motion, as in a cartoon. Why it's important: *Animation is a component of multimedia; it is used in online video games as well as in moving banners displaying sports scores or stock prices.*

applets (p. 66, KQ 2.4) Small programs that can be quickly downloaded and run by most browsers. Why it's important: *Web pages contain links to applets, which add multimedia capabilities.*

B2B (business-to-business) commerce (p. 74, KQ 2.5) Electronic sales or exchange of goods and services directly between companies, cutting out traditional intermediaries. Why it's important: *Expected to grow even more rapidly than other forms of e-commerce, B2B commerce covers an extremely broad range of activities, such as supplier-to-buyer display of inventories, provision of wholesale price lists, and sales of closed-out items and used materials—usually without agents, brokers, or other third parties.*

bandwidth (p. 34) Expression of how much data—text, voice, video, and so on—can be sent through a communications channel in a given amount of time. Why it's important: *Different communications systems use different bandwidths for different purposes. The wider the bandwidth, the faster data can be transmitted.*

bps (p. 35, KQ 2.1) Bits per second. Why it's important: *Data transfer speeds are measured in bits per second.*

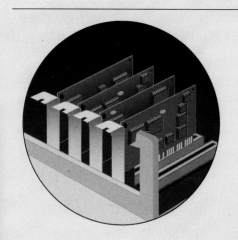

cable modem (p. 38, KQ 2.1) Device connecting a personal computer to a cable-TV system that offers an Internet connection. Why it's important: *Cable modems transmit data faster than standard modems.*

communications satellite (p. 39, KQ 2.1) Space station that transmits radio waves called *microwaves* from earth-based stations. Why it's important: *An orbiting satellite contains many communications channels and receives signals from ground microwave stations anywhere on earth.*

directory (p. 62, KQ 2.4) Search tool that provides lists of several categories of Web sites classified by topic, such as Business & Finance or Health & Fitness. Such a category is also called a *hypertext index,* and its purpose is to allow you to access information in specific categories by clicking on a hypertext link. Why it's important: *Directories are useful for browsing—looking at Web pages in a general category and finding items of interest. Search engines may be more useful for hunting specific information.*

distance learning (p. 73, KQ 2.5) Online education programs. Why it's important: *Distance learning provides educational opportunities for people who are not able to get to a campus; also distance-learning students don't have to spend time commuting, the scheduling is flexible, and they often have a greater selection of course offerings.*

domain (p. 43, KQ 2.3) A location on the Internet. Why it's important: *A domain name is necessary for sending and receiving e-mail and for many other Internet activities.*

download (p. 36, KQ 2.1) To transmit data from a remote computer to a local computer. Why it's important: *Downloading enables users to save files on their own computers for later use, which reduces the time spent online and the corresponding charges.*

DSL (digital subscriber line) (p. 38, KQ 2.1) A hardware and software technology that uses regular phone lines to transmit data in megabits per second. Why it's important: *DSL connections are much faster than regular modem connections.*

e-commerce (electronic commerce) (p. 73, KQ 2.5) Conducting business activities online. Why it's important: *E-commerce is not only widening consumers' choice of products and services but is also creating new businesses and compelling established businesses to develop Internet strategies.*

emoticons (p. 49, KQ 2.3) Keyboard-produced pictorial representations of expressions. Why it's important: *Emoticons can smooth online communication.*

FAQs (Frequently Asked Questions) (p. 49, KQ 2.3) Guides that explain expected norms of online behavior for a particular group. Why it's important: *Users should read a group's/site's FAQs to know how to proceed properly.*

flaming (p. 49, KQ 2.3) Writing an online message that uses derogatory, obscene, or inappropriate language. Why it's important: *Flaming should be avoided. It is a form of public humiliation inflicted on people who have failed to read FAQs or otherwise not observed netiquette (although it can happen just because the sender has poor impulse control and needs a course in anger management).*

frame (p. 60, KQ 2.4) An independently controllable section of a Web page. Why it's important: *A Web page designer can divide a page into separate frames, each with different features or options.*

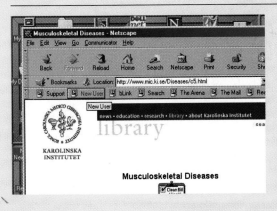

FTP (File Transfer Protocol) (p. 70, KQ 2.5) Method whereby you can connect to a remote computer called an *FTP site* and transfer publicly available files to your own microcomputer's hard disk. Why it's important: *The free files offered cover nearly anything that can be stored on a computer: software, games, photos, maps, art, music, books, statistics.*

gigabits per second (Gbps) (p. 35, KQ 2.1) 1 billion bits per second. Why it's important: *Gbps is a common measure of data transmission speed.*

home page (p. 53, KQ 2.4) Also called *welcome page;* Web page that, somewhat like the title page of a book, identifies the Web site and contains links to other pages at the site. Why it's important: *The first page you see at a Web site is the home page.*

hypertext (p. 52, KQ 2.4) System in which documents scattered across many Internet sites are directly linked, so that a word or phrase in one document becomes a connection to a document in a different place. Why it's important: *Hypertext links many documents by topics, allowing users to find information on topics they are interested in.*

hypertext markup language (HTML) (p. 52, KQ 2.4) Set of special instructions (called "tags" or "markups") used to specify Web document structure, formatting, and links to other documents. Why it's important: *HTML enables the creation of Web pages.*

HyperText Transfer Protocol (HTTP) (p. 54, KQ 2.4) Communications rules that allow browsers to connect with Web servers. Why it's important: *Without HTTP, files could not be transferred over the Web.*

instant messaging (IM) (p. 45, KQ 2.3) Any user on a given e-mail system can send a message and have it pop up instantly on the screen of anyone else logged onto that system. Why it's important: *If both parties agree, they can initiate online typed conversations in real time. As they are typed, the messages appear on the display screen in a small window.*

Internet service provider (ISP) (p. 40, KQ 2.2) Company that connects you through your communications line to its server, or central computer, which connects you to the Internet. Why it's important: *Unless they subscribe to an online information service (such as AOL) or have a direct network connection (such as a T1 line), microcomputer users need an ISP to connect to the Internet.*

Internet service provider

Telephony & conferencing
Make inexpensive phone calls; have online meetings.

Internet telephony (p. 69, KQ 2.4) Using the Net to make phone calls, either one-to-one or for audioconferencing. Why it's important: *Long-distance phone calls by this means are surprisingly inexpensive.*

ISDN (Integrated Services Digital Network) (p. 38, KQ 2.1) Hardware and software that allows voice, video, and data to be communicated over traditional copper-wire telephone lines (POTS). Why it's important: *ISDN provides faster data transfer speeds than do regular modem connections.*

Java (p. 66, KQ 2.4) Complex programming language that enables programmers to create animated and interactive Web pages using applets. Why it's important: *Java applets enhance Web pages by playing music, displaying graphics and animation, and providing interactive games.*

kilobits per second (Kbps) (p. 35, KQ 2.1) 1000 bits per second. Why it's important: *Kbps is a common measure of data transfer speed. The speed of a modem that is 28,800 bps might be expressed as 28.8 Kbps.*

list-serves (p. 48, KQ 2.3) E-mail mailing lists of people who regularly participate in discussion topics. To subscribe, the user sends an e-mail to the list-serve moderator and asks to become a member, after which he or she automatically receives e-mail messages from anyone who responds to the server. Why it's important: *Anyone connected to the Internet can subscribe to list-serve services. Subscribers receive information on particular subjects and can post e-mail to other subscribers.*

log on (p. 41, KQ 2.2) To make a connection to a remote computer. Why it's important: *Users must be familiar with log-on procedures to go online.*

megabits per second (Mbps) (p. 35, KQ 2.1) 1 million bits per second. Why it's important: *Mbps is a common measure of data transmission speed.*

menu (p. 41, KQ 2.2) List of commands or options offered by software programs. Why it's important: *Menus make software easier to use; the user can choose from a list instead of having to remember specific commands and addresses.*

netiquette (p. 49, KQ 2.3) "Network etiquette"—guides to appropriate online behavior. Why it's important: *In general, netiquette has two basic rules: (a) Don't waste people's time, and (b) don't say anything to a person online that you wouldn't say to his or her face.*

newsgroup (p. 71, KQ 2.5) Giant electronic bulletin board on which users conduct written discussions about a specific subject. Why it's important: *There are more than 30,000 newsgroup forums—which charge no fee—and they cover an amazing array of topics.*

newsreader (p. 72, KQ 2.5) Program included with most browsers that allows users to access a newsgroup and read or type messages. Why it's important: *Users need a newsreader to participate in a newsgroup.*

physical connection (p. 34, KQ 2.1) The wired or wireless means of connecting to the Internet. Why it's important: *Without physical connections, the Internet would be impossible, as would telephone and other types of communication connections.*

plug-in (p. 65, KQ 2.4) Also called a *player* or a *viewer;* program that adds a specific feature to a browser, allowing it to play or view certain files. Why it's important: *To fully experience the content of many Web sites, you will need to use plug-ins.*

protocol (p. 54, KQ 2.4) Set of communication rules for exchanging information. Why it's important: *HyperText Transfer Protocol (HTTP) provides the communications rules that allow browsers to connect with Web servers.*

The push technology from Entrypoint.com periodi[cally] delivers news that you preset according to your sp[ecifications]

push technology (p. 67, KQ 2.4) Software that automatically downloads information to your computer (as opposed to "pull" technology, in which you go to a Web site and pull down the information you want). Why it's important: *With little effort, users can obtain information that is important to them.*

real-time chat (RTC) (p. 72, KQ 2.5) Typed discussion ("chat") among participants who are online at the same time; it is like a telephone conversation except that messages are typed rather than spoken. Why it's important: *RTC provides a means of immediate electronic communication.*

site (p. 53, KQ 2.4) Computer with a domain name. Why it's important: *Sites provide Internet and Web content.*

scroll arrows (p. 60, KQ 2.4) Small up/down and left/right arrows located at the bottom and side of your screen display. Why it's important: *Clicking on scroll arrows with your mouse pointer moves the screen so that you can see the rest of the Web page, or the content displayed on the screen.*

scrolling (p. 60, KQ 2.4) Moving quickly upward or downward through text or other screen display, using the mouse and scroll arrows (or the arrow keys on the keyboard). Why it's important: *Normally a computer screen displays only part of, for example, a Web page. Scrolling enables users to view an entire document, no matter how long.*

search engine (p. 62, KQ 2.4) Search tool that allows you to find specific documents through keyword searches and menu choices, in contrast to directories, which are lists of Web sites classified by topic. Why it's important: *Search engines enable users to find Web sites of specific interest to them.*

spam (p. 50, KQ 2.3) Unsolicited e-mail in the form of advertising or chain letters. Why it's important: *Spam filters are available that can spare users the annoyance of receiving junk mail, ads, and other unwanted e-mail.*

streaming audio (p. 67, KQ 2.4) Process of downloading audio in which you can listen to the file while the data is still being downloaded to your computer. Why it's important: *Users don't have to wait until the entire audio is downloaded to hard disk before listening to it.*

streaming video (p. 67, KQ 2.4) Process of downloading video in which the data is transferred in a continuous flow so that you can begin viewing a file even before the end of the file is sent. Why it's important: *Users don't have to wait until the entire video is downloaded to hard disk before watching it.*

T1 line (p. 38, KQ 2.1) Traditional trunk line that carries 24 normal telephone circuits and has a transmission rate of 1.5 Mbps. Why it's important: *High-capacity T1 lines are used at many corporate, government, and academic sites; these lines provide greater data transmission speeds than do regular modem connections.*

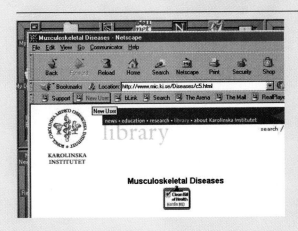

Telnet (p. 71, KQ 2.5) Program or command that allows you to connect to remote computers on the Internet. Why it's important: *This feature, which allows microcomputers to communicate successfully with mainframes, enables users to tap into Internet computers and access public files as though they were connected directly instead of, for example, through an ISP site.*

upload (p. 36, KQ 2.1) To transmit data from a local computer to a remote computer. Why it's important: *Uploading allows users to easily exchange files over networks.*

URL (Universal Resource Locator) (p. 54, KQ 2.4) String of characters that points to a specific piece of information anywhere on the Web. A URL consists of (1) the Web protocol, (2) the name of the Web server, (3) the directory (or folder) on that server, and (4) the file within that directory (perhaps with an extension such as html or htm). Why it's important: *URLs are necessary to distinguish among Web sites.*

Usenet (p. 71, KQ 2.5) Worldwide network of servers that can be accessed through the Internet. Why it's important: *Newsgroups take place on Usenet.*

Web browser (browser) (p. 53, KQ 2.4) Software that enables users to view Web pages and to jump from one page to another. Why it's important: *Users can't surf the Web without a browser. The two best known browsers are Microsoft's Internet Explorer and Netscape Communicator.*

webcasting (p. 67, KQ 2.4) Services, based on push technology, in which customized text, video, and audio are sent to the user automatically on a regular basis. Why it's important: *Users choose the categories, or the channels, of Web sites that will automatically send updated information. Thus, it saves time because users don't have to go out searching for the information.*

Web page (p. 53, KQ 2.4) Document on the World Wide Web that can include text, pictures, sound, and video. Why it's important: *A Web site's content is provided on Web pages. The starting page is the home page.*

Web portal (p. 61, KQ 2.4) Web site that groups popular features such as search tools, e-mail, electronic commerce, and discussion groups. The most popular portals are America Online, Yahoo!, Microsoft Network, Netscape, Lycos, Go Network, Excite Network, AltaVista, and WebCrawler. Why it's important: *Web portals provide an easy way to access the Web.*

Web site (website) (p. 53, KQ 2.4) Location of a Web domain name in a computer somewhere on the Internet. Why it's important: *Web sites provide multimedia content to users.*

window (p. 45, KQ 2.3) Rectangular area containing a document or activity. Why it's important: *This feature enables different outputs to be displayed at the same time on the screen. For example, users can exchange messages almost instantaneously while operating other programs.*

Chapter Review

Self-Test Questions

1. Today's data-transmission speeds are measured in _____, _____, _____, and _____ per second.

2. A(n) _____ connects a personal computer to a cable-TV system that offers an Internet connection.

3. A space station that transmits data as microwaves is a _____.

4. A company that connects you through your communications line to its server, which connects you to the Internet is a(n) _____.

5. A small rectangular area on the computer screen that contains a document or displays an activity is called a(n) _____.

6. _____ is writing an online message that uses derogatory, obscene, or inappropriate language.

7. A(n) _____ is software that enables users to view Web pages and to jump from one page to another.

8. A computer with a domain name is called a(n) _____.

9. _____ comprises the communications rules that allow browsers to connect with Web servers.

10. A(n) _____ is a program that adds a specific feature to a browser, allowing it to play or view certain files.

Multiple-Choice Questions

1. Kbps means _____ bits per second.
   a. 1 billion
   b. 1 thousand
   c. 1 million
   d. 1 hundred
   e. 1 trillion

2. A location on the Internet is called a(n) _____.
   a. network
   b. user ID
   c. domain
   d. browser
   e. web

3. In the e-mail address *Kim_Lee@earthlink.net.us*, *Kim_Lee* is the _____.
   a. domain
   b. URL
   c. site
   d. user ID
   e. location

4. Which of the following is not one of the four components of a URL?
   a. Web protocol
   b. name of the Web server
   c. name of the browser
   d. name of the directory on the Web server
   e. name of the file within the directory

5. Which of the following is the fastest method of data transmission?
   a. ISDN
   b. DSL
   c. modem
   d. T1 line
   e. cable modem

6. Which of the following is not a netiquette rule?
   a. consult FAQs
   b. flame only when necessary
   c. don't shout
   d. avoid huge file attachments
   e. avoid sloppiness and errors

True/False Questions

T F 1. AOL, EarthLink, and Prodigy are full-service ISPs.

T F 2. Users can do nothing to stop spam.

T F 3. A Web page is a document on the World Wide Web.

T F 4. Bookmarks let you store the URLs of Web pages you frequently visit.

T F 5. Search engines do not allow you to use keywords to initiate information searches.

Short-Answer Questions

1. Briefly define *bandwidth*.

2. What are three methods of data transmission that are faster than via a regular modem connection?

3. What does *log on* mean?

4. What is netiquette, and why is it important?

5. On the Web, many documents are "linked." What does that mean?

Concept Mapping

On a separate sheet of paper, draw a concept map, or visual diagram, linking concepts. Show how the following terms are related.

Cable modem	Physical connection
Communications satellite	Protocol
Directory	Real-time chat (RTC)
Domain	Site
Download	Search engine
DSL	T1 line
Home page	Upload
Hypertext	URL
Instant Messaging	Usenet
Internet service provider	Web browser
ISDN	Webcasting
List-serves	Web page
Newsgroup	Web portal
Newsreader	Web site

Knowledge in Action

1. Identify the four parts of the following URL:
    http://www.nps.gov/yell/swimming.htm

2. Some Web sites go overboard with multimedia effects, while others don't include enough. Locate a Web site that you think makes effective use of multimedia. What is the purpose of the site? Why is the site's use of multimedia effective? Take notes, and repeat the exercise for a site with too much multimedia and one with too little.

3. Visit the following job-hunting Web sites:

    www.monster.com

    www.occ.com

    www.careerpath.com

    www.cweb.com

    Investigate job offerings in a field you are interested in. Which is the easiest site to use? Why? What recommendations would you make for improving the site?

4. Distance learning uses electronic links to extend college campuses to people who otherwise would not be able to take college courses. Is your school or someone you know involved in distance learning? If so, research the system's components and uses. What hardware and software do students need in order to communicate with the instructor and classmates? What courses are offered? Prepare a short report on this topic.

5. It's difficult to conceive how much information is available on the Internet and the Web. One method you can use to find information among the millions of documents is to use a search engine, which helps you find Web pages based on typed keywords or phrases. Use your browser to visit the following search sites: *www.yahoo.com* and *www.goto.com*. Click in the Search box and then type the phrase "personal computers," then hit "Go," "Find it!", or hit the Enter key. When you locate a topic that interests you, print it out by choosing File, Print from the menu bar or by clicking the Print button on the toolbar. Report on your findings.

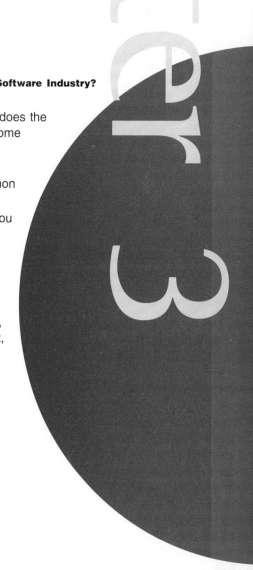

Software

The Power behind the Power

Key Questions

You should be able to answer the following questions.

3.1 **Online Software & Application Software Providers: Turning Point for the Software Industry?** What are some recent trends in online software?

3.2 **System Software** What are three components of system software; what does the operating system (OS) do; what is on the OS interface; and what are some common desktop, network, and portable OSs?

3.3 **Application Software: Getting Started** What are four ways of obtaining application software, tools available to help you learn them, three common types of files, and the types of software?

3.4 **Word Processing** What can you do with word processing software that you can't do with pencil and paper?

3.5 **Spreadsheets** What can you do with an electronic spreadsheet that you can't do with pencil and paper and a standard calculator?

3.6 **Database Software** What is database software, and what is personal information management software?

3.7 **Specialty Software** What are the principal uses of presentation graphics, financial, desktop publishing, drawing and painting, project management, and computer-aided design software?

A Visual Overview of This Chapter

1 Online Software & Application Software Providers. Application service providers **(ASPs)** are firms that lease software over the Internet. ASPs fit the strategy of users of **network computers**—thin clients, or inexpensive, stripped-down computers that connect to networks and run applications tied to servers. ASPs were anticipated by **enterprise resource planning (ERP) software,** which consists of large client/server software applications that help companies organize and operate their businesses. With ASPs, however, clients can rent instead of buy software to run off of servers.

2 System Software. Three basic types of system software are operating systems, device drivers, and utility programs. The first, the **operating system (OS),** consists of the master system of programs that manage the basic operations of the computer. Features of the OS are (1) **booting,** when it is loaded into the computer's main memory; (2) the **supervisor** or **kernel's** management of the CPU; (3) file management; (4) task management, sometimes including **multitasking,** when a computer works on several processes at a time; and (5) **formatting** or **initializing,** preparing a disk to store data or programs. The second, **device drivers,** allow input/output devices to communicate with the rest of the computer system. The third, **utility programs,** provide functions (such as data recovery) not supplied by other system software.

The OS **user interface** is the display screen that enables user interaction with the computer via keyboard or via the mouse, which has an onscreen **pointer.** Today's OSs have a **graphical user interface (GUI),** which consists of a **desktop,** or main interface screen, that allows you to select from **icons** (little pictorial symbols) or **menus** (lists of activities). Most icons have a **rollover** feature that pops up an explanation when a mouse rolls over it. Menus may be **pull-down** (from the screen top), **fly-out** (explode out to the right), **pull-up** (from screen bottom), and **pop-up** (anywhere on screen). **Toolbars** (top of screen) and **taskbars** (bottom) display frequently used icons and menus. The data and programs appear in a frame called a **window,** which can be resized or repositioned on screen. Most toolbars contain a **Help command** to provide answers to questions; for some tasks, **context-sensitive help** is available about that specific task.

There are three categories of **platforms,** or particular combinations of processors and OSs—for desktops/laptops, for networks, and for handhelds. (1) Principal desktop/laptop OSs are DOS, Macintosh OS, and the Microsoft Windows series. **DOS** was Microsoft's original OS. **Macintosh operating system** runs only on Apple Macintoshes. **Microsoft Windows** (versions 3.1, **95, 98,** and most recently **Me**) is the most popular OS for desktops and portables. (2) Principal network server OSs are **NetWare** from Novell; **Windows NT** and its successor **Windows 2000** from Microsoft; **Unix,** available in several versions, including Sun's **Solaris;** and **Linux,** a free version of Unix and a kind of **open-source software** modifiable by anyone. (3) Principal OSs for handhelds are **Palm OS,** which runs the Palm and the Visor, and **Windows CE** (a slimmed-down version of Windows 95), which became **Pocket PC,** a simpler version.

3 Application Software. There are five ways of obtaining application software. (1) You can use commercial software, which is **copyrighted** and is available for a fee under **software license,** meaning it may not be duplicated without permission. (2) You can freely duplicate **public-domain software,** which is not copyrighted. (3) You can use copyrighted **shareware,** which is distributed free but requires a fee for continued use. (4) You can use **freeware,** which is copyrighted but distributed free. (5) You can lease **rentalware,**

the concept behind ASPs. To learn software you can use step-by-step **tutorials** (lessons) or **documentation** (reference guides).

The purpose of application software is to manipulate raw data into files of information. A **file** is a named collection of data or a program existing in secondary storage. Three types are **document files** (created by word processing), **worksheet files** (created by spread-sheets), and **database files** (created by database management programs). Files can be **imported** or acquired from other programs and **exported** or sent to other programs.

Productivity software, which is designed to make users more productive, may exist in stand-alone form, such as word processing or spreadsheet programs. Or several programs may be combined in an **office suite.** Some productivity software exists as **groupware,** which several users may share online.

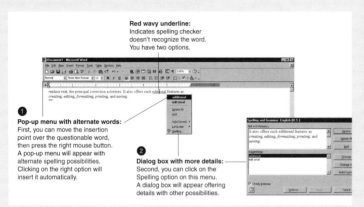

Red wavy underline: Indicates spelling checker doesn't recognize the word. You have two options.

❶ Pop-up menu with alternate words: First, you can move the insertion point over the questionable word, then press the right mouse button. A pop-up menu will appear with alternate spelling possibilities. Clicking on the right option will insert it automatically.

❷ Dialog box with more details: Second, you can click on the Spelling option on this menu. A dialog box will appear offering details with other possibilities.

4 Word Processing. Perhaps the most useful productivity program is **word processing** (Microsoft Word, Corel WordPerfect), which allows you to create, edit, format, print, and store text material, using mouse and keyboard. A computer keyboard contains **special-purpose keys** to enter, delete, and edit data and execute commands and **function keys** (F1, F2, etc.) for executing commands specific to the software. Several keystrokes may be combined in one or two keystroke commands, or a **macro.**

Three features that help you create documents are the **cursor,** the movable symbol on the display screen; **scrolling,** the ability to move up, down, or sideways through the text; and **word wrap,** which continues text on the next line automatically when you reach the end of a line. An **outline feature** enables you to show the hierarchy of headings within a document. Features for editing documents are insert and delete, undelete, find and replace, cut/copy and paste, **spelling checker, grammar checker,** and **thesaurus** (for presenting alternate words).

Formatting, or determining the appearance of a document, is helped with the aid of **templates,** preformatted documents, and **wizards,** which answers your questions and formats a document. Aspects of formatting are **fonts,** or typefaces and type sizes; spacing and columns; margins and **justification** (spacing of words in a line); page numbers and page **headers/footers** (repeated text at top/bottom); and other formatting such as use of **clip art** (ready-made pictures). Most formatting matters come from the manufacturer with automatically standardized settings, or **default settings.** Most programs give several options for printing out documents. Documents may be **saved,** or preserved, in secondary storage, such as hard disk.

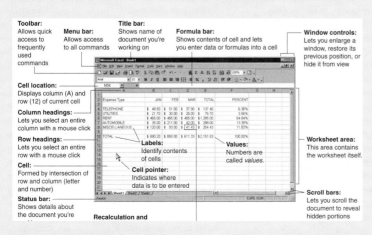

Toolbar: Allows quick access to frequently used commands

Menu bar: Allows access to all commands

Title bar: Shows name of document you're working on

Formula bar: Shows contents of cell and lets you enter data or formulas into a cell

Window controls: Lets you enlarge a window, restore its previous position, or hide it from view

Cell location: Displays column (A) and row (12) of current cell

Column headings: Lets you select an entire column with a mouse click

Row headings: Lets you select an entire row with a mouse click

Cell: Formed by intersection of row and column (letter and number)

Status bar: Shows details about the document you're

Labels: Identify contents of cells

Cell pointer: Indicates where data is to be entered

Recalculation and

Values: Numbers are called *values.*

Worksheet area: This area contains the worksheet itself.

Scroll bars: Lets you scroll the document to reveal hidden portions

5 Spreadsheets. The **electronic spreadsheet** (Microsoft Excel, Corel Quattro Pro, Lotus 1-2-3) allows you to create tables and financial schedules by entering data and formulas into rows and columns arranged as a grid. A spreadsheet file contains **worksheets,** or single tables, with several related worksheets collected into a **workbook.**

Spreadsheet grids are organized with **column headings** across the top, **row headings** down the left side, and various **labels** or descriptive text. Columns and rows intersect in a **cell,** and its position is called a **cell address;** several adjacent cells constitute a **range.** A number entered in the cell is called a **value,** and its location is indicated by a **cell pointer** or **spreadsheet cursor. Formulas,** or instructions for calculations, are used to manipulate data; built-in formulas are called **functions.** Values can be changed and then recomputed, the process of **recalculation,** an important reason for the popularity of the spreadsheet, since it allows you to do **what-if analysis**—see how changing numbers can change outcomes. For specialized needs, **worksheet templates,** custom-designed forms, are available. A nice feature of spreadsheets is the ability to create **analytical graphics,** graphical forms that make numeric data easier to analyze.

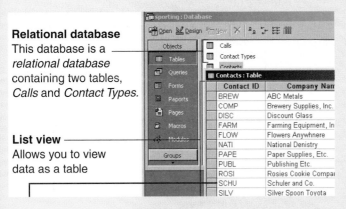

Relational database
This database is a *relational database* containing two tables, *Calls* and *Contact Types.*

List view
Allows you to view data as a table

6 **Database Software.** A **database** is a collection of interrelated files, and **database software** (Microsoft Access, Corel Paradox, Lotus Approach) is a program that controls the structure of a database and access to the data. The most widely used form is the **relational database,** in which data is organized into related tables. Each **table** contains **records** (rows) and **fields** (columns). The records within the various tables in a database are linked by a **key field,** a common identifier that is unique, such as a Social Security number.

You can find what you want with a **query,** locating and displaying records. Records can also be **sorted,** as alphabetically, numerically, or geographically. Search results can then be saved, or they can be put into different formats, printed out, copied and placed in other documents, or transmitted (as via e-mail) to someone else. A specialized type of database software is a **personal information manager (PIM),** which helps you manage addresses, appointments, and to-do lists.

7 **Specialty Software.** Among the many thousands of specialized productivity programs available, the following are important.

Presentation graphics software (Microsoft PowerPoint, Corel Presentations, Lotus Freelance Graphics) uses graphics, animation, sound, and data to make visual presentations, commonly called *slide shows.* Design templates and content templates are available to help users get started. Material may be viewed from several perspectives (Outline, Slide, Notes Page, and Slide Show views). Presentations may be dressed up with clip art, textures, audio clips, and the like.

Financial software includes personal-finance managers, entry-level accounting programs, and business financial-management packages. **Personal-finance managers** (Quicken, Microsoft Money) help track income and expenses, write checks, do online banking, and plan financial goals. Tax programs help with tax preparation and filing. Some financial programs automate bookkeeping and payroll tasks. Others help in business start-ups or in making investments.

Desktop publishing (DTP) software (QuarkXPress, PageMaker, Microsoft Publisher) enables users to mix text and graphics to produce high-quality output for commercial printing. Users can choose various type and layout styles and can use files (text, graphics) from other programs. **Drawing programs** allows users to design and illustrate objects and products; **painting programs** allow them to simulate painting on screen.

Project management software helps users plan and schedule people, costs, and resources required to complete a project on time.

Computer-aided design (CAD) programs are used to design products, structures, engineering drawings, and maps. **CAD/CAM software**—for computer-aided design/computer-aided manufacturing—allows products designed with CAD to be input automatically into an automated manufacturing system that makes the products.

uestion: "I've heard a lot about the idea of renting software programs over the Internet, rather than buying and installing them on a PC. How well do you think this is going to work, when will it start to show up in the market, and how much is it going to cost?"

Answer: "I don't know, I don't know, and I don't know. But I do know that it's coming fast."

So wrote David Einstein, columnist and West Coast bureau chief of *Forbes Digital Tool,* in reply to a reader's inquiry.[1] Actually, by now we do know: It's already here, and it's cheap or free. How well it works, however, remains to be seen.

Everything else seems to be going on the Internet. Why not word processors, spreadsheets, and other software? Indeed, even now, if you're tired of using a paper calendar or datebook, you can go online to jump.com, yahoo.com, or when.com and start using a Web calendar to keep track of appointments, deadlines, and birthdays. (Cautionary note: If, however, some day the Web site doesn't come up on your computer screen, as happened to *Time* magazine technology writer Anita Hamilton, you'll be back to jotting down your appointments on Post-It notes.)[2] Other Web-based software has also been offered free for some time—browsers, e-mail, and address books, for example—designed to keep users coming back to a particular portal such as Alta Vista or Yahoo!

But for businesses in particular—and ultimately probably for you—it does make sense to rent software rather than buy it. Let's take a look at this concept as we begin our discussion of software.

3.1 Online Software & Application Software Providers: Turning Point for the Software Industry?

KEY QUESTION

What are some recent trends in online software?

Software, or *programs,* consists of the instructions that tell the computer how to perform a task. As we said in Chapter 1, there are two types of software: (1) *Application software* enables you to perform specific tasks—solve problems, perform work, entertain yourself. Examples are word processors, spreadsheets, and video games. (2) *System software* enables the computer to perform essential operating tasks and run the application software. Examples you may have heard of are Windows 98, Windows 2000, DOS, Unix, and Mac OS. We discuss both types of software in this chapter.

Software has been part and parcel of stand-alone computer systems since the early days. Even now, all those people lugging their laptops across campus or through airports are packing most of their software with them, stored on the hard drive. But the rise of the Internet has made online software a likely next step.

Online Software & the Application Service Provider

Jim Obsitnik, 30, is a software salesman and avid jogger—so avid, in fact, that he keeps track of his miles and running times on an Excel spreadsheet. Business travel used to be inconvenient because he had to record his jogging data on slips of paper (which could get lost), then enter it into his PC when he got home. No more. Now when he's out of town he can log onto a free, password-protected runner's Web site called Desktop.com and download his spreadsheet. He doesn't even have to have his own computer; he can use someone else's.[3]

Is this a picture of the future? Every month you pay the phone bill, the electric bill, maybe the ISP (Internet service provider) bill. What about paying an ASP—an application software provider—as well as an ISP? Or even getting both ISP and ASP for free?

An **_ASP (application service provider)_ is a firm that leases software over the Internet.** You no longer have to buy software in a store, in shrink-wrapped packages. Instead, you can simply download a particular program when you need it, for as long as you need it.

Could this be a major turning point for computers and communications? "The [ASP] trend is rocking the $8 billion software industry," says one analysis, "threatening titans such as Microsoft, IBM," and others.[4] Among the purveyors of Web-based software are Damango, Desktop.com, Halfbrain.com, iAmaze.com, Mi8, myWebOS, Sun Microsystems, ThinkFree.com, and Visto.com. *(See* ● *Panel 3.1.)* Mi8, for example, offers Microsoft Office and Lotus Notes—which are well-known business application programs—for $21.95 a month. The ASP called myWebOS offers a free word processor, e-mail, and calendar; free spreadsheet and presentation software is in the works. Sun Microsystems *(www.sun.com)* offers StarOffice and Star-Portal. Microsoft also plans to provide its Office program as a monthly service. In addition, there are ASPs that make available online specialized software to handle tasks for small businesses, such as office services (HotOffice.com), expense tracking services (TimeBills.com), or human resources (Employease).[5]

● **PANEL 3.1**

Web pages of some application service providers

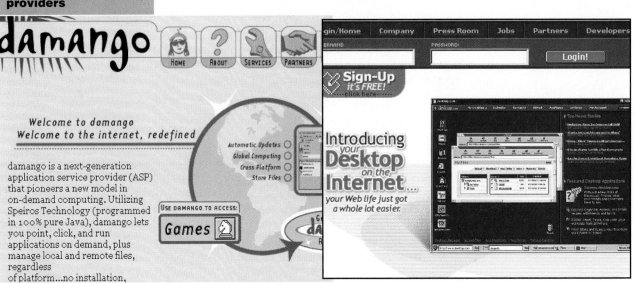

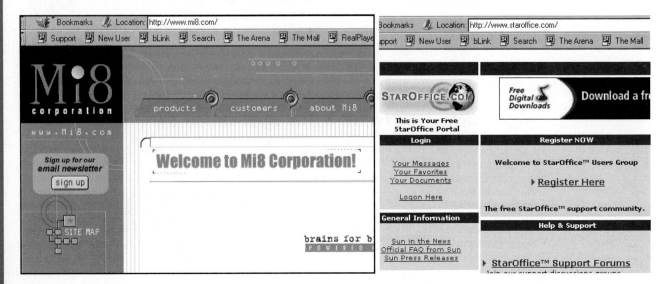

Network Computers Revisited: "Thin Clients" versus "Fat Clients"

We can link the growth in ASPs to the concept of *network computer*, proposed a few years ago by Larry Ellison, CEO of database-software maker Oracle Corp. A **_network computer_ is an inexpensive, stripped-down computer that connects people to networks and runs applications tied to servers.**

Out of this grew a distinction between "thin clients" and "fat clients." A *client,* you'll recall from Chapter 1, is one part of a client/server network. The *server* is the central computer that holds collections of data and programs; the clients are the PCs, workstations, and other computers on the network that use the data and programs from the server.

A *fat client* is a regular computer, perhaps a PC, that is on a network. Often it contains software with a great many features. Such "bloatware" is the result of software makers constantly trying to top themselves when they issue new versions. To run these programs efficiently, a computer requires lots of main memory and hard-disk storage capacity, as well as a powerful microprocessor. A *thin client,* by contrast, is a network computer—a stripped-down computer without hefty microprocessors or even much storage—that is supposed to operate as an inexpensive terminal tied to a server. (See ● *Panel 3.2.*)

● **PANEL 3.2**
Old way and new way of getting software

Fat client

Users provide their own software and are usually responsible for any upgrades of hardware and software. Data can be input or downloaded from online sources.

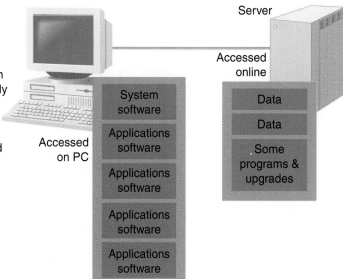

Thin client

Users download not only data but also different kinds of applications software from an online source.

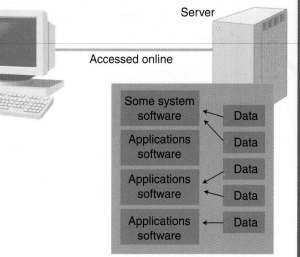

Contrary to Ellison's expectations, the widespread appearance of network computers/thin clients was delayed because the prices of regular, fat-client PCs dropped sharply, below $1000. But in the meantime, another movement advanced the cause of online software—the development of enterprise resource planning software.

From ERP to ASP: The Evolution of "Rentalware"

About 10 years ago, businesses began to adopt client/server arrangements. The tasks once performed by mainframes and minicomputers were being divided between desktop computers and servers. To ease this transition, companies relied on business and accounting programs called ERP software. ___ERP (enterprise resource planning) software___ consists of large client/server software applications that help companies organize and operate their businesses. The makers of such software include SAP, Oracle, and PeopleSoft, as well as Microsoft.

ERP is expensive—a corporation might well spend $30 million on such a system. And the risk was all on the buyer. Even if the system didn't work as well as it should, the buyer would be inclined to keep it, because it represented such a huge investment.

Enter the ASP model. Instead of buying an ERP system, a company can "rent" the same thing, and the installation and management of all the equipment and software becomes someone else's headache.[6] "The Internet creates an opportunity to change the way people manage information technology," says Gary Bloom, an executive with Oracle Corp. "You will buy software as a service, just the way you buy telephone service today."[7]

Whatever the future of Web-based programs, not many experts believe that today's principal arrangement—software installed on the hard disk of a desktop or portable computer—will ever entirely disappear. And while a network terminal won't require system software, the traditional PC still does. Let us therefore discuss this next.

CONCEPT CHECK

What is an application software provider?

Explain the concept of thin clients versus fat clients.

What is enterprise resource software?

3.2 System Software

As we've said, software is of two types. *Application software* is software that can perform useful work on general-purpose tasks, such as word processing or spreadsheets, or that is used for entertainment. There are hundreds of application software packages available for personal computers. *System software*, which you will find already installed if you buy a new computer, enables the application software to interact with the computer and helps the computer manage its internal and external resources. There are only a handful of systems software packages for personal computers.

There are three basic types of system software that you need to know about. *(See ● Panel 3.3.)*

- **Operating systems:** An *operating system* is the principal component of system software in any computing system.
- **Device drivers:** *Device drivers* help the computer control a peripheral device.

● PANEL 3.3
Three components of system software
An operating system is required for application software to run on your computer.

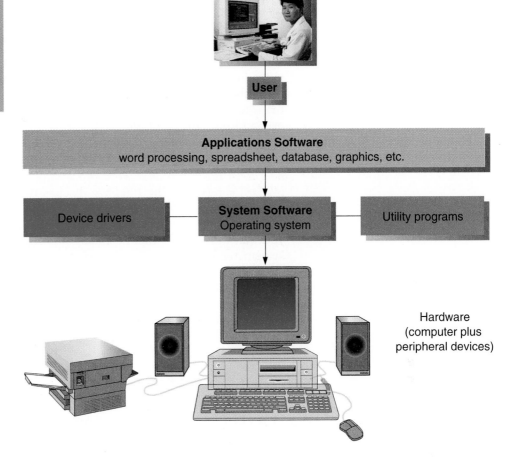

User

Applications Software
word processing, spreadsheet, database, graphics, etc.

Device drivers — **System Software**
Operating system — Utility programs

Hardware
(computer plus peripheral devices)

- **Utility programs:** *Utility programs* are generally used to support, enhance, or expand existing programs in a computer system.

Let's consider these briefly, then return to the operating system for a more extended discussion, since it is the most important of the three.

The Operating System: What It Does

The _**operating system (OS)**_ **consists of the master system of programs that manage the basic operations of the computer.** These programs provide resource management services of many kinds. In particular, they handle the control and use of hardware resources, including disk space, memory, CPU time allocation, and peripheral devices. The operating system allows you to concentrate on your own tasks or applications rather than on the complexities of managing the computer.

Some important functions of the operating system are as follows:

- Booting: The work of the operating system begins as soon as you turn on, or "boot," the computer. _**Booting**_ **is the process of loading an operating system into a computer's main memory.** This loading is accomplished by a program that is stored permanently in the computer's electronic circuitry. When you turn on the machine, programs called *diagnostic routines* test the main memory, the central processing unit, and other parts of the system to make sure they are running properly. Next, BIOS (for basic input/output system) programs are copied to main memory and help the computer interpret keyboard characters or transmit characters to the display screen or to a diskette. Then the boot program obtains the operating system, usually from hard disk, and loads it into the computer's main memory, where it remains until you turn the computer off.

- **CPU management:** The central component of the operating system is the *supervisor*. Like a police officer directing traffic, the <u>**supervisor,**</u> **or kernel, manages the CPU. It remains in main memory while the computer is running and directs other "nonresident" programs to perform tasks that support application programs.** The operating system also manages memory—it keeps track of the locations within main memory (and on hard disk) where the programs and data are stored.

- **File management:** A *file* is a named collection of related information. A file can be a *program*, such as a word processing program. Or it can be a *data file*, such as a word processing document, a spreadsheet, images, songs, and the like. We discuss files in more detail later in the chapter.

  Files containing programs and data are located in many places on your hard disk and other secondary-storage devices. The operating system records the storage location of all files. If you move, rename, or delete a file, the operating system manages such changes and helps you locate and gain access to it. For example, you can *copy*, or duplicate, files and programs from one disk to another. You can *back up*, or make a duplicate copy of, the contents of a disk. You can *erase*, or remove, from a disk any files or programs that are no longer useful. You can *rename*, or give new file names to, the files on a disk.

- **Task management:** In word processing, the operating system accepts input data, stores the data on a disk, and prints out a document— seemingly simultaneously. Some operating systems can handle more than one program—word processing, spreadsheet, database—at the same time; the programs are displayed in separate windows on the screen. Others can accommodate the needs of several users at the same time. All these examples illustrate process, or task, management. A task is an operation such as storing, printing, or calculating.

  <u>**Multitasking**</u> **is the activity in which a computer works on more than one process at a time.** For instance, you might write on one word-processing document while printing out another and downloading a third from a Web site. (Note, however, that unless a computer has more than one processor, it can perform only a single task at a time. Most PCs have only one processor.)

- **Formatting:** <u>**Formatting,**</u> **or initializing, a disk is the process of preparing that disk so that it can store data or programs.** Today it is easier to buy preformatted diskettes, which bear the label "Formatted IBM" (for floppies designed to run on PCs) or "Formatted Macintosh." However, it's useful to know how to format a blank floppy disk or reformat a diskette that wasn't intended for your machine.

Device Drivers: Running Peripheral Hardware

<u>*Device drivers*</u> **are specialized software programs that allow input and output devices to communicate with the rest of the computer system.** Many basic device drivers come with the system software when you buy a computer. If, however, you buy a new peripheral device, such as a mouse, scanner, or printer, the package will include a device driver (probably on a CD-ROM). You'll need to install the driver on your computer's hard-disk drive (by following the manufacturer's instructions) before the device will operate.

Utilities: Service Programs

<u>*Utility programs,*</u> **also known as service programs, perform tasks related to the control and allocation of computer resources. They enhance existing**

functions or provide services not supplied by other system software programs. Most computers come with utilities built in as part of the system software. (Windows 95/98 offers several of them.) However, they may also be bought separately as external utility programs (such as Norton Desktop, 911 Utilities, and PC Tools).

Among the tasks performed by utilities are the backing up of data, recovery of lost data, and identification of hardware problems.

CONCEPT CHECK

What distinguishes the three types of system software?

Explain the important features of the operating system.

Now that we've described the three components of system software—the operating system, device drivers, and utility programs—let's take a more detailed look at the operating system, which is the most important of the three.

The Operating System's User Interface: Your Computer's Dashboard

When you turn on the power switch of your personal computer and it goes through the boot process, it will display a starting screen. From the screen, you choose the application programs you want to run or the files of data you want to open. This part of the operating system is called the ***user interface***, **the user-controllable display screen that allows you to communicate, or interact, with the computer.** Like the dashboard on a car, the user interface has gauges that show you what's going on and switches and buttons for controlling what you want to do.

You can interact with this display screen using the keys on your keyboard, but you will also frequently use your mouse. The mouse allows you to direct an on-screen pointer to perform any number of activities. The ***pointer* usually appears as an arrow, although it changes shape depending on the application.** The mouse is used to move the pointer to a particular place on the display screen or to point to little symbols, or icons. You can activate the function corresponding to the symbol by pressing ("clicking") buttons on the mouse. Using the mouse, you can pick up and slide ("drag") an image from one side of the screen to the other or change its size. *(See ● Panel 3.4, next page.)*

In the beginning, personal computers had *command-driven* interfaces, which required you to type in strange-looking instructions (such as "copy a:\filename c:\" to copy a file from a floppy disk to a hard disk). In the next version, they had *menu-driven interfaces,* in which you could use the arrow keys on your keyboard (or a mouse) to choose a command from a menu, or list of activities. Today the computer's "dashboard" is usually a ***graphical user interface (GUI)*, pronounced "gooey," that allows you to use a mouse or keystrokes to select icons (little symbols) and commands from menus (lists of activities).** The GUIs on the PC and on the Apple Macintosh (which was the first easy-to-use personal computer available on a wide scale) are somewhat similar. Once you learn one version, it's fairly easy to learn the other. However, the best-known GUI is that of Microsoft Windows, the latest version of which is shown on the opposite page. *(See ● Panel 3.5, page 95.)*

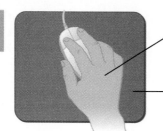

Resting your hand on the mouse, use your thumb and outside two fingers to move the mouse on your desk or mouse pad. Use your first two fingers to press the mouse buttons.

Mouse pad provides smooth surface.

Left button. Click once on an item on screen to select it. Click twice to perform an action.

Right button. Click on an object to display a shortcut list of options.

Term	Action	Purpose
Point	Move mouse across desk to guide pointer to desired spot on screen. The pointer assumes different shapes, such as arrow, hand, or I-beam, depending on the task you're performing.	To execute commands, move objects, insert data, or similar actions on screen
Click	Press and quickly release left mouse button.	To select an item on the screen
Double-click	Quickly press and release left mouse button twice.	To open a document or start a program
Drag and drop	Position pointer over item on screen, press and hold down left mouse button while moving pointer to location in which you want to place item, then release.	To move an item on the screen
Right-click	Press and release right mouse button.	To display a shortcut list of commands, such as a pop-up menu of options

Outlook Express: Part of Microsoft's browser, Internet Explorer, that enables you to use e-mail.

Microsoft Network: Click here to connect to Microsoft Network (MSN), the company's online service.

Norton Protected: Click here to activate anti-virus software.

Network Neighborhood: If your PC is linked to a network, click here to get a glimpse of everything on the network.

My Documents: Where your documents are stored unless you specify otherwise.

My Computer: Gives you a quick overview of all the files and programs on your PC.

Documents: Multitasking capabilities allow users to smoothly run more than one program at once.

Start menu: After clicking on the start button, a menu appears, giving you a quick way to handle common tasks. You can launch programs, call up documents, change system settings, get help, and shut down your PC.

Start button: Click for an easy way to start using the computer.

Taskbar: Gives you a log of all programs you have opened. To switch programs, click on the icon buttons on the taskbar.

Multimedia: Windows 98 features sharper graphics and improved video capabilites compared to Windows 95.

PANEL 3.5
A graphical user interface
This is for Windows 98.

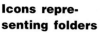
**Icons repre-
senting folders**

To-do list

Class
schedules

Term paper

Finances

Letters

Three features of a GUI are the *desktop, icons,* and *menus.*

- **Desktop:** After the boot-up process is completed, the first screen you will encounter is the *desktop,* a term that embodies the idea of folders of work (memos, schedules, to-do lists) on a businessperson's desk. The **_desktop_, which is the operating system's main interface screen, displays pictures (icons) that provide quick access to programs and information.**

- **Icons and rollovers:** We're now ready to give a formal definition: **_Icons_ are small pictorial figures that represent programs, data files, or procedures.** For example, a trash can represents a place to dispose of a file you no longer want. If you click your mouse pointer on a little picture of a printer, you can print out a document. One of the most important icons is the *folder,* a representation of a manila folder; folders are the collections of files in which you store your documents and other data.

  Of course, you can't always be expected to know what an icon or graphic means. Most icons have a **_rollover_ feature, in which a small text box explaining the icon's function appears when you roll the mouse pointer over the icon.** A rollover may also produce an animated graphic.

Icon: Symbol representing a program, data file, or procedure. Icons are designed to communicate their function, such as a floppy disk for saving.

Rollover

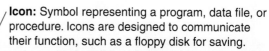

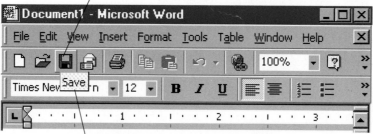

Rollover: When you roll your mouse cursor over an icon or graphic, a small box with text appears that briefly explains its function.

- **Menus:** Like a restaurant menu, a **_menu_ offers you a list of options to choose from**—in this case, a list of commands for manipulating data, such as Print or Edit. Menus are of several types. Resembling a pull-down window shade, a **_pull-down menu_, also called a drop-down menu, is a list of options that pulls down from the top of the screen.** For example, if you use the mouse to "click on" (activate) a command (for example, File) on the menu bar, you will see a pull-down menu offering further commands. Choosing one of these options may produce further menus called **_fly-out menus_, menus that seem to explode out to the right.**

  A **_pull-up menu_ is a list of options that pulls up from the bottom of the screen.** In Windows 98, a pull-up menu appears in the lower left-hand corner when you click on the Start button.

  A **_pop-up menu_ is a list of command options that can "pop up" anywhere on the screen when you click the right mouse button.** Pop-up menus are not connected to a toolbar (explained later).

Suppose you want to go to a document—say, a term paper you've been working on. There are two ways to begin working from the Windows 98

Types of menus

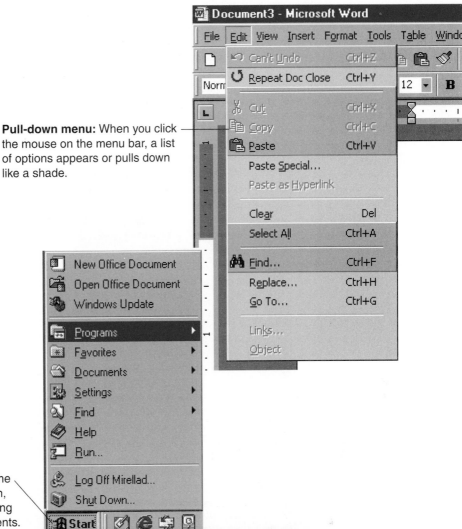

Pull-down menu: When you click the mouse on the menu bar, a list of options appears or pulls down like a shade.

Pull-up menu: When you click the mouse cursor on the *Start* button, it produces a pull-up menu offering access to programs and documents.

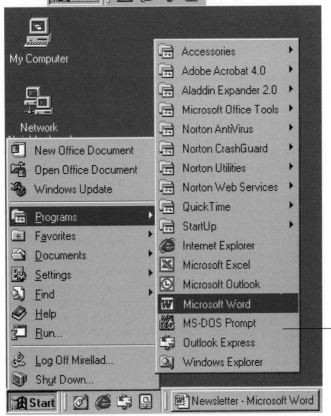

Fly-out menu: Moving the mouse cursor to an option on the pull-up menu produces a fly-out menu with more options.

desktop: (1) You can click on the Start button at lower left and then make a selection from the pull-up menu that appears. Or (2) you can click on one of the icons on the desktop, probably the most important of which is the *My Computer* icon, and pursue the choices offered there. Either way, the result is the same: The document will be displayed on the window. *(See ● Panel 3.6.)*

From Start menu

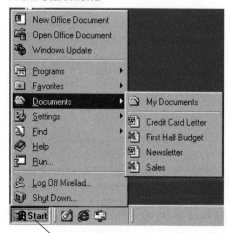

Click on *Start* button to produce Start menu, then go to *Documents* option.

Click on C, which opens a window that provides access to information stored on your hard disk.

From My Computer icon

Click on *My Computer* icon, which opens a window that provides access to information on your computer.

Click on *My Documents* icon, which opens a window providing access to document files and folders.

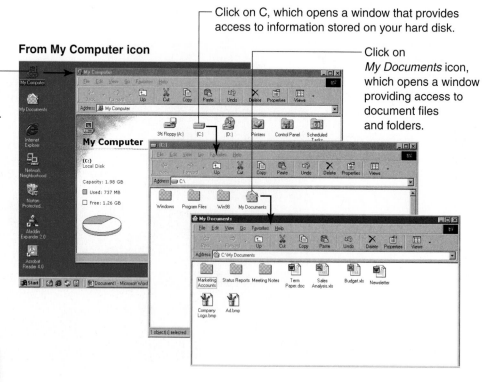

● PANEL 3.6

Two ways to go to a document in Windows 98

Once you're past the desktop, the GUI's opening screen, if you click on the *My Computer* icon you will encounter toolbars and windows.

Partial toolbar

- **Toolbar:** A ***toolbar*** **is a bar across the top of the display window. It displays menus and icons representing frequently used options or commands.** Examples of menus are File, Edit, View, Favorites, and Help. An example of an icon is the picture of a printer, which issues a print command. **A graphic toolbar appearing at the bottom of the Windows screen, showing the application running, is called a *taskbar*.**
- **Windows:** When spelled with a capital "W," Windows is the name of Microsoft's system software (Windows 95, 98, Me, and so on). When spelled with a lowercase "w," a ***window*** **is a rectangular frame on the**

Some windows

Minimize: Click here to shrink window so it collapses to an icon on the desktop.

Maximize: Click here to enlarge window so it fills the screen.

Close: Click here to exit a file and remove its window from your display.

computer display screen. Through this frame you can view a file of data—such as a document, spreadsheet, or database—or an application program.

In the right-hand corner of the Windows 98 toolbar are three icons that represent *Minimize, Maximize,* and *Close.* By clicking on these icons, you can *minimize* the window (shrink it down to an icon at the bottom of the screen), *maximize* it (enlarge it), or *close* it (exit the file and make the window disappear).

You can also *move* the window around the desktop, using the mouse.

Finally, you can create *multiple windows* to show operations going on concurrently. For example, one window might show the text of a paper you're working on, another might show the reference section for the paper, and a third might show something you're downloading from the Internet.

● **The Help command:** Don't understand how to do something? Forgotten a command? Accidentally pressed some keys that messed up your screen layout and you want to undo it? Most toolbars contain a **_Help command_**, **which leads to a table of contents, an index, and a search feature that can help you locate answers. In addition, many applications have context-sensitive help,** which leads you to information about the task you're performing. *(See ● Panel 3.7.)*

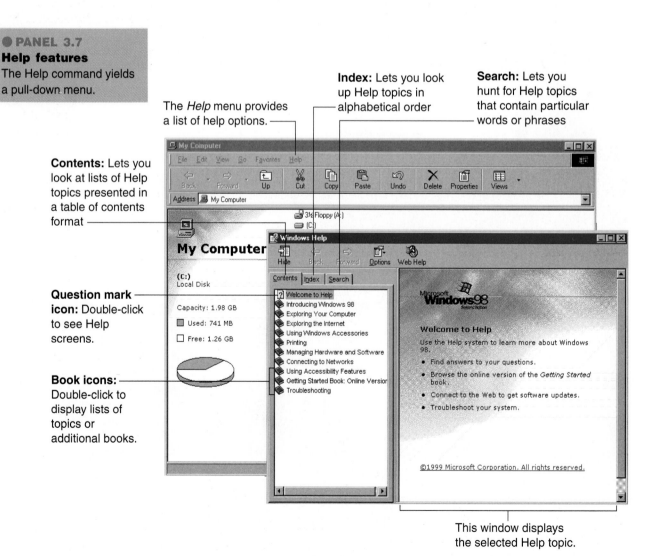

● PANEL 3.7
Help features
The Help command yields a pull-down menu.

Contents: Lets you look at lists of Help topics presented in a table of contents format

Question mark icon: Double-click to see Help screens.

Book icons: Double-click to display lists of topics or additional books.

The *Help* menu provides a list of help options.

Index: Lets you look up Help topics in alphabetical order

Search: Lets you hunt for Help topics that contain particular words or phrases

This window displays the selected Help topic.

Describe the features of the GUI: desktop, icons, and the various kinds of menus.

What do toolbars and windows enable you to do?

What is the Help command?

Common Desktop Operating Systems: From DOS to Macintosh to Windows 3.1, 95, 98, & Me

The _**platform**_ **is the particular processor model and operating system on which a computer system is based.** For example, there are "Mac platforms" (Apple Macintosh) and "Windows platforms" or "PC platforms" (for personal computers such as Dell, Compaq, Gateway, Hewlett-Packard, or IBM that run Microsoft Windows). Sometimes the latter are called "_Wintel_ platforms," for "Windows + Intel," because they often combine the Windows operating system with the Intel processor chip. (We discuss processors in Chapter 4.)

Despite the dominance of the Windows platform, many so-called _legacy systems_ are still in use. A legacy system is an older, outdated, yet still functional technology, such as the DOS operating system. You may find yourself having to use DOS at some point.

Let us quickly describe the principal platforms used on desktop computers: DOS, the Macintosh OS, and the Windows series (3.1, 95, 98, and Me). Desktop operating systems are used mainly on single-user computers (both desktops and laptops) rather than on mainframes or servers. We discuss operating systems for servers and for portable information appliances shortly.

- DOS—the old-timer: _**DOS**_ (rhymes with "boss")—_**Disk Operating System**_—**was the original operating system produced by Microsoft and had a hard-to-use command-driven user interface.** Its initial 1982 version was designed to run on the IBM PC as PC-DOS. Later Microsoft licensed the same system to other computer makers as MS-DOS. With the growing popularity of cheaper PCs produced by these companies, MS-DOS came to dominate the industry. Two years before the advent of Windows 95, which eventually grew out of DOS, there were reportedly more than 100 million users of DOS, which at that time made it the most popular software ever adopted—of any sort.

- The Macintosh operating system—for the love of Mac: The _**Macintosh operating system (Mac OS),**_ **which runs only on Apple Macintosh computers, set the standard for icon-oriented, easy-to-use graphical user interfaces.** The software generated a strong legion of fans shortly after its launch in 1984 and inspired Microsoft to upgrade DOS to the more user-friendly Windows operating systems. Much later, in 1998, Apple introduced its iMac computer (the "i" stands for Internet), which added other capabilities such as small-scale networking. (See ● Panel 3.8.)

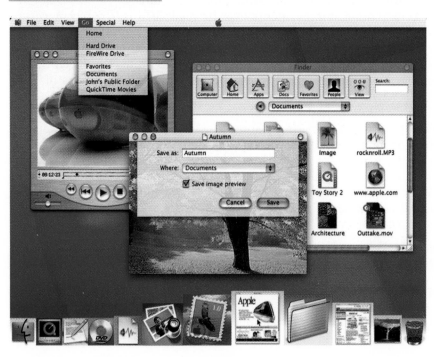

The Apple iMac

Can you run Windows-based application software on a Macintosh? Indeed, you can, using special add-on software (such as Virtual PC). However, it runs more slowly, and most Mac users prefer to stick with application programs written specifically for the Apple machine. Macintosh is still considered king in areas such as desktop publishing, and programs for games or for common business uses such as word processing and spreadsheets are widely available. For more specialized applications, however, most programs are written for the Windows platform.

Even so, the newest version of the operating system, Mac OS X (called "ten"), offers some superior features. The user interface known as Aqua has Hollywood-like tricks (thanks to Apple chairman Steve Jobs's experience running Pixar Animation Studios, makers of *Toy Story* and other animated films). "Jelly-colored onscreen buttons pulse as if alive," says one description. "Menu borders are translucent, allowing you to see the documents under them. Sliders glow luminously."[8] In addition, Apple claims that OS X won't allow software conflicts, a frequent headache with Microsoft's Windows operating systems. For example, you might install a game and find that it interferes with the device driver for a sound card. Then, when you uninstall the game, the problem persists. With Mac OS X, when you try to install an application program that conflicts with any other program, the Mac simply won't allow you to run it.

Ultimately, Apple's strategy is to marry its easy-to-use operating system with its iMac to offer free Web services for everything from your photos to your home page.[9]

- **Microsoft Windows for desktops and laptops—Windows 3.1, 95, 98, and Me:** Taking its cue from the popularity of Mac's easy-to-use GUI, Microsoft began working on Windows—trying to make DOS more user-friendly. Early attempts (Windows 1.0, 2.0, 3.0) did not catch on. However, in 1992, Windows 3.1 emerged as the preferred system among PC users. (Technically, Windows 3.1 wasn't a full operating system; it was simply a layer or shell over DOS.) Later, this version evolved into the Windows 95 operating system, which was succeeded by Windows 98. **_Microsoft Windows 95/98_ is today's most popular operating system for desktop and portable microcomputers, supporting the most hardware and the most application software. Its most recent version is Windows Me (for Millennium Edition).**

Among other improvements over its predecessors, the recent versions of Windows adhere to a standard called Universal Plug and Play, which promises to let a variety of electronics seamlessly network with each other. More importantly, Me claims to have reduced the problem of frequent "crashes," notorious with earlier versions of Windows. When your computer crashes, all the elements on the screen simply freeze up, and you're obliged to turn off the computer and reboot. Now the operating system tries to protect core system files from being altered when you load third-party programs. If they do go haywire, it provides you with a "system restore" tool to roll back the computer to the time when it was running smoothly. In addition, Windows Me makes it easier for users to download pictures from a digital camera and share them with others online.[10]

Network Operating Systems: NetWare, Windows NT/2000, Unix/Solaris, & Linux

The operating systems described so far were principally designed for use with stand-alone desktop machines, not large systems of networked computers. Now let us consider the important operating systems designed to work with networks: NetWare, Windows NT/2000, Unix/Solaris, and Linux.

1999 Share of Server (Non-Mainframe) Market	
Netware	19%
Windows NT	38%
Unix	15%
Linux	25%
Other	3%

- **Novell's NetWare—PC Networking Software:** **_NetWare_ has long been a popular network operating system for coordinating microcomputer-based local area networks (LANs) throughout a company or a campus.** LANs allow PCs to share programs, data files, and printers and other devices. Novell, the maker of NetWare, thrived as corporate data managers realized that networks of PCs could exchange information more cheaply than the previous generation of mainframes and midrange computers.

  In 1999, NetWare's share of the server market was 19%, down from earlier times, owing to the rise of the Internet and competition from Microsoft, Unix/Solaris, and Linux. Nevertheless, Novell's directory software (it runs on Windows 2000, Solaris, and Linux servers as well as on NetWare), which helps keep track of computers, programs, and people on a network, has been important to Ford, Wal-Mart, and similar big companies, and this innovation has kept it ahead.[11]

- **Windows NT and 2000—the challenge from Microsoft:** Windows 95 and 98 can be used to link PCs in small networks in homes and offices. However, something more powerful was needed to run the huge networks linking a variety of computers—PCs, workstations, mainframes—used by many companies, universities, and other organizations, which previously were served principally by Unix and Netware operating systems. **_Windows NT_ (the NT stands for New Technology), later upgraded to _Windows 2000_, is Microsoft's multi-tasking operating system designed to run on network servers. It allows multiple users to share resources such as data and programs.**

  When it first appeared, in 1993, the system came in two versions. The *Windows NT Workstation* version enabled graphic artists, engineers, and others using stand-alone workstations to do intensive computing at their desks. The *Windows NT Server* version was designed to benefit multiple users tied together in client/server networks. In 1999, Windows NT had 38% of the server market. In early 2000, Microsoft rolled out its updated version, *Windows 2000*, to replace Windows NT version 4.0.

  The name Windows 2000 is somewhat unfortunate. This is not the successor to Windows 95 and 98 for home and non-network use. If you're interested in arcade-style games, for instance, 2000 won't work, since it was specifically designed to eliminate crashes and reboots and as a result is incompatible with a whole class of programs and devices (though they will work with Windows 98 and Me). Windows 2000 has better Plug-and-Play hardware installation and works better for portables, giving slightly better battery life than either NT or 98. One weakness is that Windows 2000 still can't run the most powerful servers.[12]

- **Unix and Solaris—first to exploit the Internet:** Unix (pronounced "*Yu*-niks") was developed at AT&T's Bell Laboratories in 1969 as an operating system for minicomputers. Today **_Unix_ is a multitasking operating system for multiple users that has built-in networking capability and versions that can run on all kinds of computers.** Government agencies, universities, research institutions, large corporations, and banks all use Unix for everything from designing airplane parts to currency trading. Unix is also used for Web-site management. Indeed, the developers of the Internet built their communications system around Unix because it has the ability to keep large systems (with hundreds of processors) churning out transactions day in and day out for years without fail. In 1999, Unix had 15% of the server market (down from 19% in 1998, probably because of inroads made by Linux, as we'll discuss).

  Sun Microsystem's *Solaris* is a super-reliable version of Unix that seems to be most popular for handling large e-commerce servers and large Web sites. Solaris can handle servers with as many as 64 micro-processors, compared with 32 for Windows 2000. In addition, eight computers can be clustered together to work as one, compared with four for Windows 2000.[13] As we mentioned, Sun is also an ASP, offering over-the-Net application software StarOffice and StarPortal.

- **Linux—software built by a community:** It began in 1991 when programmer Linus Torvalds, a graduate student in Finland, posted his free Linux operating system on the Internet. With 25% of the server market in 1999 (up from 16% the year before), Linux (pronounced "*Linn-uks*") is the rising star of network software. ***Linux* is a free version of Unix, and its continual improvements result from the efforts of tens of thousands of volunteer programmers.** Whereas Windows is Microsoft's proprietary product, Linux is ***open-source software*—meaning any programmer can download it from the Internet for free and modify it with suggested improvements.** The only qualification is that changes can't be copyrighted; they must be made available to all and remain in the public domain. From these beginnings, Linux has attained cult-like status. "What makes Linux different is that it's part of the Internet culture," says an IBM general manager. "It's essentially being built by a community."[14] (The People's Republic of China announced in July 2000 that it was adopting Linux as a national standard for operating systems because it feared being dominated by the OS of a company of a foreign power—namely, Microsoft.)

  If Linux belongs to everyone, how do companies like Red Hat Software and VA Linux Systems—two companies that base their business on Linux—make money? Their strategy is to give away the software but then sell services and support. Red Hat, for example, makes available an inexpensive ($149) application-software package that offers word processing, spreadsheets, and the like.

  Because it was built for use on the Internet, Linux is more reliable than Windows for online applications. Hence, it is better suited to run Web sites and e-commerce software. Its real growth, however, may come as it reaches outward to other applications. IBM, Red Hat, and 45 other companies have formed the Embedded Linux Consortium, which envisions a time when microchips running Linux are built into everything from Internet-linked microwave ovens and refrigerators to wireless phones and TV set-top cable boxes.[15]

Operating Systems for Handhelds: Palm OS & Windows CE/Pocket PC

Maybe you're not one of the estimated (at the end of 2000) 7.7 million owners of a handheld computer or personal digital assistant (PDA). But perhaps you've seen people poking through calendars and address books, beeping through games, or (in the latest versions) checking e-mail or the Web on these palm-size devices. Handhelds have gained headway in the corporate world and on college campuses. Because of their small size, they rely on specialized operating systems, including the Palm OS and Windows CE.

- **Palm OS—the dominant OS for handhelds:** In 1994, Jeff Hawkins took blocks of mahogany and plywood into his garage and emerged with a prototype for the PalmPilot. Two years later, Hawkins and business brain Donna Dubinsky pulled one of the most successful new-product launches in history. Today the company, Palm Computing, with 80% of the handheld market, sells the popular Palm III, V, and VII. However, Hawkins and Dubinsky left to form another company, Handspring, whose product, Visor, acts like a Palm but costs less and has an expansion slot that can transform the device into a cell phone, MP3 music player, or two-way pager.

Visor handheld computer
This runs on the Palm operating system.

The **_Palm OS_, which runs the Palm and Visor, is the dominant operating system for handhelds,** with about 85% of the market, and is being licensed to competitors, such as IBM and Nokia. With a Palm OS, you have access to thousands of application programs over the Internet, many of them free. (For instance, one $10 program, PayUp, lets you track 19 people's dinner orders and calculate an exact bill for each, including tax and tip.)[16]

- Windows CE/Pocket PC—Microsoft Windows for handhelds: In 1996, Microsoft released **_Windows CE_, a greatly slimmed-down version of Windows 95 for handheld computing devices,** such as those made by Casio, Compaq, and Hewlett-Packard. Windows CE had some of the familiar Windows look and feel and included rudimentary word processing, spreadsheet, e-mail, Web browsing, and other software.

But whereas Palm concentrated on simplicity, reliability, compactness, and long battery life, Microsoft and its allies focused on what had worked in the conventional-PC marketplace: piling on new features and options. Consumers rejected it. "Windows CE is a bear to use," wrote one reviewer about the Compaq Aero, the leader in the CE market, which accounted for only 7% of handheld dollar sales in 1999. "Microsoft is essentially cramming Windows down our throats."[17]

Microsoft abandoned Windows CE because too many buyers and developers regarded it as a loser. Its latest version, **_Pocket PC_ for handhelds, is simpler and less cluttered than CE and looks and feels a lot less like desktop Windows.** Based on research findings that customers want their handhelds to be more than just electronic organizers, however, Pocket PC offers pocket versions of Word and Excel that let users read standard word processing and spreadsheet files sent as e-mail attachments from their PCs. Microsoft was also planning to offer a special version of Internet Explorer that would automatically reformat Web pages so they could be read on the small screens of handhelds.[18]

In what is coming to be called "the post-PC era," companies are rushing to design not only handhelds but also *intelligent linking technology,* which will connect embedded chips in all kinds of household appliances. Besides Palm OS and Windows CE/Pocket PC, some companies, the British Psion Revo, a handheld, runs a swift Java-language compatible operating system.[19] (We describe Java in the Appendix.) Another operating system called BeIA, an open-source system from Be Inc. in Menlo Park, California, forms the basis of a number of Internet appliances.[20]

CONCEPT CHECK

What are the principal desktop operating systems and their features?

What are the common network operating systems and their features?

Describe the leading operating systems for handhelds.

3.3 Application Software: Getting Started

KEY QUESTIONS
What are four ways of obtaining application software, tools available to help you learn them, three common types of files, and the types of software?

As we've stated, *application software* is software that has been developed to solve a particular problem, to perform useful work on specific tasks, or to provide entertainment. New microcomputers are usually equipped not only with system software but also with some application software.

Software

Application Software: For Sale, for Free, or for Rent?

At one time, just about everyone paid for their PC application software. You bought it as part of a computer or in a software store, or you downloaded it online with a credit card charge. Now, as we suggested in describing ASPs earlier, this business model may be changing. Let's look at the alternatives: *commercial software, public-domain software, shareware, freeware,* and *rentalware. (See ● Panel 3.9.)*

- Commercial software: *Commercial software,* also called *proprietary software,* is software that's offered for sale, such as Microsoft Word or Office 2000. Although such software may not show up on the bill of sale when you buy a new PC, you've paid for it as part of the purchase. And, most likely, whenever you order a new game or other commercial program, you'll have to pay for it. This software is copyrighted. A **_copyright_ is the exclusive legal right that prohibits copying of intellectual property without the permission of the copyright holder.**

    Software manufacturers don't sell you their software; rather, they sell you a license to become an authorized user of it. What's the difference? In paying for a **_software license_, you sign a contract in which you agree not to make copies of the software to give away or for resale.** That is, you have bought only the company's permission to use the software and not the software itself. This legal nicety allows the company to retain its rights to the program and limits the way its customers can use it. The small print in the licensing agreement usually allows you to make one copy (backup copy or archival copy) for your own use. (Each software company has a different license; there is no industry standard.)

    Every year or so, software developers find ways to enhance their products and put forth new versions or new releases. A *version* is a major upgrade in a software product, traditionally indicated by numbers such as 1.0, 2.0, 3.0. More recently, other notations have been used, as in Microsoft's Office 97, Office 2000, and so forth. A *release,* which now may be called an "add" or "addition," is a minor upgrade. Often this is indicated by a change in number after the decimal point. (For instance, 3.0 may become 3.1, 3.11, 3.2, and so on.) Some releases are now also indicated by the year in which they are marketed. And, unfortunately, some releases are not clearly indicated at all. (These are "patches," which may be downloaded from the software maker's Web site.)

- Public-domain software: **_Public-domain software_ is not protected by copyright and thus may be duplicated by anyone at will.** Public domain programs—usually developed at taxpayer expense by government agencies—have been donated to the public by their creators. They are often available through sites on the Internet. You can duplicate public domain software without fear of legal prosecution.

● PANEL 3.9
Choices among application software

Types	Definition
Commercial software	Copyrighted. If you don't pay for it, you can be prosecuted
Public-domain software	Not copyrighted. You can copy it for free without fear of legal prosecution
Shareware	Copyrighted. Available free, but you should pay to continue using it
Freeware	Copyrighted. Available free
Rentalware	Copyrighted. Lease for a fee

- **Shareware:** <u>**Shareware**</u> **is copyrighted software that is distributed free of charge but requires users to make a monetary contribution in order to continue using it.** Shareware is distributed primarily through the Internet, but because it is copyrighted, you cannot use it as the basis for developing your own program that would compete with the original product.

- **Freeware:** <u>**Freeware**</u> **is copyrighted software that is distributed free of charge,** today most often over the Internet. Why would any software creator let his or her product go for free? Sometimes developers want to see how users respond, so they can make improvements in a later version. Sometimes it is to further some scholarly or humanitarian purpose—for instance, to create a standard for software on which people are apt to agree. (Linux is such a program.) In its most recent form, freeware is made available by companies trying to make money some other way—actually, by attracting viewers to their advertising. (The Web browsers Internet Explorer and Netscape Navigator are of this type.) Freeware developers generally retain all rights to their programs; technically, you are not supposed to duplicate and distribute them further.

- **Rentalware:** <u>**Rentalware**</u> **is software that users lease for a fee.** This is the concept behind application services providers. AT&T Corp., for instance, is joining with ASPs to allow corporations to run software from AT&T's network rather than buying software of their own.

Tutorials & Documentation

How are you going to learn a given software program? Most commercial packages come with tutorials and documentation.

- **Tutorials:** A <u>**tutorial**</u> **is an instruction book or program that helps you learn to use the product by taking you through a prescribed series of steps.** For instance, our publisher offers several how-to books, known as the *Advantage Series,* that enable you to learn different kinds of software. Tutorials may also form part of the software package.

- **Documentation:** <u>**Documentation**</u> **is a user guide or reference manual that provides a narrative and graphical description of a program.** While documentation may be print-based, today it is usually available on CD-ROM, as well as via the Internet. Documentation may be instructional, but features and functions are usually grouped by category for reference purposes. For example, in word processing documentation, all features related to printing are grouped together so you can easily look them up.

Files of Data—& the Usefulness of Importing & Exporting

There is only one reason for having application software: to take raw data and manipulate it into useful files of information. A <u>**file**</u> **is a named collection of data or a program that exists in a computer's secondary storage,** such as a floppy disk, hard disk, or CD-ROM.

Three common types of files are as follows:

- **Document files:** <u>**Document files**</u> **are created by word processing programs and consist of documents such as reports, letters, memos, and term papers.**

- **Worksheet files:** <u>**Worksheet files**</u> **are created by electronic spreadsheets and consist of collections of (usually) numerical data such as budgets, sales forecasts, and schedules.**

PRACTICAL ACTION BOX
When Software Causes Problems

In Issaquah, Washington, a man was coaxed out of his home by police officers "after he pulled a gun and shot several times at his personal computer, apparently in frustration," according to an Associated Press story.[a]

Experienced computer users will recognize the phrase "apparently in frustration" as an understatement. Fortunately, newcomers will likely be spared these agonies with software because they will be operating at rather basic levels (and can seek help in computer classrooms from their instructors or staff members). There are also a number of sources of user assistance—although even here, forewarned is forearmed.

- **Instruction manuals:** User guides or instruction manuals printed on paper have traditionally accompanied a box of application software diskettes or CD-ROMs. Sometimes these are just what you need. Sometimes, however, you need to know what you need help with, and it may be difficult to define the problem. In addition, these guides may be a puzzle to anyone not trained as a programmer because they are often written by the very people who developed the software. And experts are not always able to anticipate the problems of the amateur.

- **Help software:** In place of paper manuals, software publishers are now relying more on Help programs on a diskette or CD-ROM accompanying the software, which contain a series of Help menus. However, these, too, may feature technospeak that is baffling to newcomers. Also, like paper manuals, help programs may suffer the drawbacks of having been rushed out at the last minute.

    Help programs may also be available through the Internet. The problem with trying to use the Net or the Web, of course, is: How do you go online to solve the problem of your computer not working if your computer isn't working? Or how do you access the Internet to get advice on how to fix your software and simultaneously fix your software? (It helps to have two computers.)

- **Telephone help lines:** If you encounter a software glitch, it's possible you could use a help line to call a customer-support technician at the program's manufacturer. The technical advisor would then try to talk you through the problem or might refer you to a Web site from which you could obtain a file that would remedy your problem.

    Both software and hardware makers have long offered telephone help lines for customers. If your computer breaks down, you may find that your hardware company still offers emergency advice for free (although you may have to pay the phone company's long-distance tolls). However, software manufacturers now invariably charge for help calls if you've had the software longer than 30–90 days.

- **Commercial how-to books:** Because of the inadequacies in user support by the software developers themselves, an entire industry has sprung up devoted to publishing how-to books. These are the kind of books found both in computer stores

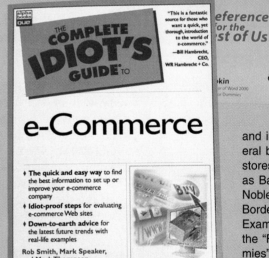

and in general bookstores such as Barnes & Noble or Borders. Examples are the "For Dummies" or "Complete Idiot's" books (such as *PCs for Dummies* and *The Complete Idiot's Guide to Microsoft Office*).

- **Knowledgeable friends:** Believe it or not, nothing beats having a knowledgeable friend: your instructor, a student more advanced than you, or someone with a technical interest in computers. We can't stress enough how important it is to get to know people—from your classes, from computer user groups (including online Internet groups), from family friends, or whatever—who can lend aid and expertise when your computer software gives you trouble.

- **Database files:** **_Database files_ are created by database management programs and consist of organized data that can be analyzed and displayed in various useful ways.** Examples are student names and addresses that can be displayed according to age, grade-point average, or home state.

Other common types of files (such as graphics, audio, and video files) are discussed in Chapter 6.

It's useful to know that often files can be exchanged—that is, *imported* and *exported*—between programs.

- **Importing:** **_Importing_ is defined as getting data from another source and then converting it into a format that can be used in the program in which you are currently working.** For example, you might write a letter in your word processing program and include in it—that is, import—a column of numbers from your spreadsheet program.
- **Exporting:** **_Exporting_ is defined as transforming data into a format that can be used in another program and then transmitting it.** For example, you might work up a list of names and addresses in your database program, then send it—export it—to a document you are writing in your word processing program.

CONCEPT CHECK

Distinguish among the following kinds of software: commercial, public-domain, shareware, freeware, and rentalware.

Describe tutorials and documentation, and discuss three common types of files.

Distinguish importing from exporting.

The Types of Software

Software can be classified in many ways—for entertainment, personal, education/reference, productivity, and specialized uses. *(See ● Panel 3.10.)*

● PANEL 3.10
Types of software

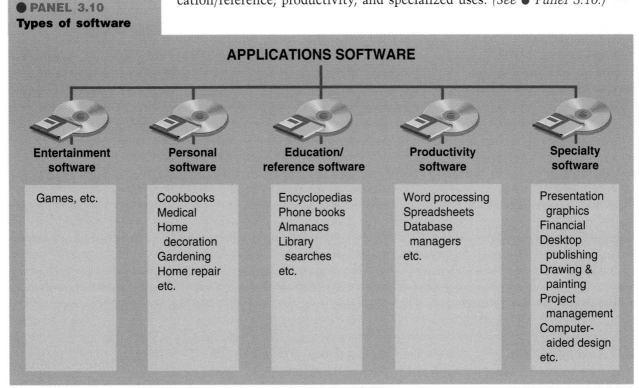

APPLICATIONS SOFTWARE

Entertainment software	Personal software	Education/ reference software	Productivity software	Specialty software
Games, etc.	Cookbooks Medical Home decoration Gardening Home repair etc.	Encyclopedias Phone books Almanacs Library searches etc.	Word processing Spreadsheets Database managers etc.	Presentation graphics Financial Desktop publishing Drawing & painting Project management Computer-aided design etc.

In the rest of this chapter we will discuss types of ***productivity software***—such as **word processing, spreadsheets, and database managers—whose purpose is to make users more productive at particular tasks.** Some productivity software comes in the form of an ***office suite,* which bundles several applications together into a single large package.** Microsoft Office 2000, for example, includes (among other things) Word, Excel, and Access—word processing, spreadsheet, and database programs, respectively. Other productivity software, such as Lotus Notes, is sold as ***groupware*—online software that allows several people to collaborate on the same project.**

Let us now consider the three most important types of productivity software: word processing, spreadsheets, and database software (including personal information managers). We will then discuss more specialized software: presentation graphics, financial software, desktop publishing, drawing and painting, project management, and computer-aided design software.

3.4 Word Processing

KEY QUESTION
What can you do with word processing software that you can't do with pencil and paper?

After a long and productive life, the typewriter has gone to its reward. Indeed, it is practically as difficult today to get a manual typewriter repaired as to find a blacksmith. Word processing software offers a much-improved way of dealing with documents.

***Word processing software* allows you to use computers to format, create, edit, print, and store text material,** among other things. The best-known word processing program is probably Microsoft Word but there are others such as Corel WordPerfect. Word processing software allows users to maneuver through a document and *delete, insert,* and *replace* text, the principal correction activities. It also offers such additional features as *creating, editing, formatting, printing,* and *saving.*

● **PANEL 3.11**
Common keyboard layout

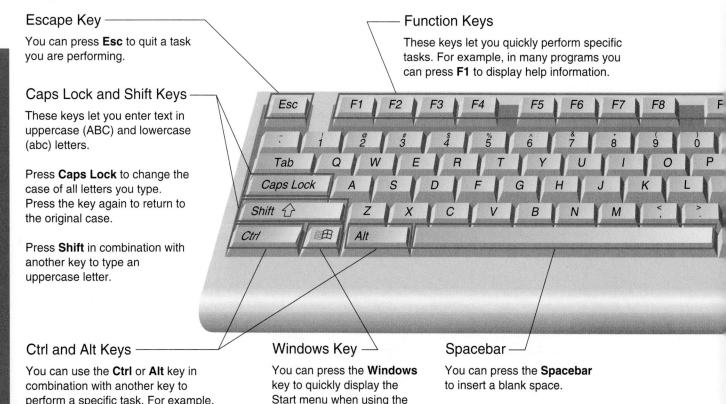

Escape Key
You can press **Esc** to quit a task you are performing.

Function Keys
These keys let you quickly perform specific tasks. For example, in many programs you can press **F1** to display help information.

Caps Lock and Shift Keys
These keys let you enter text in uppercase (ABC) and lowercase (abc) letters.

Press **Caps Lock** to change the case of all letters you type. Press the key again to return to the original case.

Press **Shift** in combination with another key to type an uppercase letter.

Ctrl and Alt Keys
You can use the **Ctrl** or **Alt** key in combination with another key to perform a specific task. For example, in some programs, you can press **Ctrl** and **S** to save a document.

Windows Key
You can press the **Windows** key to quickly display the Start menu when using the Windows 95, 98, or NT operating systems.

Spacebar
You can press the **Spacebar** to insert a blank space.

Features of the Keyboard

Besides the mouse, the principal tool of word processing is the keyboard. As well as letter, number, and punctuation keys and often a calculator-style numeric keypad, computer keyboards have special-purpose and function keys. *(See ● Panel 3.11.)* Sometimes, for the sake of convenience, strings of keystrokes are combined in combinations called *macros.*

- **Special-purpose keys:** ***Special-purpose keys* are used to enter, delete, and edit data and to execute commands.** An example is the *Esc* (for "Escape") key, which tells the computer to cancel an operation or leave ("escape from") the current mode of operation. The *Enter,* or *Return,* key, which you will use often, tells the computer to execute certain commands and to start new paragraphs in a document. Commands are instructions that cause the software to perform specific actions.

  Special-purpose keys are generally used the same way regardless of the application software package being used. Most keyboards include the following special-purpose keys: *Esc, Ctrl, Alt, Del, Ins, Home, End, PgUp, PgDn, Num Lock,* and a few others. (*Ctrl* means Control, *Del* means Delete, *Ins* means Insert, for example.)

- **Function keys:** ***Function keys,* labeled F1, F2, and so on, are positioned along the top or left side of the keyboard. They are used to execute commands specific to the software being used.** For example, one application software package may use *F6* to exit a file, whereas another may use *F6* to underline a word.

- **Macros:** Sometimes you may wish to reduce the number of keystrokes required to execute a command. To do this, you use

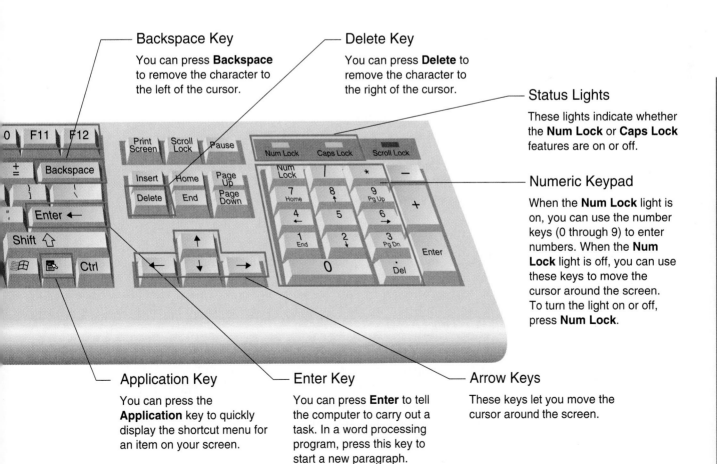

Backspace Key

You can press **Backspace** to remove the character to the left of the cursor.

Delete Key

You can press **Delete** to remove the character to the right of the cursor.

Status Lights

These lights indicate whether the **Num Lock** or **Caps Lock** features are on or off.

Numeric Keypad

When the **Num Lock** light is on, you can use the number keys (0 through 9) to enter numbers. When the **Num Lock** light is off, you can use these keys to move the cursor around the screen. To turn the light on or off, press **Num Lock**.

Application Key

You can press the **Application** key to quickly display the shortcut menu for an item on your screen.

Enter Key

You can press **Enter** to tell the computer to carry out a task. In a word processing program, press this key to start a new paragraph.

Arrow Keys

These keys let you move the cursor around the screen.

a macro. A **_macro_, also called a keyboard shortcut, is a single keystroke or command—or a series of keystrokes or commands—used to automatically issue a longer, predetermined series of keystrokes or commands.** Thus, you can consolidate several activities into only one or two keystrokes. The user names the macro and stores the corresponding command sequence; once this is done, the macro can be used repeatedly. (To set up a macro, pull down the Help menu and type in "macro.")

Although many people have no need for macros, others who find themselves continually repeating complicated patterns of keystrokes say they are quite useful.

Creating Documents

Creating a document means entering text using the keyboard. Word processing software has three features that affect this process—the *cursor, scrolling,* and *word wrap.*

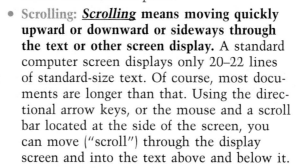

- Cursor: The **_cursor_ is the movable symbol on the display screen that shows you where you may next enter data or commands.** The symbol is often a blinking rectangle or I-beam. You can move the cursor on the screen using the keyboard's directional arrow keys or a mouse. The point where the cursor is located is called the *insertion point.*

- Scrolling: **_Scrolling_ means moving quickly upward or downward or sideways through the text or other screen display.** A standard computer screen displays only 20–22 lines of standard-size text. Of course, most documents are longer than that. Using the directional arrow keys, or the mouse and a scroll bar located at the side of the screen, you can move ("scroll") through the display screen and into the text above and below it.

- Word wrap: *Word wrap* automatically continues text on the next line when you reach the right margin. That is, the text "wraps around" to the next line. You don't have to hit a "carriage return" key or Enter key, as you do with a typewriter.

Scrolling

One feature that can help you organize term papers and reports is an *outline feature,* which enables you to put tags on various headings to show the hierarchy of heads—for example, main head, subhead, and sub-subhead.

Editing Documents

Editing is the act of making alterations in the content of your document. Some features of editing are *insert and delete, undelete, find and replace, cut/copy and paste, spelling checker, grammar checker,* and *thesaurus.* Some of these commands are in the Edit pull-down menu and icons on the toolbar. *(See ● Panel 3.12.)*

- Insert and delete: *Inserting* is the act of adding to the document. You simply place the cursor wherever you want to add text and start typing; the existing characters will be pushed along. The *Insert* key toggles between inserting and writing over.

  Deleting is the act of removing text, usually using the *Delete* or *Backspace* keys.

Toolbar:
Allows quick access to frequently used commands

Menu bar:
Allows access to all commands

Title bar:
Shows name of document you're working on

Spelling and Grammar button:
Click on to check for misspelled words and incorrect grammar.

Text alignment buttons:
Click on to align text to be left, right, center, or full justified.

Style button:
Click on to access variety of format styles.

Ruler:
Shows tabs and margins

Insertion point:
Blinking symbol shows where the next character you type will appear.

Window controls:
Lets you enlarge a window, restore its previous position, or hide it from view

Mouse pointer:
Use the mouse to move the insertion point, to click on icons, or to select text for editing.

Scroll bars:
Lets you scroll the document to reveal hidden portions

Status bar:
Shows details about the document you're working on

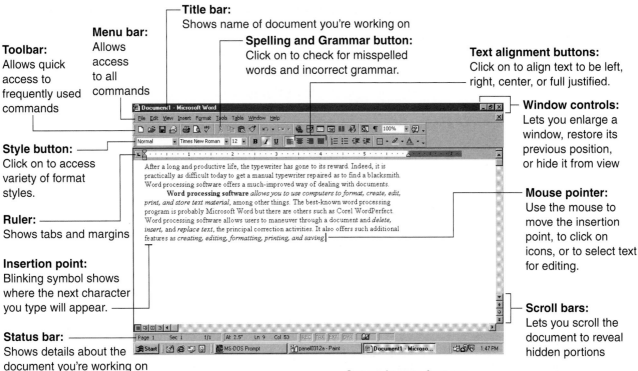

Cut-and-paste feature:
Enables you to move blocks of text. First highlight the text. On Edit menu, select Cut option. Then, on Edit Menu, select Paste option. (You can also use the icons in the toolbar.)

Outline feature:
Enables you to view headings in your document

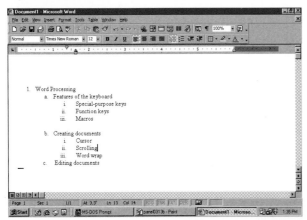

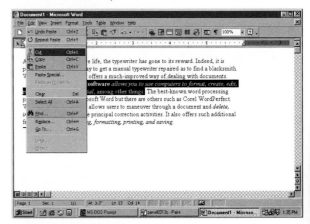

Formatting feature:
Enables you to change type font and size. First highlight the text. Then click next to Font button for pull-down menu of fonts. Click next to Font Size button for menu of type sizes.

Font button **Font size button**

To format text, first highlight it.

Then select formatting.

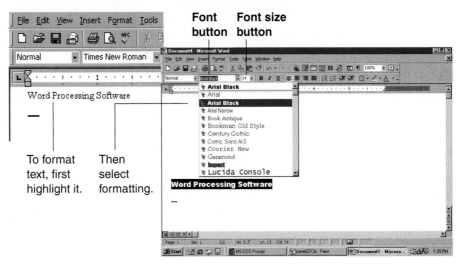

● PANEL 3.12
Word processing
Some basic functions of Microsoft Word.

Software

The *Undelete command* allows you to change your mind and restore text that you have deleted. Some word processing programs offer as many as 100 layers of "undo," so that users who delete several paragraphs of text, but then change their minds, can reinstate the material.

- **Find and replace:** The *Find,* or *Search, command* allows you to find any word, phrase, or number that exists in your document. The *Replace command* allows you to automatically replace it with something else.

- **Cut/Copy and paste:** Typewriter users who wanted to remove a paragraph or block of text from one place to another in a manuscript used scissors and glue to "cut and paste." With word processing, it only takes a few keystrokes. You select (highlight) the portion of text you want to copy or move. Then you use the *Copy* or *Cut command* to move it to the *clipboard,* a special holding area in the computer's memory. From there, you can "paste," or transfer, the material to any point (indicated with the cursor) in the existing document or in a new document.

- **Spelling checker:** Most word processors have a ***spelling checker,*** **which tests for incorrectly spelled words.** As you type, the spelling checker indicates (perhaps with a squiggly line) words that aren't in its dictionary and thus may be misspelled. *(See ● Panel 3.13.)* Special add-on dictionaries are available for medical, engineering, and legal terms. In addition, programs such as Microsoft Word have an Auto Correct function that automatically fixes such common mistakes as transposed letters—replacing "teh" with "the," for instance.

● PANEL 3.13
Spelling checker
How a word processing program checks for misspelled words and offers alternatives

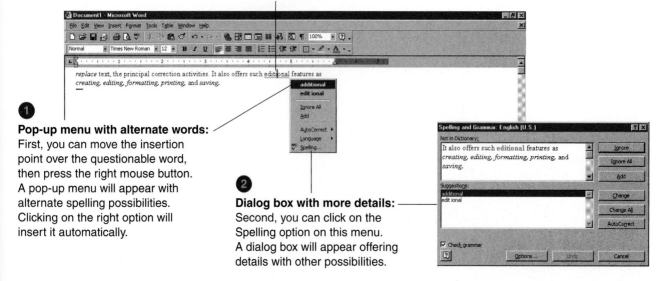

Red wavy underline:
Indicates spelling checker doesn't recognize the word. You have two options.

① Pop-up menu with alternate words:
First, you can move the insertion point over the questionable word, then press the right mouse button. A pop-up menu will appear with alternate spelling possibilities. Clicking on the right option will insert it automatically.

② Dialog box with more details:
Second, you can click on the Spelling option on this menu. A dialog box will appear offering details with other possibilities.

- **Grammar checker:** A ***grammar checker*** **highlights poor grammar, wordiness, incomplete sentences, and awkward phrases.** The grammar checker won't fix things automatically, but it will flag (perhaps with a different color of squiggly line) possible incorrect word usage and sentence structure. *(See ● Panel 3.14.)*

- **Thesaurus:** If you find yourself stuck for the right word while you're writing, you can call up an on-screen ***thesaurus,*** **which will present you with the appropriate word or alternative words.**

Formatting Documents with the Help of Templates & Wizards

In the sense in which we use it here, ***formatting*** **means determining the appearance of a document.** Word processing programs help you do this with two helpful devices—templates and wizards. A ***template*** **is a preformatted document that provides basic tools for shaping a final document**—the text, layout, and style for a letter, for example. A ***wizard*** **answers your questions and uses the answers to lay out and format a document.** In Word, you can use the Memo Wizard to create professional-looking memos or the Resume Wizard to create a resume.

● PANEL 3.14
Grammar checker
This program points out possible errors in sentence structure and word usage and suggests alternatives.

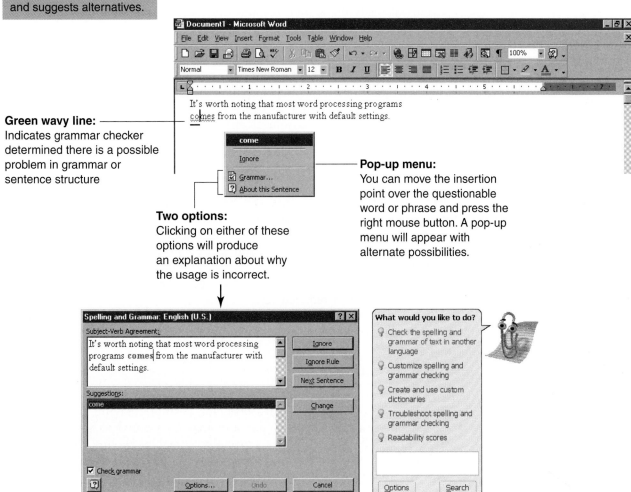

Green wavy line:
Indicates grammar checker determined there is a possible problem in grammar or sentence structure

Pop-up menu:
You can move the insertion point over the questionable word or phrase and press the right mouse button. A pop-up menu will appear with alternate possibilities.

Two options:
Clicking on either of these options will produce an explanation about why the usage is incorrect.

Software

Among the many aspects of formatting are the following:

- **Font:** You can decide what __*font*__—**typeface and type size**—you wish to use. For instance, you can specify whether it should be Arial, Courier, or Times New Roman. You can indicate whether the text should be, say, 10 points or 12 points in size and the headings should be 14 points or 16 points. (There are 72 points in an inch.) You can specify what parts of it should be underlined, *italic,* or **boldface.**
- **Spacing and columns:** You can choose whether you want the lines to be *single-spaced* or *double-spaced* (or something else). You can specify whether you want text to be *one column* (like this page), *two columns* (like many magazines and books), or *several columns* (like newspapers).
- **Margins and justification:** You can indicate the dimensions of the *margins*—left, right, top, and bottom—around the text. You can specify the text *justification*—how the letters and words are spaced in each line. To *justify* means to align text evenly between left and right margins, as is done with most newspaper columns and this text. To *left-justify* means to align text evenly on the left. (It is "ragged right," like many business letters or this paragraph.)
- **Pages, headers, footers:** You can indicate page numbers and headers or footers. A *header* is common text (such as a date or document name) that is printed at the top of every page. A *footer* is the same thing printed at the bottom of every page.
- **Other formatting:** You can specify *borders* or other decorative lines, *shading, tables,* and *footnotes.* You can even import *graphics* or drawings from files in other software programs, including *clip art*—collections of ready-made pictures and illustrations available online or on CD-ROM disks.

It's worth noting that word processing programs (and indeed most forms of application software) come from the manufacturer with *default settings.* __*Default settings*__ **are the settings automatically used by a program unless the user specifies otherwise, thereby overriding them.** Thus, for example, a word processing program may automatically prepare a document single-spaced, left-justified, with 1-inch right and left margins, unless you alter these default settings.

Printing, Faxing, or E-Mailing Documents

Most word processing software gives you several options for printing. For example, you can print *several copies* of a document. You can print *individual pages* or a *range of pages.* You can even preview a document before printing it out. *Previewing (print previewing)* means viewing a document on screen to see what it will look like in printed form before it's printed. Whole pages are displayed in reduced size.

You can also send your document off to someone else by fax or e-mail attachment, if your computer has the appropriate communications link.

Saving Documents

__*Saving*__ **means storing, or preserving, a document as an electronic file permanently**—on floppy disk or hard disk, for example. Saving is a feature of nearly all application software, but anyone accustomed to writing with a typewriter will find this activity especially valuable. Having the document stored in electronic form spares you the tiresome chore of retyping it from scratch whenever you want to make changes. You need only retrieve it from the storage medium and make the changes you want. Then you can print it out again.

CONCEPT CHECK

What are the important features of the keyboard?

Describe the role of the cursor, scrolling, and word wrap in creating documents.

What word processing features are available to help edit documents?

What assistance is available to help you format documents, and what aspects of formatting should be of concern to you?

3.5 Spreadsheets

KEY QUESTION

What can you do with an electronic spreadsheet that you can't do with pencil and paper and a standard calculator?

What is a spreadsheet? Traditionally, it was simply a grid of rows and columns, printed on special light-green paper, that was used to produce financial projections and reports. A person making up a spreadsheet spent long days and weekends at the office penciling tiny numbers into countless tiny rectangles. When one figure changed, all others numbers on the spreadsheet had to be recomputed. Ultimately, there might be wastebaskets full of jettisoned worksheets.

In the late 1970s, Daniel Bricklin was a student at the Harvard Business School. One day he was staring at columns of numbers on a blackboard when he got the idea for computerizing the spreadsheet. The first **electronic spreadsheet, now called simply a _spreadsheet_, allowed users to create tables and financial schedules by entering data and formulas into rows and columns arranged as a grid on a display screen.** Before long, the electronic spreadsheet was the most popular small-business program. Unfortunately for Bricklin, his version (called VisiCalc) was quickly surpassed by others. Today the principal spreadsheets are Microsoft Excel, Corel Quattro Pro, and Lotus 1-2-3.

You can put your checkbook register on a spreadsheet, and then use it to compute totals and compare income and expenses from one month to the next. In addition, within a spreadsheet file you may have workbooks containing worksheets. A *worksheet* is a single table. A *workbook* is a collection of related worksheets. Thus, within your Microsoft Excel spreadsheet file, you might have one workbook headed *Checkbook*, which would contain worksheets with the history of your checking account for the years 2000, 2001, and so on. You might have another workbook headed *Credit cards*, containing worksheets for each year.

The Basics: How Spreadsheets Work

The arrangement of a spreadsheet is as follows. *(See ● Panel 3.15, next page.)*

- **How a spreadsheet is organized—column headings, row headings, and labels:** In the worksheet's frame area (work area), lettered *column headings* appear across the top ("A" is the name of the first column, "B" the second, and so on). Numbered *row headings* appear down the left side ("1" is the name of the first row, "2" the second, and so forth). *Labels* are any descriptive text, such as APRIL, RENT, or GROSS SALES. You use your computer's keyboard to type in the various headings and labels.

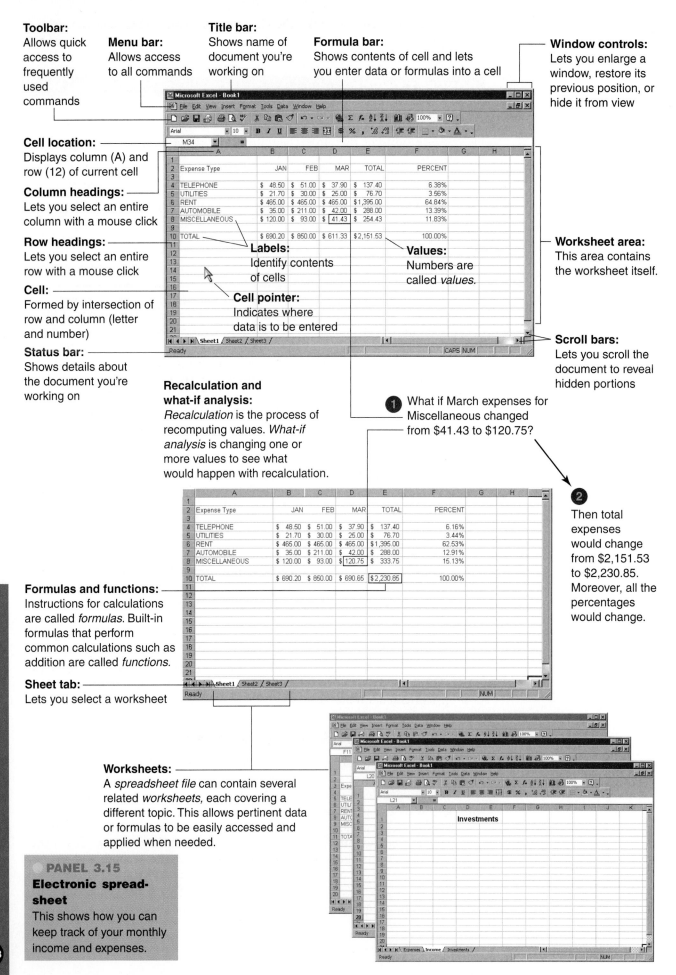

Toolbar:
Allows quick access to frequently used commands

Menu bar:
Allows access to all commands

Title bar:
Shows name of document you're working on

Formula bar:
Shows contents of cell and lets you enter data or formulas into a cell

Window controls:
Lets you enlarge a window, restore its previous position, or hide it from view

Cell location:
Displays column (A) and row (12) of current cell

Column headings:
Lets you select an entire column with a mouse click

Row headings:
Lets you select an entire row with a mouse click

Cell:
Formed by intersection of row and column (letter and number)

Status bar:
Shows details about the document you're working on

Labels:
Identify contents of cells

Values:
Numbers are called *values.*

Worksheet area:
This area contains the worksheet itself.

Cell pointer:
Indicates where data is to be entered

Scroll bars:
Lets you scroll the document to reveal hidden portions

Recalculation and what-if analysis:
Recalculation is the process of recomputing values. *What-if analysis* is changing one or more values to see what would happen with recalculation.

① What if March expenses for Miscellaneous changed from $41.43 to $120.75?

② Then total expenses would change from $2,151.53 to $2,230.85. Moreover, all the percentages would change.

Formulas and functions:
Instructions for calculations are called *formulas.* Built-in formulas that perform common calculations such as addition are called *functions.*

Sheet tab:
Lets you select a worksheet

Worksheets:
A *spreadsheet file* can contain several related *worksheets,* each covering a different topic. This allows pertinent data or formulas to be easily accessed and applied when needed.

● PANEL 3.15
Electronic spreadsheet
This shows how you can keep track of your monthly income and expenses.

Chapter 3

118

- **Where columns and rows meet—cells, cell addresses, ranges, and values:** A _**cell**_ **is the place where a row and a column intersect; its position is called a cell address.** For example, "A1" is the cell address for the top left cell, where column A and row 1 intersect. A _range_ is a group of adjacent cells—for example, A1 to A5. **A number or date entered in a cell is called a** _**value.**_ The values are the actual numbers used in the spreadsheet—dollars, percentages, grade points, temperatures, or whatever. Headings and labels also go into cells. A _cell pointer,_ or _spreadsheet cursor,_ indicates where data is to be entered. The cell pointer can be moved around like a cursor in a word processing program.

- **Why the spreadsheet has become so popular—formulas, functions, recalculation, and what-if analysis:** Now we come to the reason the electronic spreadsheet has taken offices by storm. _**Formulas**_ **are instructions for calculations.** For example, a formula might be @SUM(A5:A15), meaning _Sum_ (that is, add) _all the numbers in the cells with cell addresses A5 through A15._

  Functions are built-in formulas that perform common calculations. For instance, a function might average a range of numbers or round off a number to two decimal places.

  After the values have been entered into the worksheet, the formulas and functions can be used to calculate outcomes. However, what was revolutionary about the electronic spreadsheet was its ability to easily do recalculation. _**Recalculation**_ **is the process of recomputing values,** either as an ongoing process as data is being entered or afterward, with the press of a key. With this simple feature, the hours of mind-numbing work required to manually rework paper spreadsheets became a thing of the past.

  The recalculation feature has opened up whole new possibilities for decision making. In particular, _**what-if analysis**_ **allows the user to see how changing one or more numbers changes the outcome of the recalculation.** That is, you can create a worksheet, putting in formulas and numbers, and then ask, "What would happen if we change that detail?"—and immediately see the effect on the bottom line.

- **Using worksheet templates—pre-arranged forms for specific tasks:** You may find that your spreadsheet software makes worksheet templates available for specific tasks. _Worksheet templates_ are forms containing formats and formulas custom-designed for particular kinds of work. Examples are templates for calculating loan payments, tracking travel expenses, monitoring personal budgets, and keeping track of time worked on projects. Templates are also available for a variety of business needs—providing sales quotations, invoicing customers, creating purchase orders, and writing a business plan.

Analytical Graphics: Creating Charts

A nice feature of spreadsheet packages is the ability to create analytical graphics, or charts. _**Analytical graphics**_**, or business graphics, are graphical forms that make numeric data easier to analyze** than when it is organized as rows and columns of numbers. Whether viewed on a monitor or printed out, analytical graphics help make sales figures, economic trends, and the like easier to comprehend and analyze.

The principal examples of analytical graphics are _bar charts, line graphs,_ and _pie charts._ (See ● _Panel 3.16, next page._) If you have a color printer, these charts can appear in color. In addition, they can be displayed or printed out so that they look three-dimensional. Spreadsheets can even be linked to more exciting graphics, such as digitized maps.

● PANEL 3.16

Analytical graphics

Bar charts, line graphs, and pie charts are used to display numbers in graphical form.

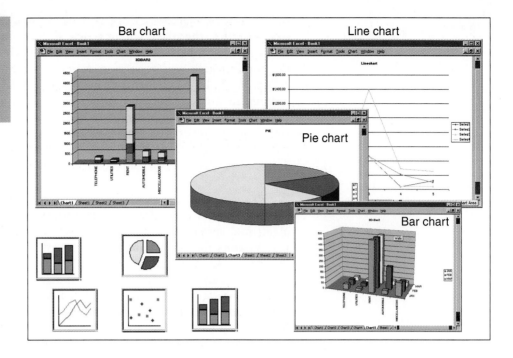

Bar chart

Line chart

Pie chart

Bar chart

CONCEPT CHECK

What is a spreadsheet? A worksheet? A workbook?

What are the components of a spreadsheet?

What is the significance of recalculation and what-if analysis?

What's useful about worksheet templates and analytical graphics?

3.6 Database Software

In its most general sense, a database is any electronically stored collection of data in a computer system. In its more specific sense, a **_database_ is a collection of interrelated files** in a computer system. These computer-based files are organized according to their common elements, so that they can be retrieved easily. (Databases are covered in detail in Chapter 6.) Sometimes called a *database manager* or *database management system (DBMS)*, **_database software_ is a program that sets up and controls the structure of a database and access to the data.**

The Benefits of Database Software

When data is stored in separate files, the same data will be repeated in many files. In the old days, each college administrative office—registrar, financial aid, housing, and so on—might have a separate file on you. Thus, there was *redundancy*—your address, for example, was repeated over and over. The advantage of database software is that data is not in separate files. Rather, it is *integrated*. Thus, your address need only be listed once, and all the separate administrative offices will have access to the same information. For that reason, information in databases is considered to have more *integrity*. That is, the information is more likely to be accurate and up to date.

Databases are a lot more interesting than they used to be. Once they included only text. Now they can also include pictures, sound, and animation. It's likely, for instance, that your personnel record in a future company database will include a picture of you and perhaps even a clip of your voice. If you go looking for a house to buy, you will be able to view a real estate agent's database of video clips of homes and properties without leaving the realtor's office.

Today the principal microcomputer database programs are Microsoft Access, Corel Paradox, and Lotus Approach. (In larger systems, Oracle is a major player.) These programs also allow users to attach multimedia—sound, motion, and graphics—to forms.

The Basics: How Databases Work

Let's consider some basic features of databases.

- **How a relational database is organized—tables, records, and fields:** The most widely used form of database, especially on PCs, is the **_relational database_, in which data is organized into related tables.** Each *table* contains rows and columns; the rows are called *records*, and the columns are called *fields*. An example of a record is a person's address—name, street address, city, and so on. An example of a field is that person's last name; another field would be that person's first name, a third field would be that person's street address, and so on. *(See ● Panel 3.17, next page.)*

  Just as a spreadsheet may include a workbook with several worksheets, so a relational database might include a database with several tables. For instance, if you're running a small company, you might have one database headed *Employees*, containing three tables—*Addresses, Payroll,* and *Benefits.* You might have another database headed *Customers*, with *Addresses, Orders,* and *Invoices* tables.

- **How various records can be linked—the key field:** The records within the various tables in a database are linked by a **_key field_, a field that can be used as a common identifier because it is unique.** The most frequent key field used in the United States is the Social Security number, but any unique identifier could be used, such as employee number.

- **Finding what you want—querying and displaying records:** The beauty of database software is that you can locate records quickly. For example, several offices at your college may need access to your records, but for different reasons: the registrar, financial aid, student housing, and so on. Any of these offices can *query records—locate and display records*—by calling them up on a computer screen for viewing and updating. Thus, if you move, your address field will need to be corrected for all relevant offices of the college. A person making a search might make the query, "Display the address of [your name]." Once a record is displayed, the address field can be changed. Thereafter, any office calling up your file will see the new address.

- **Sorting and analyzing records and applying formulas:** With database software you can easily find and change the order of records in a table. Normally, records are displayed in a database in the same order in which they are entered—for instance, by the date a person registered to attend college. However, all these records can be *sorted* in different ways—arranged alphabetically, numerically, geographically, or in some other order. For example, they can be rearranged by state, by age, or by Social Security number.

  In addition, database programs contain built-in mathematical formulas so that you can analyze data. This feature can be used, for example, to find the grade-point averages for students in different majors or in different classes.

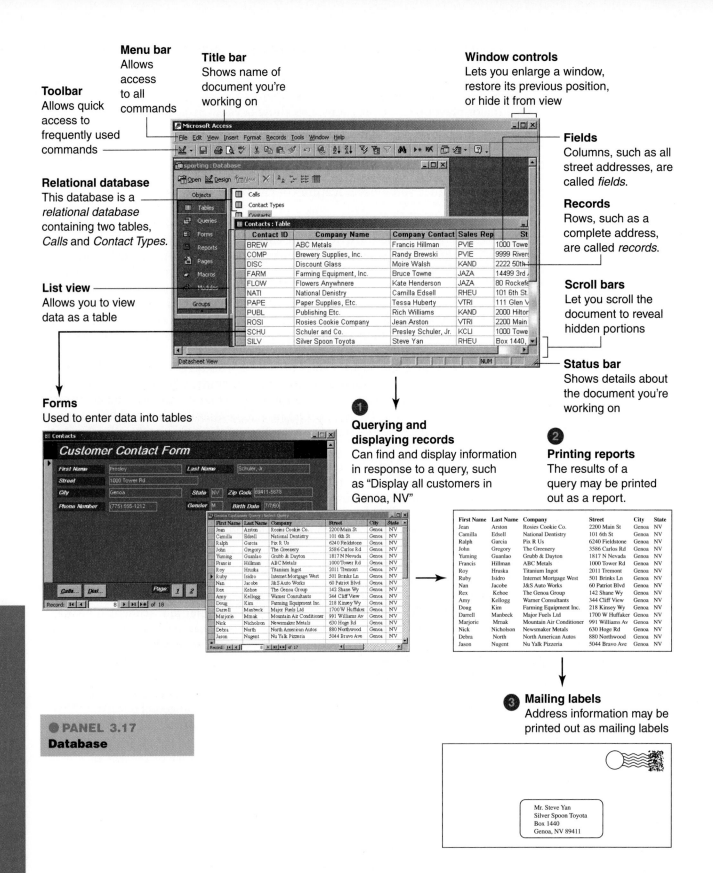

Menu bar
Allows access to all commands

Title bar
Shows name of document you're working on

Window controls
Lets you enlarge a window, restore its previous position, or hide it from view

Toolbar
Allows quick access to frequently used commands

Fields
Columns, such as all street addresses, are called *fields*.

Relational database
This database is a *relational database* containing two tables, *Calls* and *Contact Types*.

Records
Rows, such as a complete address, are called *records*.

List view
Allows you to view data as a table

Scroll bars
Let you scroll the document to reveal hidden portions

Status bar
Shows details about the document you're working on

Forms
Used to enter data into tables

Querying and displaying records
Can find and display information in response to a query, such as "Display all customers in Genoa, NV"

Printing reports
The results of a query may be printed out as a report.

Mailing labels
Address information may be printed out as mailing labels

● PANEL 3.17
Database

- **Putting search results to use—saving, formatting, printing, copying, or transmitting:** Once you've queried, sorted, and analyzed the records and fields, you can simply save them to your hard disk or to a floppy disk. You can format them in different ways, altering headings and type styles. You can print them out on paper as reports, such as an employee list with up-to-date addresses and phone numbers. A common use is to print out the results as names and addresses on mailing labels—adhesive-backed stickers that can be run through your printer, then stuck on envelopes. You can use the copy command to copy your search results and then paste them into a paper produced on your word processor. You can also cut and paste data into an e-mail message or make the data an attachment file to an e-mail, so that it can be transmitted to someone else.

Personal Information Managers

Pretend you are sitting at a desk in an old-fashioned office. You have a calendar, a Rolodex-type address file, and a notepad. Most of these items could also be found on a student's desk. How would a computer and software improve on this arrangement?

Many people find ready uses for specialized types of database software known as personal information managers. A ***personal information manager (PIM)* is software to help you keep track of and manage information you use on a daily basis, such as addresses, telephone numbers, appointments, to-do lists, and miscellaneous notes.** Some programs feature phone dialers, outliners (for roughing out ideas in outline form), and ticklers (or reminders). With a PIM, you can key in notes in any way you like and then retrieve them later based on any of the words you typed.

Popular PIMs are Microsoft Outlook, Lotus Organizer, and Act. Microsoft Outlook, for example, has sections labeled Inbox, Calendar, Contacts, Tasks (to-do list), Journal (to record interactions with people), Notes (scratchpad), and Files. *(See ● Panel 3.18.)*

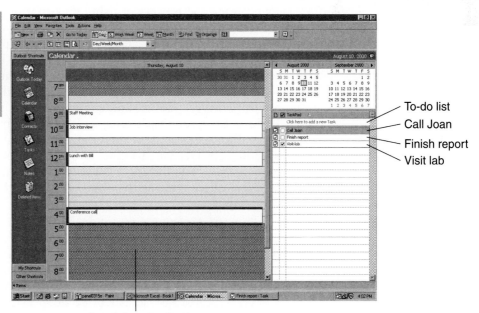

● PANEL 3.18
Personal information manager
This shows the calendar available with Microsoft Outlook.

To-do list
Call Joan
Finish report
Visit lab

Appointment calendar

What is a database, and what are its benefits?

Describe the basic features of a relational database.

How might a PIM help you?

3.7 Specialty Software

KEY QUESTION

What are the principal uses of presentation graphics, financial, desktop publishing, drawing and painting, project management, and computer-aided design software?

After learning some of the productivity software just described, you may wish to become familiar with more specialized programs. For example, you might first learn word processing and then move on to desktop publishing, or first learn spreadsheets, and then learn personal-finance software. We will consider the following kinds of software, although they are but a handful of the thousands of specialized programs available: *presentation graphics, financial, desktop publishing, drawing and painting, project management,* and *computer-aided design* software.

Presentation Graphics Software

You may already be accustomed to seeing presentation graphics because many college instructors now use such software to accompany their lectures. **_Presentation graphics software_ uses graphics, animation, sound and data or information to make visual presentations.** Well-known presentation graphics packages include Microsoft PowerPoint, Corel Presentations, and Lotus Freelance Graphics. *(See ● Panel 3.19.)*

Visual presentations are commonly called *slide shows,* although they can consist not only of 35-mm slides but also of paper copies, overhead transparencies, video, animation, and sound. Presentation graphics packages often come with slide sorters, which group together a dozen or so slides in miniature. The person making the presentation can use a mouse or keyboard to bring the slides up for viewing or even start a self-running electronic slide show.

Let's examine the process of using presentation software.

- **Using templates to get started:** Just as word processing programs offer templates for faxes, business letters, and the like, presentation-graphics programs offer templates to help you organize your presentation, whether it's for a roomful of people or over the Internet. Templates are of two types: design and content. *Design templates* offer formats, layouts, background patterns, and color schemes that can apply to general forms of content material. *Content templates* offer formats for specific subjects; for instance, PowerPoint offers templates for "Selling Your Ideas," "Facilitating a Meeting," and "Motivating a Team." The software offers wizards that walk you through the process of filling in the template.

- **Getting assistance on content development and organization:** To provide assistance as you're building your presentation, PowerPoint displays three windows on your screen at the same time—the *Outline View,* the *Slide View,* and the *Notes Page View.* The purpose of this is to enable you to add new slides, create and edit the text on the slides, and create notes (to use as lecture or speech notes) while developing your presentation.

  The *Outline View* helps you organize the content of your material in standard outline form. The text you enter into the outline is automatically formatted into slides according to the template you selected. If you wish, you can pull in (import) your outline from a word processing document.

1 Outline View
This view helps you organize the content of your material in standard outline form.

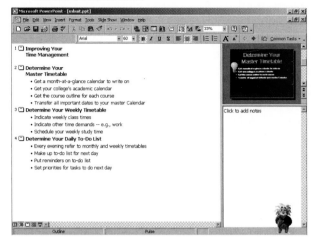

2 Dressing up your presentation
PowerPoint offers professional design templates of text format, background, and borders. You place your text for each slide into one of these templates. You can also import a graphic from the clip art that comes with the program.

3 Slide View
This view allows you to see what a single slide will look like. You can use this view to edit the content and looks of each slide.

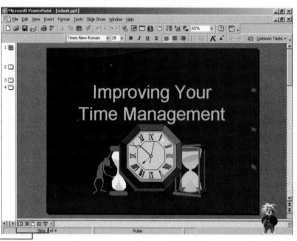

4 Notes Page View
This view displays a small version of the slide plus the notes you will be using as speaker notes.

View icons
Clicking on these offers different views: *Slide, Outline, Slide Sorter, Notes Page,* and *Slide Show*

5 Slide Sorter View
This view displays miniatures of each slide, enabling you to adjust the order of your presentation.

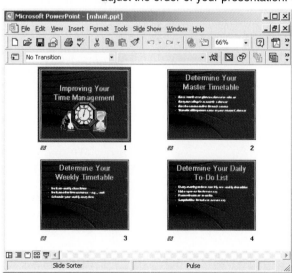

● **PANEL 3.19**
Presentation graphics
Microsoft PowerPoint 2000 helps you prepare and make visual presentations.

Software

The *Slide View* helps you see what a single slide will look like. The outline text appears as slide titles and subtitles in subordinate order.

The *Notes Page View* displays the notes you will be using as speaker notes. It includes a small version of the slide.

Two other views are helpful in organizing and practicing. The *Slide Sorter View* allows you to view a number of slides (12 or more) at once, so you can see how to order and reorder them. The *Slide Show View* presents the slides in the order in which your audience will view them, so you can practice your presentation.

- **Dressing up your presentation:** Presentation software makes it easy to dress up each visual page ("slide") with artwork by pulling in clip art from other sources. Although presentations may make use of some basic analytical graphics—bar, line, and pie charts—they usually look much more sophisticated. For instance, they may utilize different texture (speckled, solid, cross-hatched), color, and three-dimensionality. In addition, you can add audio clips, special visual effects (such as blinking text), animation, and video clips.

CONCEPT CHECK

What are the benefits of presentation graphics software?

What kind of help is available with a presentation graphics package?

Financial Software

Financial software is a growing category that ranges from personal-finance managers to entry-level accounting programs to business financial-management packages.

Consider the first of these, which you may find particularly useful. **_Personal-finance managers_ let you keep track of income and expenses, write checks, do online banking, and plan financial goals.** Such programs don't promise to make you rich, but they can help you manage your money. They may even get you out of trouble. Many personal-finance programs, such as Quicken and Microsoft Money, include a calendar and a calculator, but the principal features are the following:

- **Tracking of income and expenses:** The programs allow you to set up various account categories for recording income and expenses, including credit card expenses.
- **Checkbook management:** All programs feature checkbook management, with an on-screen check writing form and check register that look like the ones in your checkbook. Checks can be purchased to use with your computer printer.
- **Reporting:** All programs compare your actual expenses with your budgeted expenses. Some will compare this year's expenses to last year's.
- **Income tax:** All programs offer tax categories, for indicating types of income and expenses that are important when you're filling out your tax return.
- **Other:** Some of the more versatile personal-finance programs also offer financial-planning and portfolio-management features.

Besides personal-finance managers, financial software includes small business accounting and tax software programs, which provide virtually all the forms you need for filing income taxes. Tax programs such as TaxCut and Turbo Tax make complex calculations, check for mistakes, and even unearth deductions you didn't know existed. Tax programs can be linked to personal finance software to form an integrated tool.

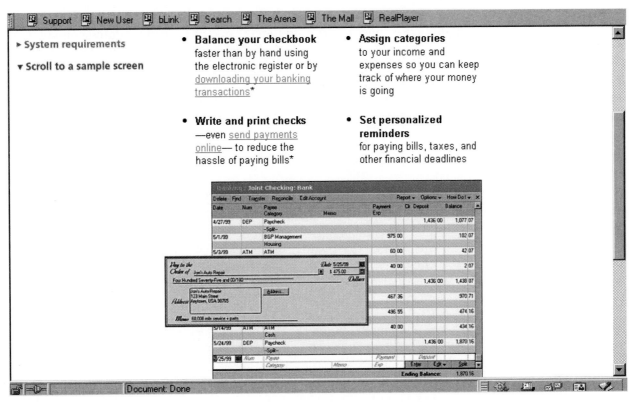

▸ System requirements

▾ Scroll to a sample screen

Balance your checkbook
faster than by hand using
the electronic register or by
downloading your banking
transactions*

Write and print checks
—even send payments
online— to reduce the
hassle of paying bills*

Assign categories
to your income and
expenses so you can keep
track of where your money
is going

**Set personalized
reminders**
for paying bills, taxes, and
other financial deadlines

Document: Done

Quicken

This is one of the most popular personal-finance programs.

A lot of financial software is of a general sort and may be used in all kinds of enterprises. For instance, accounting software automates bookkeeping tasks, while payroll software keeps records of employee hours and produces reports for tax purposes.

Some programs go beyond financial management and tax and accounting management. For example, Business Plan Pro, Management Pro, and Performance Now can help you set up your own business from scratch.

Finally, there are investment software packages, such as StreetSmart from Charles Schwab and Online Xpress from Fidelity, as well as various retirement planning programs.

CONCEPT CHECK

What is financial software?

What various things can you do with financial software?

Desktop Publishing

When Margaret Trejo, then 36, was laid off from her job because her boss couldn't meet the payroll, she was stunned. "Nothing like that had ever happened to me before," she said later. "But I knew it wasn't a reflection on my work. And I saw it as an opportunity."[22]

Today Trejo Production is a successful desktop-publishing company in Princeton, New Jersey, using Macintosh equipment to produce scores of books, brochures, and newsletters. "I'm making twice what I ever made in management positions," says Trejo, "and my business has increased by 25% every year."

Not everyone can set up a successful desktop-publishing business, because many complex layouts require experience, skill, and knowledge of graphic design. Indeed, use of these programs by nonprofessional users can lead to rather unprofessional-looking results. Nevertheless, the availability of microcomputers and reasonably inexpensive software has opened up a career area formerly reserved for professional typographers and printers.

Software

Desktop publishing (DTP) **involves mixing text and graphics to produce high-quality output for commercial printing, using a microcomputer and mouse, scanner, laser or ink-jet printer, and DTP software.** Often the printer is used primarily to get an advance look before the completed job is sent to a typesetter for even higher-quality output. Professional DTP programs are QuarkXPress and PageMaker. Microsoft Publisher is a "low-end," consumer-oriented DTP package. Some word processing programs, such as Word and WordPerfect, also have many DTP features, though at nowhere near the level of the specialized DTP packages.

Desktop publishing has the following characteristics:

- **Mix of text with graphics:** Desktop-publishing software allows you to precisely manage and merge text with graphics. As you lay out a page on-screen, you can make the text "flow," liquid-like, around graphics such as photographs. You can resize art, silhouette it, change the colors, change the texture, flip it upside down, and make it look like a photo negative.

- **Varied type and layout styles:** As do word processing programs, DTP programs provide a variety of fonts, or typestyles, from readable Times Roman to staid Tribune to wild Jester and Scribble. Additional fonts can be purchased on disk or downloaded online. You can also create all kinds of rules, borders, columns, and page numbering styles.

- **Use of files from other programs:** It's usually not efficient to do word processing, drawing, and painting with DTP software. As a rule, text is composed on a word processor, artwork is created with drawing and painting software, and photographs are scanned in using a scanner and then manipulated and stored using photo-manipulation software. Prefabricated art to illustrate DTP documents may be obtained from disks of clip art, or "canned" images. The DTP program is used to integrate all these files. You can look at your work on the display screen as one page or as two facing pages (in reduced size). Then you can see it again after it has been printed out. *(See ● Panel 3.20.)*

Drawing & Painting Programs

John Ennis was trained in realistic oil painting, and for years he used his skill creating illustrations for book covers and dust jackets. Now he "paints" using a computer, software, and mouse. The greatest advantage, he says, is that if "I do a brush stroke in oil and it's not right, I have to take a rag and wipe it off. With the computer, I just hit the 'undo' command."[23]

It may be no surprise to learn that commercial artists and fine artists have begun to abandon the paintbox and pen and ink for software versions of palettes, brushes, and pens. The surprise, however, is that an artist can use mouse and pen-like stylus to create computer-generated art as good as that achievable with conventional artist's tools. More surprising, even *nonartists* can produce good-looking work with these programs.

There are two types of computer art programs: drawing and painting.

- **Drawing programs:** A *drawing program* **is graphics software that allows users to design and illustrate objects and products.** Some drawing programs are CorelDRAW, Adobe Illustrator, Macromedia Freehand, and Sketcher. Drawing programs create *vector images*—images created from mathematical calculations.

- **Painting programs:** *Painting programs* **are graphics programs that allow users to simulate painting on screen.** A mouse or a tablet stylus is used to simulate a paintbrush. The program allows you to select "brush" sizes, as well as colors from a color palette. Examples of painting programs are MetaCreations' Painter, Adobe Photoshop, Corel PhotoPaint, and JASC's PaintShop Pro. Painting programs produce *raster images* made up of little dots.

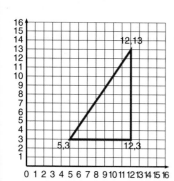
Vector image

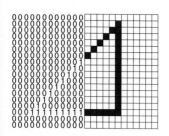
Raster image

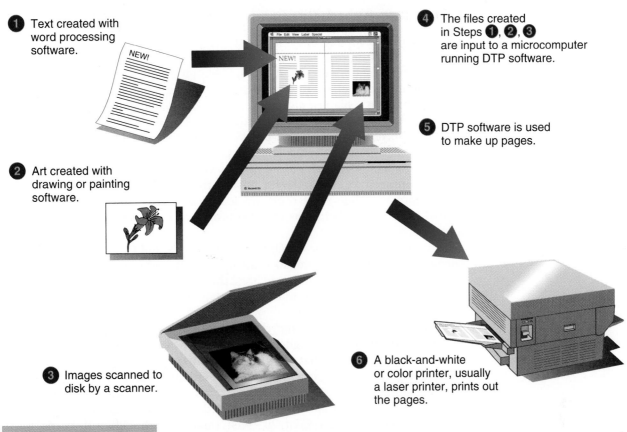

1 Text created with word processing software.

NEW!

2 Art created with drawing or painting software.

3 Images scanned to disk by a scanner.

4 The files created in Steps **1**, **2**, **3** are input to a microcomputer running DTP software.

5 DTP software is used to make up pages.

6 A black-and-white or color printer, usually a laser printer, prints out the pages.

● **PANEL 3.20**
How desktop publishing uses other files

CONCEPT CHECK

What can you do with desktop publishing software?

What are some characteristics of DTP software?

What are some features of drawing programs? Of painting programs?

Project Management Software

The kind of database program we called a personal information manager (PIM) can help you schedule your appointments and do some planning. That is, it can help you manage your own life. But what if you need to manage the lives of others in order to accomplish a full-blown project, such as steering a political campaign or handling a nationwide road tour for a band? Strictly defined, a *project* is a one-time operation involving several tasks and multiple resources that must be organized toward completing a specific goal within a given period of time. The project can be small, such as an advertising campaign for an in-house advertising department, or large, such as construction of an office tower or a jetliner.

__Project management software__ **is a program used to plan and schedule the people, costs, and resources required to complete a project on time.** For instance, the associate producer on a feature film might use such software to keep track of the locations, cast and crew, materials, dollars, and schedules needed to complete the picture on time and within budget. The software would show the scheduled beginning and ending dates for a particular task—such as shooting all scenes on a certain set—and then the date that task was actually completed. Examples of project management software are Harvard Project Manager, Microsoft Project, Suretrack Project Manager, and ManagerPro.

Computer-Aided Design

Computers have long been used in engineering design. ***Computer-aided design (CAD) programs* are intended for the design of products, structures, civil engineering drawings, and maps.** CAD programs, which are available for microcomputers, help architects design buildings and workspaces and engineers design cars, planes, electronic devices, roadways, bridges, and subdivisions. CAD and drawing programs are similar. However, CAD programs provide precise dimensioning and positioning of the elements being drawn, so that they can be transferred later to CAM programs. Also, CAD programs lack the special effects for illustrations that come with drawing programs. One advantage of CAD software is that the product can be drawn in three dimensions and then rotated on the screen so the designer can see all sides. *(See ● Panel 3.21.)* Examples of CAD programs for beginners are Autosketch and CorelCAD.

A variant on CAD is *CADD*, for *computer-aided design and drafting*, software that helps people do drafting. CADD programs include symbols (points, circles, straight lines, and arcs) that help the user put together graphic elements, such as the floor plan of a house. An example is Autodesk's Auto-CAD.

CAD/CAM (computer-aided design/computer-aided manufacturing) software allows products designed with CAD to be input into an automated manufacturing system that makes the products.** For example, CAD and its companion CAM brought a whirlwind of enhanced creativity and efficiency to the fashion industry. Some CAD systems, says one writer, "allow designers to electronically drape digital-generated mannequins in flowing gowns or tailored suits that don't exist, or twist imaginary threads into yarns, yarns into weaves, weaves into sweaters without once touching needle to garment."[24] The designs and specifications are then input into CAM systems that enable robot pattern-cutters to automatically cut thousands of patterns from fabric with only minimal waste. Whereas previously the fashion industry worked about a year in advance of delivery, CAD/CAM has cut that time to 8 months—a competitive edge for a field that feeds on fads.

CONCEPT CHECK

Describe what project management software can do.

What is the purpose of CAD software? Of CAD/CAM software?

● **PANEL 3.21**
CAD
Example of computer-aided design

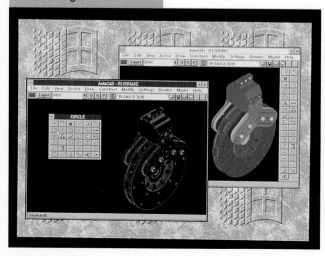

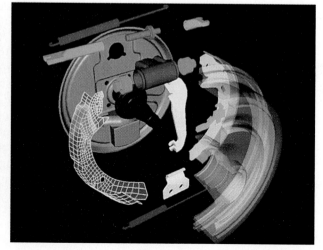

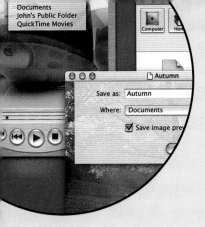

Visual Summary

analytical graphics (p. 119, KQ 3.5) Also called *business graphics;* graphical forms that make numeric data easier to analyze than when it is organized as rows and columns of numbers. The principal examples of analytical graphics are bar charts, line graphs, and pie charts. Why it's important: *Whether viewed on a monitor or printed out, analytical graphics help make sales figures, economic trends, and the like easier to comprehend and analyze.*

ASP (application service provider) (p. 88, KQ 3.1) Firm that leases software over the Internet. Why it's important: *You no longer have to buy software in a store, in shrink-wrapped packages. Instead, you can simply download a particular program when you need it, for as long as you need it.*

booting (p. 91, KQ 3.2) Process of loading an operating system into a computer's main memory. Why it's important: *This loading is accomplished by a program that is stored permanently in the computer's electronic circuitry. When you turn on the machine, programs called diagnostic routines test the main memory, the central processing unit, and other parts of the system to make sure they are running properly. Next, BIOS (for basic input/output system) programs are copied to main memory and help the computer interpret keyboard characters or transmit characters to the display screen or to a diskette. Then the boot program obtains the operating system, usually from hard disk, and loads it into the computer's main memory, where it remains until you turn the computer off.*

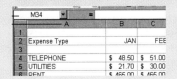

cell (p. 119, KQ 3.5) Place where a row and a column intersect in a spreadsheet worksheet; its position is called a *cell address.* Why it's important: *The cell is the smallest working unit in a spreadsheet. Data and formulas are entered into cells. Cell addresses provide location references for spreadsheet users.*

computer-aided design (CAD) programs (p. 130, KQ 3.7) Programs intended for the design of products, structures, civil engineering drawings, and maps. Why it's important: *CAD programs, which are available for microcomputers, help architects design buildings and workspaces and engineers design cars, planes, electronic devices, roadways, bridges, and subdivisions. CAD and drawing programs are similar. However, CAD programs provide precise dimensioning and positioning of the elements being drawn, so that they can be transferred later to CAM programs. Also, CAD programs lack the special effects for illustrations that come with drawing programs. One advantage of CAD software is that the product can be drawn in three dimensions and then rotated on the screen so the designer can see all sides.*

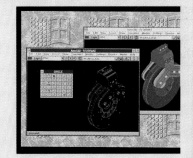

computer-aided design/computer-aided manufacturing (CAD/CAM) software (p. 130, KQ 3.7) Programs that allow products designed with CAD to be input into an automated manufacturing system that makes the products. Why it's important: *CAM systems have greatly enhanced efficiency in many industries.*

● PANEL 3.9
Choices among application softwar

Types	Definition
Commercial software	Copyrighted. I
Public-domain software	Not copyright
Shareware	Copyrighted.

copyright (p. 106, KQ 3.3) Exclusive legal right that prohibits copying of intellectual property without the permission of the copyright holder. Why it's important: *Copyright law aims to prevent people from taking credit for and profiting from other people's work.*

cursor (p. 112, KQ 3.4) Movable symbol on the display screen that shows you where you may next enter data or commands. The symbol is often a blinking rectangle or an I-beam. You can move the cursor on the screen using the keyboard's directional arrow keys or a mouse. The point where the cursor is located is called the *insertion point.* Why it's important: *All application software packages use cursors to show the current work location on the screen.*

database (p. 120, KQ 3.6) Collection of interrelated files in a computer system. These computer-based files are organized according to their common elements, so that they can be retrieved easily. *Why it's important: Businesses and organizations build databases to help them keep track of and manage their affairs. In addition, online database services put enormous resources at the user's disposal.*

database file (p. 109, KQ 3.3) File created by database management programs and consist of organized data that can be analyzed and displayed in various useful ways. *Why it's important: Database files make up a database.*

database software (p. 120, KQ 3.6) Application software that sets up and controls the structure of a database and access to the data. *Why it's important: Database software allows users to organize and manage huge amounts of data.*

default settings (p. 116, KQ 3.4) Settings automatically used by a program unless the user specifies otherwise, thereby overriding them. *Why it's important: Users need to know how to change default settings in order to customize documents.*

desktop (p. 96, KQ 3.2) The operating system's main interface screen. *Why it's important: The desktop displays pictures (icons) that provide quick access to programs and information.*

desktop publishing (DTP) (p. 128, KQ 3.7) Application software and hardware system that involves mixing text and graphics to produce high-quality output for commercial printing, using a microcomputer and mouse, scanner, laser or ink-jet printer, and DTP software. Often the printer is used primarily to get an advance look before the completed job is sent to a typesetter for even higher-quality output. Professional DTP programs are QuarkXPress and PageMaker. Microsoft Publisher is a "low-end," consumer-oriented DTP package. Some word processing programs, such as Word and WordPerfect, have rudimentary DTP features. *Why it's important: Desktop publishing has reduced the number of steps, the time, and the money required to produce professional-looking printed projects.*

device drivers (p. 92, KQ 3.2) Specialized software programs—usually components of system software—that allow input and output devices to communicate with the rest of the computer system. *Why it's important: Drivers are needed so that the computer's operating system can recognize and run peripheral hardware.*

document file (p. 107, KQ 3.3) File created by word processing programs; consists of documents such as reports, letters, memos, and term papers. *Why it's important: Document files are probably the most common type of file users will deal with.*

documentation (p. 107, KQ 3.3) User guide or reference manual that provides a narrative and graphical description of a program. While documentation may be print-based, today it is usually available on CD-ROM, as well as via the Internet. *Why it's important: Documentation helps users learn software commands and use of function keys, solve problems, and find information about system specifications.*

DOS (Disk Operating System) (p. 101, KQ 3.2) Original operating system produced by Microsoft, with a hard-to-use command-driven user interface. Its initial 1982 version was designed to run on the IBM PC as PC-DOS. Later Microsoft licensed the same system to other computer makers as MS-DOS. *Why it's important: DOS used to be the most common microcomputer operating system, and it is still used on many microcomputers. Today the most popular operating systems use GUIs.*

drawing program (p. 128, KQ 3.7) Graphics software that allows users to design and illustrate objects and products. *Why it's important: Drawing programs are vector-based and are best used for straightforward illustrations based on geometric shapes.*

ERP (enterprise resource planning) software (p. 90, KQ 3.1) Large client/server software applications that help companies organize and operate their businesses. *Why it's important: ERP software coordinates a company's entire business and moves data speedily from one department to another.*

export (p. 109, KQ 3.3) To transform data into a format that can be used in another program and then transmii it. *Why it's important: Users need to know how to export many types of files.*

file (p. 107, KQ 3.3) A named collection of data or a program that exists in a computer's secondary storage, such as on a floppy disk, hard disk, or CD-ROM disk. *Why it's important: Dealing with files is an inescapable part of working with computers. Users need to be familiar with the different types of files.*

financial software (p. 126, KQ 3.7) Application software that ranges from personal-finance managers to entry-level accounting programs to business financial-management packages. *Why it's important: Financial software provides users with powerful management tools (personal finance managers), as well as small business accounting programs. Moreover, tax programs provide virtually all the forms needed for filing income taxes, make complex calculations, check for mistakes, and even unearth deductions you didn't know existed. Tax programs can also be integrated with personal finance software. Accounting software automates bookkeeping tasks, while payroll software keeps records of employee hours and produces reports for tax purposes. Some programs allow users to set up a business from scratch. Financial software also includes investment software packages and various retirement planning programs.*

fly-out menu (p. 96, KQ 3.2) Menu that seems to explode out to the right. *Why it's important: Menus make software easier to use.*

font (p. 116, KQ 3.4) A particular typeface and type size. *Why it's important: Fonts influence the appearance and effectiveness of documents, brochures, and other types of publications.*

formatting (initializing) a disk (p. 92, KQ 3.2) Process of preparing floppy disks so that they can store data or programs. *Why it's important: Today it is easier to buy preformatted disks. However, it's useful to know how to format a blank floppy disk or re-format a disk that wasn't intended for your machine.*

formatting (p. 115, KQ 3.4) In word processing and desktop publishing, determining the appearance of a document. *Why it's important: Ways to format a document include using different fonts, boldface, italics, variable spacing, columns, and margins. The document's format should match the needs of its users.*

formulas (p. 119, KQ 3.5) In a spreadsheet, instructions for calculations entered into designated cells. *Why it's important: When spreadsheet users change data in one cell, all the cells linked to it by formulas automatically recalculate their values.*

freeware (p. 107, KQ 3.3) Copyrighted software that is distributed free of charge, today most often over the Internet. *Why it's important: Freeware saves users money.*

function keys (p. 111, KQ 3.4) Keys labeled F1, F2, and so on, positioned along the top or left side of the keyboard. *Why it's important: They are used to execute commands specific to the software being used.*

grammar checker (p. 115, KQ 3.4) Word processing feature that highlights poor grammar, wordiness, incomplete sentences, and awkward phrases. The grammar checker won't fix things automatically, but it will flag (perhaps with a different color of squiggly line) possible incorrect word usage and sentence structure. *Why it's important: Grammar checkers help users produce better-written documents.*

graphical user interface (GUI) (p. 93, KQ 3.2) User interface in which icons and commands from menus may be selected by means of a mouse or keystrokes. *Why it's important: GUIs are easier to use than command-driven interfaces.*

groupware (p. 110, KQ 3.3) Online software that allows several people to collaborate on the same project. *Why it's important: Groupware improves productivity by keeping users continually notified about what their colleagues are thinking and doing.*

Help command (p. 100, KQ 3.2) Command generating a table of contents, an index, and a search feature that can help users locate answers to questions about using the software. *Why it's important: Help features provide a built-in electronic instruction manual.*

icons (p. 96, KQ 3.2) Small pictorial figures that represent programs, data files, or procedures. *Why it's important: Icons have simplified the use of software. The feature represented by the icon can be activated by clicking on the icon.*

import (p. 109, KQ 3.3) To get data from another source and then convert it into a format compatible with the program in which you are currently working. Why it's important: *Users will often have to import files.*

key field (p. 121, KQ 3.6) Field that can be used as a common identifier because it is unique. The most frequent key field used in the United States is the Social Security number, but any unique identifier could be used, such as an employee number. Why it's important: *Key fields are needed to identify and retrieve specific items in a database.*

Linux (p. 104, KQ 3.2) Free version of Unix. Why it's important: *Linux is an inexpensive, open-source operating system useful for online applications and to PC users who have to maintain a Web server or a network server.*

Macintosh operating system (Mac OS) (p. 101, KQ 3.2) System software that runs only on Apple Macintosh computers. Why it's important: *Although Macs are not as common as PCs, many people believe they are easier to use. Macs are often used for graphics and desktop publishing.*

macro (p. 112, KQ 3.4) Also called a keyboard shortcut; a single keystroke or command—or a series of keystrokes or commands—used to automatically issue a longer, predetermined series of keystrokes or commands. Why it's important: *Users can consolidate several activities into only one or two keystrokes. The user names the macro and stores the corresponding command sequence; once this is done, the macro can be used repeatedly.*

---

menu (p. 96, KQ 3.2) Displayed list of options—such as commands—to choose from. Why it's important: *Menus are a feature of GUIs to make software easier to use.*

Microsoft Windows 95/98 (p. 102, KQ 3.2) Operating system for desktop and portable microcomputers, supporting the most hardware and the most application software. Why it's important: *Windows has become the most common system software used on microcomputers.*

multitasking (p. 92, KQ 3.2) Operating system software feature that allows the execution of more than one process at a time. For instance, you might write on one word-processing document while printing out another and downloading a third from a Web site. Why it's important: *Multitasking allows the computer to switch rapidly back and forth among different tasks. The user is generally unaware of the switching process.*

netware (p. 103, KQ 3.2) Popular network operating system for orchestrating microcomputer-based local area networks (LANs). Why it's important: *Netware allows PCs to share files, printers, and servers.*

network computer (p. 89, KQ 3.1) Stripped-down computer that connects people to networks and runs applications tied to servers. Why it's important: *It's less expensive than a personal computer, and, since the software is stored on a server, the user theoretically has fewer software headaches.*

office suite (p. 110, KQ 3.3) A single large software package that bundles several applications together. Why it's important: *Office suites cost less than do the applications purchased separately.*

open-source software (p. 104, KQ 3.2) Software that any programmer can download from the Internet for free and modify with suggested improvements. The only qualification is that changes can't be copyrighted; they must be made available to all and remain in the public domain. Why it's important: *Because this software is not proprietary, any programmer can make improvements, which can result in better quality software.*

operating system (p. 91, KQ 3.2) Master system of programs that manage the basic operations of the computer. Why it's important: *These programs provide resource management services of many kinds. In particular, they handle the control and use of hardware resources, including disk space, memory, CPU time allocation, and peripheral devices. The operating system allows users to concentrate on their own tasks or applications rather than on the complexities of managing the computer.*

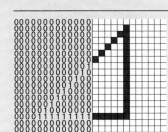

painting program (p. 128, KQ 3.7) Graphics program that allows users to simulate painting on screen. A mouse or a tablet stylus is used to simulate a paintbrush. The program allows you to select "brush" sizes, as well as colors from a color palette. Why it's important: *Painting programs, which produce raster images made up of little dots, are good for creating art with soft edges and many colors.*

Palm OS (p. 105, KQ 3.2) Operating system for the Palm and Visor; the dominant operating system for handhelds. Why it's important: *Because it not a Windows derivative but was specifically designed for handhelds, Palm OS is a smoother-running operating system.*

personal-finance manager (p. 126, KQ 3.7) Application software that lets you keep track of income and expenses, write checks, do online banking, and plan financial goals. Why it's important: *Personal finance software can help people manage their money more effectively.*

personal information manager (PIM) (p. 123, KQ 3.6) Software to help you keep track of and manage information you use on a daily basis, such as addresses, telephone numbers, appointments, to-do lists, and miscellaneous notes. Some programs feature phone dialers, outliners (for roughing out ideas in outline form), and ticklers (or reminders). Why it's important: *PIMs can help users better organize and manage daily business activities.*

platform (p. 101, KQ 3.2) Particular processor model and operating system on which a computer system is based. Why it's important: *User should be aware that there are Mac platforms (Apple Macintosh) and Windows platforms, or "PC platforms (for personal computers such as Dell, Compaq, Gateway, Hewlett-Packard, or IBM that run Microsoft Windows). Sometimes the latter are called Wintel platforms, for "Windows + Intel," because they often combine the Windows operating system with the Intel processor chip.*

Pocket PC (p. 105, KQ 3.2) Operating system for handhelds, is simpler and less cluttered than CE and looks and feels a lot less like desktop Windows. Why it's important: *Pocket PC offers pocket versions of Word and Excel that let users read standard word processing and spreadsheet files sent as e-mail attachments from their PCs.*

pointer (p. 93, KQ 3.2) Indicator that usually appears as an arrow, although it changes shape depending on the application. The mouse is used to move the pointer to a particular place on the display screen or to point to little symbols, or icons. Why it's important: *It is often easier to manipulate the pointer on the screen by means of the mouse than to type commands on a keyboard.*

pop-up menu (p. 96, KQ 3.2) List of command options that can "pop up" anywhere on the screen. In contrast to pull-down or pull-up menus, pop-up menus are not connected to a toolbar. Why it's important: *Pop-up menus make programs easier to use.*

presentation graphics software (p. 124, KQ 3.7) Software that uses graphics, animation, sound and data or information to make visual presentations. Why it's important: *Presentation graphics software provides a means to produce sophisticated graphics.*

productivity software (p. 110, KQ 3.3) Application software such as word processing, spreadsheets, and database managers. Why it's important: *Productivity software makes users more productive at particular tasks.*

project management software (p. 129, KQ 3.7) Program used to plan and schedule the people, costs, and resources required to complete a project on time. Why it's important: *Project management software increases the ease and speed of planning and managing complex projects.*

public-domain software (p. 106, KQ 3.3) Software, often available on the Internet, that is not protected by copyright and thus may be duplicated by anyone at will. Why it's important: *Public domain software offers lots of software options to users who may not be able to afford much commercial software. Users may download such software from the Internet for free and make as many copies as they wish.*

pull-down menu (p. 96, KQ 3.2) Also called a *drop-down menu;* list of options that pulls down from the menu bar at the top of the screen. Why it's important: *Like other menu-based and GUI features, pull-down menus make software easier to use.*

pull-up menu (p. 96, KQ 3.2) List of options that pulls up from the menu bar at the bottom of the screen. *Why it's important: See* pull-down menu.

recalculation (p. 119, KQ 3.5) Recomputing values in a spreadsheet, either as an ongoing process as data is entered or afterward, with the press of a key. *Why it's important: With this simple feature, the hours of mind-numbing work required to manually rework paper spreadsheets became a thing of the past.*

relational database (p. 121, KQ 3.6) Database in which data is organized into related tables. Each table contains rows and columns; the rows are called records, and the columns are called fields. An example of a record is a person's address—name, street address, city, and so on. An example of a field is that person's last name; another field would be that person's first name; a third field would be that person's street address, and so on. *Why it's important: The relational database a common type of database.*

rentalware (p. 107, KQ 3.3) Software that users lease for a fee. *Why it's important: This is the concept behind application services providers (ASPs).*

rollover (p. 96, KQ 3.2) Icon feature in which a small text box explaining the icon's function appears when you roll the mouse pointer over the icon. A rollover may also produce an animated graphic. *Why it's important: The rollover gives the user an immediate explanation of an icon's meaning.*

save (p. 116, KQ 3.4) To store, or preserve, a document as an electronic file permanently— on diskette, hard disk, or CD-ROM, for example. *Why it's important: Saving is a feature of nearly all application software. Having the document stored in electronic form spares users the tiresome chore of retyping it from scratch whenever they want to make changes. Users need only retrieve it from the storage medium and make the changes, then resave it and print it out again.*

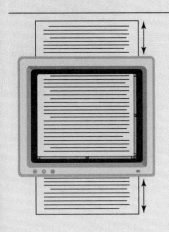

scrolling (p. 112, KQ 3.4) Moving quickly upward, downward, or sideways through the text or other screen display. *Why it's important: A standard computer screen displays only 20–22 lines of standard-size text; however, most documents are longer than that. Using the directional arrow keys, or the mouse and a scroll bar located at the side of the screen, users can move ("scroll") through the display screen and into the text above and below it.*

shareware (p. 107, KQ 3.3) Copyrighted software that is distributed free of charge but requires users to make a monetary contribution in order to continue using it. Shareware is distributed primarily through the Internet. Because it is copyrighted, you cannot use it to develop your own program that would compete with the original product. *Why it's important: Like public domain software and freeware, shareware offers an inexpensive way to obtain new software.*

software license (p. 106, KQ 3.3) Contract by which users agree not to make copies of software to give away or for resale. *Why it's important: Software manufacturers don't sell people software so much as licenses to become authorized users of the software.*

special-purpose keys (p. 111, KQ 3.4) Keys used to enter, delete, and edit data and to execute commands. For example, the Esc (for "Escape") key tells the computer to cancel an operation or leave ("escape from") the current mode of operation. The Enter, or Return, key tells the computer to execute certain commands and to start new paragraphs in a document. *Why it's important: Special-purpose keys are essential to the use of software.*

spelling checker (p. 114, KQ 3.4) Word processing feature that tests for incorrectly spelled words. As you type, the spelling checker indicates (perhaps with a squiggly line) words that aren't in its dictionary and thus may be misspelled. Special add-on dictionaries are available for medical, engineering, and legal terms. *Why it's important: Spelling checkers help users prepare accurate documents.*

spreadsheet (p. 117, KQ 3.5) Application software that allows users to create tables and financial schedules by entering data and formulas into rows and columns arranged as a grid on a display screen. *Why it's important: When data is changed in one cell, values in other cells in the spreadsheet are automatically recalculated.*

Expense Type	JAN	FEB	MAR	TOTAL
TELEPHONE	$ 48.50	$ 51.00	$ 37.90	$ 137.40
UTILITIES	$ 21.70	$ 30.00	$ 25.00	$ 76.70
RENT	$ 465.00	$ 465.00	$ 465.00	$1,395.00
AUTOMOBILE	$ 35.00	$ 211.00	$ 42.00	$ 288.00
MISCELLANEOUS	$ 120.00	$ 93.00	$ 41.43	$ 254.43

supervisor (p. 92, KQ 3.2) Also called *kernel;* the central component of the operating system that manages the CPU. Why it's important: *The supervisor remains in main memory while the computer is running. As well as managing the CPU, it directs other nonresident programs to perform tasks that support application programs. The operating system also manages memory—it keeps track of the locations within main memory (and on hard disk) where the programs and data are stored.*

taskbar (p. 99, KQ 3.2) Graphic toolbar that appears at the bottom of the Windows screen. Why it's important: *The taskbar presents the applications that are running.*

template (p. 115, KQ 3.4) In word processing, a preformatted document that provides basic tools for shaping a final document—the text, layout, and style for a letter, for example. Why it's important: *Templates make it very easy for users to prepare professional-looking documents, because most of the preparatory formatting is done.*

thesaurus (p. 115, KQ 3.4) Word processing feature that will present you with the appropriate word or alternative words. Why it's important: *The thesaurus feature helps users prepare well-written documents.*

toolbar (p. 99, KQ 3.2) Bar across the top of the display window. It displays menus and icons representing frequently used options or commands. Why it's important: *Toolbars make it easier to identify and execute commands.*

tutorial (p. 107, KQ 3.3) Instruction book or program that helps you learn to use the product by taking you through a prescribed series of steps. Why it's important: *Tutorials enable users to practice using new software in a graduated fashion and learn the software in an effective manner.*

Unix (p. 103, KQ 3.2) Multitasking operating system for multiple users that has built-in networking capability and versions that can run on all kinds of computers. Why it's important: *Government agencies, universities, research institutions, large corporations, and banks all use Unix for everything from designing airplane parts to currency trading. Unix is also used for Web-site management. The developers of the Internet built their communications system around Unix because it has the ability to keep large systems (with hundreds of processors) churning out transactions day in and day out for years without fail.*

user interface (p. 93, KQ 3.2) Display screen that allows users to communicate, or interact, with the computer. The three types of user interface are command-driven, menu-driven, and graphical (GUI), now the most common interface. Why it's important: *Without user interfaces, no one could use a computer system.*

utility programs (p. 92, KQ 3.2) Also known as *service programs;* system software component that performs tasks related to the control and allocation of computer resources. Why it's important: *Utility programs enhance existing functions or provide services not supplied by other system software programs. Most computers come with utilities built in as part of the system software.*

value (p. 119, KQ 3.5) A number or date entered in a spreadsheet cell. Why it's important: *Values are the actual numbers used in the spreadsheet—dollars, percentages, grade points, temperatures, or whatever.*

what-if analysis (p. 119, KQ 3.5) Spreadsheet feature that employs the recalculation feature to investigate how changing one or more numbers changes the outcome of the calculation. Why it's important: *Users can create a worksheet, putting in formulas and numbers, and then ask, "What would happen if we change that detail?"—and immediately see the effect.*

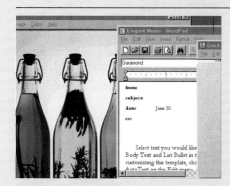

window (p. 99, KQ 3.2) Rectangular frame on the computer display screen. Through this frame you can view a file of data—such as a document, spreadsheet, or database—or an application program. Why it's important: *Using windows, users can display at the same time portions of several documents and/or programs on the screen.*

Windows CE (p. 105, KQ 3.2) Greatly slimmed-down version of Windows 95 for handheld computing devices, such as those made by Casio, Compaq, and Hewlett-Packard. Windows CE had some of the familiar Windows look and feel and included rudimentary word processing, spreadsheet, e-mail, Web browsing, and other software. Why it's important: *Windows CE was Microsoft's attempt to modify its Windows desktop operating system for use with handhelds.*

Windows NT (p. 103, KQ 3.2) Later upgraded to Windows 2000, Microsoft's multitasking operating system designed to run on network servers. Why it's important: *It allows multiple users to share resources such as data and programs.*

wizard (p. 115, KQ 3.4) Word processing software feature that answers your questions and uses the answers to lay out and format a document. Why it's important: *Wizards make it easy to prepare professional-looking memos, faxes, resumes, and other documents.*

word processing software (p. 110, KQ 3.4) Application software that allows you to use computers to format, create, edit, print, and store text material, among other things. Why it's important: *Word processing software allows users to maneuver through a document and delete, insert, and replace text, the principal correction activities. It also offers such additional features as creating, editing, formatting, printing, and saving.*

worksheet file (p. 107, KQ 3.3) File created by electronic spreadsheets; it consists of a collection of (usually) numerical data such as budgets, sales forecasts, and schedules. Why it's important: *Worksheet files are one common type of file users will have to deal with.*

Chapter Review

Self-Test Questions

1. _____ is the term for programs designed to perform specific tasks for the user.

2. _____ software allows you to create and edit documents.

3. _____ is the activity of moving upward or downward through the text or other screen display.

4. _____ enables the computer to perform essential operating tasks.

5. A(n) _____ is a firm that leases software over the Internet.

6. _____ is the activity in which a computer works on more than one process at a time.

7. Name four editing features offered by word processing programs: _____, _____, _____, _____.

8. In a spreadsheet, the place where a row and a column intersect is called a _____.

9. A(n) _____ is a keyboard shortcut used to automatically issue a longer, predetermined series of keystrokes or commands.

10. The _____ is the movable symbol on the display screen that shows you where you may next enter data or commands.

Multiple-Choice Questions

1. Which of the following is NOT an advantage of using database software?

   a. integrated data

   b. improved data integrity

   c. lack of structure

   d. elimination of data redundancy

2. Which of the following are functions of the operating system?

   a. file management

   b. CPU management

   c. task management

   d. booting

   e. all the above

3. Which of the following is not a type of menu?

   a. fly-out menu

   b. pop-in menu

   c. pop-out menu

   d. pull-down menu

   e. pull-up menu

4. Which of the following is not a feature of word processing software?

   a. spelling checker

   b. cell address

   c. formatting

   d. cut and paste

   e. find and replace

True/False Questions

T F   1. Spreadsheet software enables you to perform "what if" calculations.

T F   2. *Font* refers to a preformatted document that provides basic tools for shaping a final document.

T F   3. Rentalware is software that users lease for a fee.

T F   4. Public-domain software is protected by copyright and so is offered for sale by license only.

T F   5. The records within the various tables in a database are linked by a key field.

Short-Answer Questions

1. Briefly define *booting*.

2. What is the difference between a command-driven interface and a graphical user interface (GUI)?

3. Why can't you run your computer without system software?

4. Why is multitasking useful?

5. What is a device driver?

6. What is a utility program?

7. What are the following types of application software used for?

   a. project management software

   b. desktop publishing software

   c. database software

   d. spreadsheet software

   e. word processing software

8. What is a *platform*?

9. What are the three components of system software? What is the basic function of each?

10. Which program is more sophisticated, analytical graphics or presentation graphics? Why?

Concept Mapping

On a separate sheet of paper, draw a concept map, or visual diagram linking concepts. Show how the following terms are related.

Analytical graphics
Booting
Cell
CAD programs
Cursor
Database software
Desktop publishing
Device drivers
DOS
Drawing program
File
Financial software
Graphical user interface
Menu

Netware
Office suite
Operating system
Presentation graphics
   software
Project management
   software
Scrolling
Spreadsheet software
Toolbar
Utility programs
Windows
Word processing soft-
   ware

Knowledge in Action

1. If you were in the market for a new microcomputer today, what application software would you want to use on it? What system software? Why?

2. Several Web sites include libraries of shareware programs. Visit the *www.winfiles.com* site, click on the Windows shareware icon, and identify three shareware programs that interest you. For each program, state its name, the operating system it runs on, and its capabilities. Also, describe the contribution you must make to receive technical support.

3. Has anyone ever asked you for a copy of one of your software programs? What did you do? Write a few paragraphs on your position on software copying.

4. What operating system is used in your school's computer lab? What version? Who makes this OS? What are this OS's security features? Why was this OS chosen? Write a short report.

Hardware

How to Buy a Multimedia Computer System

Key Questions

You should be able to answer the following questions.

4.1 **The System Unit** How is data represented in a computer; what are the components of the system cabinet; what are processing speeds; how do the processor and memory work; and what are some important ports, buses, and cards?

4.2 **Secondary Storage** What are the features of floppy disks, hard disks, optical disks, magnetic tape, smart cards, and online secondary storage?

4.3 **Input Hardware** What are the three categories of input hardware, what devices do they include, and what are their features?

4.4 **Output Hardware** What are the two categories of output hardware, what devices do they include, and what are their features?

A Visual Overview of This Chapter

1 **The System Unit.** The basis of the processing part of the computer, the system unit, is the **binary system,** which has only 1 and 0. These two digits, called **bits,** correspond to the on and off states of electricity used in computers. A group of 8 bits, called a **byte,** represents one character in the computer. Storage capacities are expressed in multiples of bytes: about 1000 bytes = 1 **kilobyte;** about 1 million bytes = 1 **megabyte;** about 1 billion bytes = 1 **gigabyte;** about 1 trillion bytes = 1 **terabyte;** about 1 quadrillion bytes = 1 **petabyte.** Letters, numbers, and special characters are represented within a computer by *binary coding schemes,* such as **ASCII,** the code most widely used in microcomputers; **EBCDIC,** used with large computers; and **Unicode,** a subset of ASCII that uses 16 bits for each character.

Floppy disk drive Hard-disk drive CD-ROM drive

The *system unit,* or *case,* houses the motherboard, processor chip, memory chips, and power supply. The system unit also includes storage devices, such as disk drives, which are housed on shelves called **bays.** The *motherboard* contains sockets for **expansion**—for adding new components, such as video cards—or **upgrading**—for changing to more powerful components, such as more memory chips.

The microprocessor: The most fundamental part of the motherboard, the **microprocessor,** is the miniaturized circuitry that stores program instructions that manipulate data into information. This manipulation or processing is done by tiny on/off switches called **transistors.** Transistors and other circuitry are printed on tiny **chips** (semiconductors, integrated circuits) made of a material found in sand called *silicon.* Two designs for microprocessors are **CISC chips,** used mostly in PCs and mainframes, and **RISC chips,** used mainly in workstations. Most personal computers today use either **Intel-type chips** for PCs (made by Intel, AMD, Cyrix, and others for Compaq, Dell, Gateway, Hewlett-Packard, and IBM) and **Motorola-type chips** (made by Motorola for Apple Macintosh computers).

The speed of a microprocessor is determined by its **system clock.** For microcomputers, processing speeds are measured in **megahertz** (million cycles per second) or **gigahertz** (billion cycles). For workstations, such speeds are measured in *MIPS* (millions of instructions per second); for supercomputers, in *flops* (floating-point operations per second); and for all computers, in fractions of a second (milliseconds, microseconds, nanoseconds, and picoseconds). Another measure, **word size,** is the number of bits a computer can process at any one time, with 64 bits being faster than 32 bits.

A processor, also called a **CPU (central processing unit),** follows the instructions of the software (program) to manipulate data into information. The CPU consists of the **control unit,** which deciphers each instruction stored in it and then carries it out, and the **arithmetic/logic unit (ALU),** which performs arithmetic and logical operations. Both the control unit and the ALU contain **registers,** high-speed storage areas that temporarily store data during processing. Data within the CPU and between it and other components of the motherboard are transmitted along electrical roadways called **buses.**

Memory: Also on the motherboard are *memory chips,* of which there are four types: *RAM, ROM, CMOS,* and *flash.* (1) **RAM** (for random access memory) **chips** temporarily hold software instructions and also data before and after processing by the CPU. RAM is **volatile;** its contents are lost when the power goes off. Three types of RAM chips are **DRAM,** which must be constantly refreshed by the CPU or it will lose its contents; **SDRAM,** which is faster than DRAM; and **SRAM,** also faster than DRAM and able to retain its contents without being refreshed by the CPU. RAM chips often appear on *memory modules*—**SIMM** has chips on one side, **DIMM** has chips on both sides—which can be plugged into expansion slots on the motherboard. (2) **ROM** (for read-only memory) **chips** contain fixed start-up instructions. (**Read** means to transfer data from an input source to the CPU or memory; **write** means to transfer data from the CPU or memory to an output device.) A variant is **PROM,** a ROM chip that allows users to load read-only programs and data, although only once. (3) **CMOS chips** are battery-powered and so don't lose contents when the power is turned off; this makes them useful for holding times and dates. (4) **Flash memory chips** can be erased and reprogrammed more than once.

First level 1 cache, then level cache, then RAM, then hard disk (or CD-ROM) is the order in which the processor goes about searching for data or program instructions. **Cache** temporarily stores instructions and data that the processor is likely to use frequently, thereby speeding up processing. **Level 1 cache** is built into the processor chip; **level 2 cache** resides outside the processor chip and consists of SRAM chips. **Virtual memory** is hard disk space used to extend the capacity of RAM.

Ports and cables: A **port** is a connecting socket on the outside of the system unit into which are plugged different kinds of cables. Six types of ports are as follows: (1) **Serial ports** transmit bits one after the other, or slow data over long distances. (2) **Parallel ports** transmit 8 bits simultaneously, or fast data over close distances, as to printers. (3) **SCSI ports** transmit 32 bits simultaneously in a daisy chain of up to seven devices linked in a series. (4) **USB ports** are general-purpose ports that transmit data to up to 127 devices in a daisy chain. USB permits **Plug and Play,** which allows peripheral devices and expansion cards to be automatically configured while they are being installed. (5) Other ports, called *dedicated ports,* exist for special purposes, as for the keyboard and mouse. (6) **Infrared ports** enables cableless connection with infrared devices, as between computer and some printers.

Expandability—buses and cards: Closed architecture means a personal computer has no expansion slots; **open architecture** means it does. **Expansion slots** are sockets on the motherboard into which may be plugged **expansion cards,** circuit boards that provide more memory or that control peripheral devices. Expansion slots are connected to the CPU by *expansion buses,* such as **ISA,** the oldest and slowest at 8 or 16 bits; **PCI,** faster at 32 or 64 bits; and **AGP,** twice as fast as PCI and designed to support video and 3-D graphics. Types of expansion cards include the following: **Graphics cards** convert data into video images to display on a monitor. **Sound cards** transmit digital sounds; this includes music for video games created by **wavetable synthesis,** digitized sound samples taken from recordings of actual instruments. *Modem cards* are modems installed inside the computer. **Network interface cards** allow transmission of data over a cable network. A special kind of card is the **PC card,** used principally on laptops to expand capabilities.

2 **Secondary Storage. Secondary storage hardware**—devices that permanently hold data and programs—include floppy disks, hard disks, optical disks, magnetic tape, and smart cards, as well as online storage.

Floppy disks are removable flat pieces of mylar plastic in 3.5-inch plastic cases. Floppy disks have a **write-protect notch,** which can prevent a disk from accidentally being recorded over. Data is recorded in concentric circles called **tracks;** each track is divided into **sectors,** invisible wedge-shaped sections used for storage reference purposes. In the disk drive, the **read/write head** transfers data between the computer and the disk. Besides 3.5-inch floppy disks, which hold 1.44 megabytes, other forms of removable disks are **floppy-disk cartridges,** or higher-capacity removable disks—Zip disks, SuperDisks, and HiFD disks. **Zip disks** are special disks with a capacity of 100 or 250 megabytes. SuperDisks have a capacity of 120 megabytes, and a SuperDisk drive can also read 1.44-megabyte floppies. **HiFD disks** have a capacity of 200 megabytes; the disk drive can also read 1.44-megabyte floppies.

Hard disks are thin but rigid metal platters covered with a substance that allows data to be held in the form of magnetized spots. Disks are sealed in a hard-disk-drive unit, which can be quite sensitive, susceptible to a **head crash,** when the surface of the disk touches particles or the read/write head, resulting in loss of data on the disk. Hard disks may be nonremovable or removable. (1) A **nonremovable hard disk** is housed in the microcomputer system unit and stores most programs and data. Ads for hard disks may specify the **hard-disk controller,** a circuit board that positions the disk and read/write heads. Popular hard-disk controllers are **Ultra ATA** (also known as **EIDE**), which allows fast data transfer and high storage capacity, and **SCSI,** which supports several disk drives as well as other peripheral devices. (2) A **removable hard disk** (Iomega's Jaz, SyQuest's SparQ) consists of one or two platters enclosed with read/write heads in a hard plastic case, which is inserted into a microcomputer's cartridge drive.

Optical disks are removable disks on which data is written and read through use of laser beams. Types of optical disks are CD-ROM, CD-R, CD-RW, DVD-ROM, DVD-R, DVD-RW, and DVD-RAM. **CD-ROM** is a read-only disk that holds prerecorded text, graphics, and sound. **CD-R** disks can be written to once but can be read many times. **CD-W** disks can be written to and erased so the disk can be reused several times. **DVD-ROM** is a high-capacity CD, storing up to 17 gigabytes. **DVD-R** disks allow one-time recording by consumers. **DVD-RW** (for rewritable) and **DVD-RAM** (for random access memory) can be recorded on and erased more than once.

Magnetic tape is thin plastic tape coated with a substance that can be magnetized, with data represented by magnetized (1s) or unmagnetized (0s) spots. Large computers tend to use magnetic-tape reels, small computers tend to use **tape cartridges**, resembling audio cassettes. A **smart card,** such as a telephone debit card, contains a microprocessor and memory chip. An **optical card** is a laser-recordable, wallet-type card used with an optical card reader. A final type of storage is *online secondary storage,* with users using secure online services for backup storage.

3 Input Hardware. Input hardware—devices that translate data into a form the computer can process—is categorized as three types: *keyboards, pointing devices,* and *source-data entry devices.*

Keyboards: There are two categories of **keyboards,** devices that convert characters into electrical signals readable by the processor. The first is the *traditional computer keyboard,* which has all the keys of a typewriter plus some that are unique. A variant on the traditional keyboard is the *ergonomic* keyboard, designed to alleviate wrist injuries. The purpose of **ergonomics** is to make work and equipment safer and more efficient. The second category, *specialty keyboards and terminals,* includes three types of terminals: (1) A **dumb terminal** has screen and keyboard and can input and output but not process data. (2) An **intelligent terminal** has screen, keyboard, and its own processor and memory. One example is the **automated teller machine (ATM),** the self-service banking machine. Another is the **point-of-sale (POS) terminal,** used to record purchases in a store. (3) An **Internet terminal** provides access to the Internet. Examples are *set-top boxes* or *Web terminals,* stripped-down *network computers, online game players, PC/TVs,* and handheld *wireless pocket PCs* or *personal digital assistants (PDAs).*

Pointing Devices: Devices that control the cursor or pointer on a screen are **pointing devices.** They include the mouse and its variants, the touch screen, and various forms of pen input. (1) The *mouse,* which directs a pointer on the display screen, is moved on the desktop. Variants are the **trackball,** a movable ball mounted on a stationary device; the **pointing stick,** which protrudes from the keyboard; and the **touchpad,** a surface over which you move your finger. (2) The **touch screen** is a display screen that is sensitive to touch. (3) Pen input consists of **pen-based computer systems,** in which users write with a penlike stylus on a screen; **light pens,** light-sensitive penlike devices; and **digitizers,** which convert drawings to digital data—one example is the **digitizing tablet.**

Source Data-Entry Devices: Source-data entry devices create machine-readable data on magnetic media or paper or feed it directly into the computer's processor. They include various scanning devices—imaging systems, bar-code readers, mark- and character-recognition devices, and fax machines. They also include audio-input devices, video and photographic input, digital cameras, voice-recognition systems, sensors, radio-frequency identification devices, and human-biology input devices. (1) **Scanners** use laser beams and reflected light to translate images of text, drawings, and photographs into digital form. One type is an **imaging system,** which converts text, drawings, and photos into digital form that can be processed or stored in a computer system. This has led to the new industry of **electronic imaging,** combining of separate images using scanners.

Another scanning device is the **bar-code reader,** which reads the zebra-striped **bar codes** on products to translate them into digital code. **Magnetic-ink character recognition** reads check numbers; **optical mark recognition** reads pencil marks; **optical character recognition** reads preprinted characters, such as those on store price tags. **Fax machines,** the last type of scanner, read images and send them over phone lines. **Dedicated fax machines** only send and receive fax documents; **fax modems** are modems with fax capabilities. (2) **Audio-input devices** translate analog sounds (those with continuously variable waves) into digital 0s and 1s, either through audio boards or MIDI boards. (3) *Video-input cards* translate analog film and videotape signals into digital form, using either frame-grabber video cards or full-motion video cards. (4) **Digital cameras** use light-sensitive processor chips to capture photographic images in digital form. (5) **Voice-recognition systems** convert speech into digital signals by comparing electrical patterns produced by voices with prerecorded patterns stored in a computer. (6) **Sensors** collect data directly from the environment and transmit it to a computer. (7) **Radio-frequency identification (RF-ID) devices** consist of a tag containing a microchip that contains code numbers that can be read by radio waves of a scanner linked to a database. (8) *Human-biology input devices* include *biometric systems,* which use **biometrics,** the study of body characteristics, to identify people through biological characteristics, and *line-of-sight systems,* in which people "point" their eyes at a screen.

4 **Output Hardware. Output devices** convert machine-readable information into people-readable form. Three types of output are softcopy, hardcopy, and other.

    Softcopy: Softcopy refers to nonprinted data, such as that shown on a display screen. A **display screen (monitor, screen)** shows programming instructions and data as they are being input and information after it is processed. Screen clarity is affected by **dot pitch,** or space between **pixels,** the small units on screen that can be turned on or off; by **resolution,** which involves the number of pixels per square inch; and by **refresh rate,** the number of times per second pixels are recharged. Two types of monitors are CRT and flat-panel. A **CRT** (cathode-ray tube) is a vacuum tube. A **flat-panel display** consists of two plates of glass with a substance in between them in which light is manipulated; one technology is **liquid crystal display (LCD),** in which molecules of liquid crystal create images by transmitting or blocking light. Flat-panel screens are either **active-matrix display,** in which each pixel on screen is controlled by its own transistor and so the image is brighter and sharper, or **passive-matrix display,** in which a transistor controls a row or column of pixels. Two common color and resolution standards for monitors are **SVGA** (the most common), which can produce 16 million possible colors, and **XGA,** which can produce 65,536 possible colors.

    Hardcopy: Hardcopy refers to printed output. A **printer** prints characters or images on paper or other medium. Resolution of the image is measured by **dpi** (dots per inch), with more dots producing greater sharpness. Two types of printers are impact printers and nonimpact printers. **Impact printers** form images by striking a print hammer or wheel against an inked ribbon, leaving an image on paper; one type is the **dot-matrix printer,** which contains a print head of small pins. **Nonimpact printers** form characters or images without direct physical contact between printing mechanism and paper. Three types of nonimpact printers are laser, inkjet, and thermal. A **laser printer** creates images with dots like a photocopying machine; the printer uses a **page description language,** software that describes the images to the printer. An **ink-jet printer** sprays electrically charged droplets of ink at high speed onto paper. A **thermal printer** uses colored waxes and heat to burn dots onto special paper. Another category of printer is the **multifunction printer,** which combines printing, scanning, copying, and faxing in one device.

    Other Output: Other forms of output are sound, voice, and video. **Sound-output devices** produce digitized sound. **Voice-output devices** convert digital data into speech-like sounds. **Video** consists of photographic images, played at 15–29 frames per second; in one form of video output called **videoconferencing** people have online meetings using computers and communications devices that enable them to see and hear one another.

There's no such thing as a free PC."

So writes *Business Week* technology columnist Stephen H. Wildstrom. But, he adds, "judging from the current flood of offers for free or deeply discounted computers, you might think that the laws of economics and common sense have been repealed."[1]

Indeed, some computer users may now believe there *is* such a thing as a free lunch. Not only are free PCs being offered but also free Internet access, free long-distance calls, free voice-mail boxes, free fax services, free e-mail, even free design of your e-commerce Web site. The catch? That freebie PC may actually cost you $30–$50 a month for a three-year Internet connection that locks you in to slow dial-up access when faster methods are becoming available. Or you may have to provide personal information to advertisers (jeopardizing your privacy) and to put up with a constant stream of onscreen advertising.[2]

So maybe lunch isn't free, but can you get it cheap? The cost of a PC has declined remarkably—what would have cost you $3000 in 1994 was only $1200 in 1999.[3] Prices went down 44% over the span of three years, to an average of $844 in early 2000.[4] And if all you want is Internet access, the price is even lower. You can buy a gadget called the i-Opener, for instance—consisting basically of a keyboard, mouse, processor, display screen, microphone, and speakers—for $199 plus a monthly fee.[5] (You may even be able to do everything you need, including online banking, through a Sony PlayStation2 video-game console.[6])

The i-Opener

Thin-client computing at home

Indeed, the i-Opener represents a home version of the trend toward "thin-client" computing, mentioned in Chapter 3. Thin clients, which may be had for $350–$700, are simple terminals running a Web browser and small applications downloaded as needed from high-powered network servers.[7] Large organizations are discovering that replacing fat clients (PCs) with thin clients (terminals) produces substantial savings—not only in the upfront costs of the hardware but also in the much larger cost of installing, supporting, and updating the machines over time (as we explain further in Chapter 6).[8]

But even though some people now speak of the "post-PC era," the personal computer isn't dead yet, and no doubt at some point you'll want to buy your own. In that case, you'll need to know how to interpret a typical computer ad. (See ● *Panel 4.1.*) We'll explain how in this chapter. These days a desktop computer is apt to be a *multimedia computer*, with sound and graphics capability. As we explained in Chapter 1, the word *multimedia* means "combination of media," such as the combination of pictures, video, animation, and sound in addition to text. A multimedia computer features such equipment as a fast processor, DVD drive, sound card, graphics card, and speakers, and you may also wish to have headphones and a microphone. You may even wish to add a scanner, sound recorder, and digital camera.

4.1 The System Unit

KEY QUESTIONS

How is data represented in a computer; what are the components of the system cabinet; what are processing speeds; how do the processor and memory work; and what are some important ports, buses, and cards?

How is the information in "information processing" in fact processed? The first thing to understand is that computers run on electricity. And what is the most fundamental thing you can say about electricity? Electricity is either *on* or *off*. This two-state situation allows computers to use the *binary system* to represent data and programs.

The Binary System: Using On/Off Electrical States to Represent Data & Instructions

The decimal system that we are accustomed to has 10 digits (0, 1, 2, 3, 4, 5, 6, 7, 8, 9). By contrast, the **binary system** *has only two digits: 0 and 1.* Thus, in the computer, the 0 can be represented by the electrical current being off and the 1 by the current being on. All data and program instructions that go into the computer are represented in terms of these binary numbers.

For example, the letter G is a translation of the electronic signal 01000111, or off-on-off-off-off-on-on-on. When you press the key for G on the computer keyboard, the character is automatically converted into the series of

● **PANEL 4.1**

Advertisement for a PC

Details of this ad are explained throughout this chapter. See the little magnifying glass:

Great PC Buy!

- 7-Bay Mid-Tower Case
- Intel Pentium III Processor 733 MHz
- 64 MB 133 MHz SDRAM
- 256K Cache
- 2 USB Ports
- 3D AGP Graphics Card (8 MB)
- PCI Wavetable Sound Card
- 3.5" Floppy Drive
- Iomega 100 MB Zip Drive
- 40 GB Ultra ATA 7200 RPM Hard Drive
- 44X Max CD-ROM Drive or CD-R/RW Drive
- 104-Key Keyboard
- Microsoft IntelliMouse
- 17", .25dp Monitor (16" Display)
- HP DeskJet 970Cse Printer

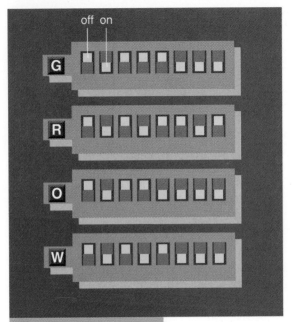

off on

G
R
O
W

electronic impulses that the computer can recognize. *(See* ● *Panel 4.2.)* Inside the computer, that character G is represented by a combination of eight *transistors* (as we will describe). Some are closed (representing the 0s), and some are open (representing the 1s).

How many representations of 0s and 1s will a computer or a storage device such as a hard disk hold? Capacity is denoted by *bits* and *bytes* and multiples thereof:

- **Bit:** In the binary system, **each 0 or 1 is called a _bit_, which is short for "binary dig*it*."**
- Byte: To represent letters, numbers, or special characters (such as ! or *), bits are combined into groups. **A group of 8 bits is called a _byte_, and a byte represents one character, digit, or other value.** (As we mentioned, in one scheme, 01000111 represents the letter *G.*) The capacity of a computer's memory or of a floppy disk is expressed in numbers of bytes or multiples such as kilobytes and megabytes.
- Kilobyte: A **_kilobyte (K, KB)_ is about 1000 bytes.** (Actually, it's precisely 1024 bytes, but the figure is commonly rounded.) The kilobyte was a common unit of measure for memory or secondary-storage capacity on older computers.
- Megabyte: A **_megabyte (M, MB)_ is about 1 million bytes** (1,048,576 bytes). Measures of microcomputer primary-storage capacity today are expressed in megabytes.
- Gigabyte: A **_gigabyte (G, GB)_ is about 1 billion bytes** (1,073,741,824 bytes). This measure was formerly used mainly with "big iron" (mainframe) types of computers, but is typical of secondary storage (hard disk) capacity of today's microcomputers.
- Terabyte: A **_terabyte (T, TB)_ represents about 1 trillion bytes** (1,009,511,627,776 bytes).
- Petabyte: A **_petabyte (P, PB)_ represents about 1 quadrillion bytes** (1,048,576 gigabytes).

Letters, numbers, and special characters are represented within a computer system by means of *binary coding schemes. (See* ● *Panel 4.3.)* That is, the off/on 0s and 1s are arranged in such a way that they can be made to represent characters, digits, or other values.

- ASCII: Pronounced *"Ask-ee," **ASCII** stands for American Standard Code for Information Interchange and is the binary code most widely used with microcomputers.* Besides more conventional characters, ASCII includes such characters as math symbols and Greek letters.
- EBCDIC: Pronounced *"Eb-see-dick," **EBCDIC** stands for Extended Binary Coded Decimal Interchange Code and is a binary code used with large computers,* such as mainframes.
- Unicode: A subset of ASCII, **_Unicode_ uses two bytes (16 bits) for each character,** rather than one byte (8 bits). Instead of the 256 character combinations of ASCII, Unicode can handle 65,536 character combinations. Thus it allows almost all the written languages of the world to be represented using a single character set.

Character	ASCII	Character	ASCII
A	0100 0001	N	0100 1110
B	0100 0010	O	0100 1111
C	0100 0011	P	0101 0000
D	0100 0100	Q	0101 0001
E	0100 0101	R	0101 0010
F	0100 0110	S	0101 0011
G	0100 0111	T	0101 0100
H	0100 1000	U	0101 0101
I	0100 1001	V	0101 0110
J	0100 1010	W	0101 0111
K	0100 1011	X	0101 1000
L	0100 1100	Y	0101 1001
M	0100 1101	Z	0101 1010
0	0011 0000	5	0011 0101
1	0011 0001	6	0011 0110
2	0011 0010	7	0011 0111
3	0011 0011	8	0011 1000
4	0011 0100	9	0011 1001
!	0010 0001	;	0011 1011

CONCEPT CHECK

What is the binary system?

Define bits, bytes, kilobytes, megabytes, gigabytes, terabytes, and petabytes.

Distinguish among ASCII, EBCDIC, and Unicode.

The Computer Case: Bays, Buttons, & Boards

What professionals tend to call the *system unit* is often called the *case* in computer ads. Whatever the term—case, system unit, or system cabinet—it is the box that houses the motherboard (including the processor chip and memory chips), the power supply, and storage devices. *(See ● Panel 4.4.)*

● PANEL 4.4

The system unit
An overhead view of an open case. It includes the motherboard, power supply, and storage devices.

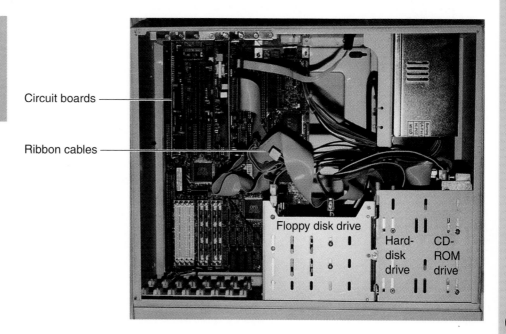

Circuit boards

Ribbon cables

Floppy disk drive

Hard-disk drive

CD-ROM drive

Hardware

For today's desktop PC, the system unit may be advertised as something like a "4-bay micro-tower case" or a "7-bay mid-tower case." A **_bay_ is a shelf or opening used for the installation of electronic equipment,** generally storage devices such as a hard drive or DVD drive. A computer may come equipped with four or seven bays. (Empty bays are covered by a panel.) A *tower* is the type of cabinet that is tall, narrow, and deep (so that it can sit on the floor beside or under a table) rather than short, wide, and deep. Originally a tower was considered to be 24 inches high. Micro- and mid-towers may be less than half that size.

The number of buttons on the outside of the case will vary, but the on/off power switch will appear somewhere, either front or back. There may also be a "sleep" switch; this allows you to suspend operations without terminating them so that you can conserve electrical power without the need for subsequent "rebooting," or restarting, the computer.

Inside the case—not visible to you unless you remove the cabinet—are various electrical circuit boards, chief of which is the motherboard, as we'll discuss.

The Motherboard & Microprocessor Chip

The *motherboard*, or *system board*, is the main circuit board in the system unit. The motherboard consists of a flat board that fills one side of the case. It contains both soldered, nonremovable components and sockets or slots for components that can be removed—microprocessor chip, RAM chips, and various expansion cards, as we explain. *(See ● Panel 4.5.)* Making some components removable allows you to expand or upgrade your system. **_Expansion_ is a way of increasing a computer's capabilities by adding hardware that performs tasks that are not part of the basic system.** For example, you might want to add video and sound cards. **_Upgrading_ means to change to newer, usually more powerful or sophisticated versions,** such as a more powerful microprocessor or more memory chips.

The most fundamental part of the motherboard is the microprocessor chip. A **_microprocessor_ is the miniaturized circuitry of a computer processor. It stores program instructions that process, or manipulate, data into information.** The key parts of the microprocessor are transistors. **_Transistors_ are tiny electronic devices that act as on/off**

Line from advertisement

🔍 Intel Pentium III Processor

● **PANEL 4.5**
The motherboard
This main board offers slots or sockets for removable components: microprocessor chip, RAM chips, and various expansion cards.

RAM (main memory) chips mounted on modules (cards)

Microprocessor chip
(with CPU)

switches, which process the on/off 1/0 bits used to represent data. According to *Moore's Law,* named for legendary Intel cofounder Gordon Moore, the number of transistors that can be packed onto a chip doubles every 18 months while the price stays the same. In 1961 a chip had only 4 transistors, in 1971 it had 2300, in 1979 it had 30,000, and in 1997 it had 7.5 million. Moore's Law actually accelerated with the debut of the 1-billion-transistor (1-gigahertz) chip in 2000.[9] (And in that year IBM announced a manufacturing breakthrough that would result in even faster chips, permitting laptop batteries, for instance, to last twice as long.[10])

All this electronic circuitry is printed (by photolithography) on a chip less than 1 centimeter square and about half a millimeter thick. **A _chip_—also called a semiconductor, or integrated circuit—consists of millions of electronic circuits printed on a tiny piece of silicon. Silicon is an element widely found in sand that has desirable electrical (or "semiconducting") properties.** Chips are mounted on carrier packages with connecting pins so that they can be plugged into sockets on the motherboard. *(See ● Panel 4.6.)*

Two principal "architectures" or designs for microprocessors are CISC and RISC. **_CISC (complex instruction set computing) chips,_ which are used mostly in PCs and in conventional mainframes,** can support a large number of instructions, but at relatively low processing speeds. In **_RISC (reduced instruction set computing) chips,_ which are used mostly in workstations,** a great many seldom-used instructions are eliminated. As a result, workstations can work up to 10 times faster then most PCs. RISC chips have been used in many Macintosh computers since 1993.

Most personal computers today use microprocessors of two kinds—those based on the model made by Intel and those based on the model made by Motorola.

- **Intel-type chips—for PCs:** About 90% of microcomputers use Intel-type microprocessors. Indeed, the Microsoft Windows operating system is designed to run on Intel chips. As a result, people in the computer industry tend to refer to the Windows/Intel joint powerhouse as *Wintel.*

    Intel-type chips for PCs are made principally by Intel Corp.—but also by Advanced Micro Devices (AMD), Cyrix, DEC, and others. They are used by manufacturers such as Compaq, Dell, Gateway 2000, Hewlett-Packard, and IBM. Since 1993, Intel has marketed its chips under the names *Pentium, Pentium Pro, Pentium MMX, Pentium II, Pentium III,* and *Celeron.* Many ads for PCs contain the logo "Intel inside" to show that the systems run an Intel microprocessor.

- **Motorola-type chips—for Macintoshes:** **_Motorola-type chips_ are made by Motorola for Apple Macintosh computers.** Since 1993, Motorola has joined forces with IBM and Apple to produce the PowerPC family of chips. With certain hardware or software add-ons, a PowerPC can run PC as well as Mac applications software.

Processing Speeds: From Megahertz to Picoseconds

Often a PC ad will carry a line that says something like "Intel Celeron processor 500 MHz," "Intel Pentium III processor 866 MHz," or "AMD Athlon processor 1-GHZ." MHz stands for *megahertz* and GHZ for *gigahertz*. These figures indicate how fast the microprocessor can process data and execute program instructions.

Every microprocessor contains a **_system clock_, which controls how fast all the operations within a computer take place.** The system clock uses fixed vibrations from a quartz crystal to deliver a steady stream of digital pulses or "ticks" to the CPU. These ticks are called *cycles*. Faster clock speeds will result in faster processing of data and execution of program instructions, as long as the computer's internal circuits can handle the increased speed.

There are four main ways in which processing speeds are measured:

- **For microcomputers—megahertz and gigahertz:** Microcomputer microprocessor speeds are usually expressed in **_megahertz (MHz)_, a measure of frequency equivalent to 1 million cycles (ticks of the system clock)**

Intel Pentium III Processor 733 MHz

  per second. The original IBM PC had a clock speed of 4.77 MHz. Today a 550-MHz Pentium III–based microcomputer processes 550 million cycles per second. The latest generation of processors operate in **_gigahertz (GHz)_—a billion cycles per second.** Since a new high-speed processor can cost many hundred dollars more than the previous generation of chip, experts often recommend that buyers fret less about the speed of the processor (since the work most people do on their PCs doesn't even tax the limits of the current hardware) and more about spending money on extra memory.[11]

- **For workstations, minicomputers, and mainframes—MIPS:** Processing speed can also be measured according to the number of instructions per second that a computer can process. MIPS stands for *millions of instructions per second*. A high-end microcomputer or workstation might perform at 100 MIPS or more, a mainframe at 200–1200 MIPS.

- **For supercomputers—flops:** The abbreviation *flops* stands for *floating-point operations per second*. A floating-point operation is a special kind of mathematical calculation. This measure, used mainly with supercomputers, is expressed as *megaflops* (mflops, or millions of floating-point operations per second), *gigaflops* (gflops, or billions), and *teraflops* (tflops, or trillions). The U.S. supercomputer known as Option Red cranks out 1.34 teraflops. (To put this in perspective, a person able to complete one arithmetic calculation every second would take about 31,000 years to do what Option Red does in a single second.)

- **For all computers—fractions of a second:** Another way to measure cycle times is in fractions of a second. A microcomputer operates in microseconds, a supercomputer in nanoseconds or picoseconds—thousands or millions of times faster. A *millisecond* is one-thousandth of a second. A *microsecond* is one-millionth of a second. A *nanosecond* is one-billionth of a second. A *picosecond* is one-trillionth of a second.

Distinguish expansion from upgrading.

Discuss the most fundamental part of the motherboard, its features, its two principal architectures, and the two principal kinds used in personal computers.

What is the system clock, and what are the various measures of processing speed?

How the Processor or CPU Works: Control Unit, ALU, & Registers

Once upon a time, the processor in a computer was measured in feet. A processing unit in the 1944 ENIAC (which had 20 such processors) was about 2 feet wide and 8 feet high. Today, computers are based on *micro*processors, less than 1 centimeter square. It may be difficult to visualize components so tiny. Yet it is necessary to understand how microprocessors work if you are to grasp what PC advertisers mean when they throw out terms such as "64 MB 133 MHz SDRAM" or "256K Advanced Transfer Cache."

Computer professionals often discuss a computer's word size. **_Word size_ is the number of bits that the processor may process at any one time.** The more bits in a word, the faster the computer. A 32-bit computer—that is, one with a 32-bit-word processor—will transfer data within each microprocessor chip in 32-bit chunks or 4 bytes at a time. (Recall there are 8 bits in a byte.) A 64-bit-word computer is faster; it transfers data in 64-bit chunks or 8 bytes at a time.

A processor is also called the *CPU,* and it works hand in hand with other circuits known as *main memory* to carry out processing. The **_CPU (central processing unit)_ is the "brain" of the computer; it follows the instructions of the software (program) to manipulate data into information. The CPU consists of two parts—(1) the control unit and (2) the arithmetic/logic unit (ALU), which both contain registers, or high-speed storage areas** (as we discuss shortly). All are linked by a kind of electronic "roadway" called a *bus.* (See ● *Panel 4.7, next page.*)

- The control unit—for directing electronic signals: The **_control unit_ deciphers each instruction stored in it and then carries out the instruction.** It directs the movement of electronic signals between main memory and the arithmetic/logic unit. It also directs these electronic signals between main memory and the input and output devices.

  For every instruction, the control unit carries out four basic operations, known as the machine cycle. In the **_machine cycle_, the CPU (1) fetches an instruction, (2) decodes the instruction, (3) executes the instruction, and (4) stores the result.**

- The arithmetic/logic unit—for arithmetic and logical operations: The **_arithmetic/logic unit (ALU)_ performs arithmetic operations and logical operations and controls the speed of those operations.**

  As you might guess, *arithmetic operations* are the fundamental math operations: addition, subtraction, multiplication, and division.

  Logical operations are comparisons. That is, the ALU compares two pieces of data to see whether one is equal to (=), greater than (>), or less than (<) the other. (The comparisons can also be combined, as in "greater than or equal to" and "less than or equal to.")

- Registers—special high-speed storage areas: The control unit and the ALU also use registers, special areas that enhance the computer's performance. **_Registers_ are high-speed storage areas that temporarily store**

The CPU and main memory

The two main CPU components on a microprocessor are the control unit and the ALU, which contain working storage areas called registers and are linked by a kind of electronic roadway called a *bus*.

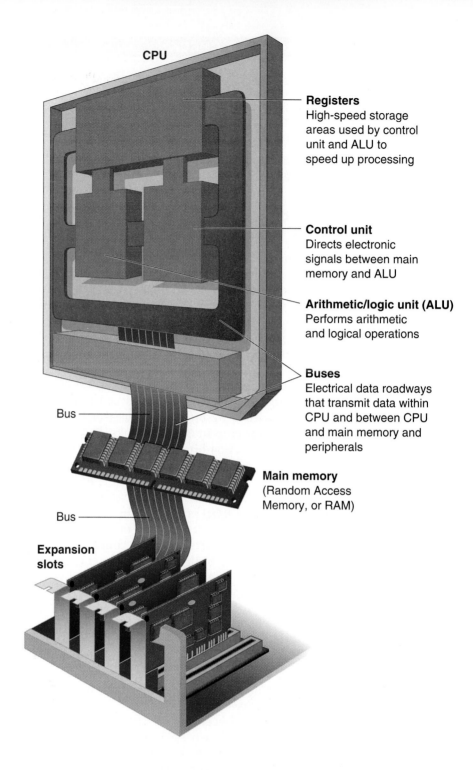

CPU

Registers
High-speed storage areas used by control unit and ALU to speed up processing

Control unit
Directs electronic signals between main memory and ALU

Arithmetic/logic unit (ALU)
Performs arithmetic and logical operations

Buses
Electrical data roadways that transmit data within CPU and between CPU and main memory and peripherals

Bus

Bus

Main memory
(Random Access Memory, or RAM)

Expansion slots

data during processing. They may store a program instruction while it is being decoded, store data while it is being processed by the ALU, or store the results of a calculation.

● Buses—data roadways: **_Buses_, or bus lines, are electrical data roadways through which bits are transmitted within the CPU and between the CPU and other components of the motherboard.** A bus resembles a multilane highway: The more lanes it has, the faster the bits can be transferred. The old-fashioned 8-bit-word bus of early microprocessors had only eight pathways. Data is transmitted four times faster in a computer with a 32-bit bus, which has 32 pathways, than in a computer with an 8-bit bus. Intel's Pentium chip is a 64-bit processor. Macintosh G4 microcomputers contain buses that are 128 bits, as do some supercomputers. Today there are several principal expansion bus standards, or "architectures," for microcomputers.

We return to a discussion of buses in a few pages.

How Memory Works: RAM, ROM, CMOS, & Flash

So far we have described only the kinds of chips known as microprocessors. But other silicon chips called *memory chips* are attached to the motherboard. The four principal types of memory chips are *RAM, ROM, CMOS,* and *flash.*

- **RAM chips—to temporarily store program instructions and data:** Recall from Chapter 1 that there are two types of storage, primary and secondary. Primary storage is temporary or working storage and is often called *memory* or *main memory;* secondary storage is relatively permanent storage (for example, on floppy disk). ___RAM chips___—**RAM stands for random access memory—are for primary storage; they temporarily hold (1) software instructions and (2) data before and after it is processed by the CPU.** Because its contents are temporary, RAM is said to be ___volatile___—**the contents are lost when the power is turned off.** This is why you should *frequently*—every 5–10 minutes, say— transfer (save) your work to a secondary-storage medium such as your hard disk, in case the electricity goes off while you're working. (However, there is one kind of RAM, called flash RAM, that is not temporary, as we'll discuss shortly.)

64 MB 133 MHz SDRAM

Three types of RAM chips are used in personal computers: *DRAM, SDRAM,* and *SRAM.* Pronounced *"dee-*ram," *DRAM (dynamic RAM)* must be constantly refreshed by the CPU or it will lose its contents. The type of dynamic RAM used in most PCs today is *SDRAM (synchronous dynamic RAM),* which is synchronized by the system clock and is much faster than DRAM. Often in computer ads the speed of SDRAM is expressed in megahertz. The third type, pronounced *"ess-*ram," *SRAM (static RAM)* is faster than any DRAM and will retain its contents without having to be refreshed by the CPU.

Microcomputers come with different amounts of RAM, which is usually measured in megabytes. An ad may list "64 MB SDRAM," but you can also get 128 megabytes of RAM. The more RAM you have, the faster the computer operates, and the better your software performs. *Having enough RAM is a critical matter.* Before you buy a software package, look at the outside of the box or check the manufacturer's Web site to see how much RAM is required. Microsoft Office 2000, for instance, states that a minimum of 16 megabytes of RAM is required.

SIMM
Single inline memory module

If you're short on memory capacity, you can usually add more RAM chips by plugging them into the motherboard. Chips can be bought single or in so-called *memory modules,* circuit boards that can be plugged into expansion slots on the motherboard. There are two types of such modules: SIMMs and DIMMS, both of which use DRAM chips. A *SIMM (single inline memory module)* has RAM chips on only one side. A *DIMM (dual inline memory module)* has RAM chips on both sides.

DIMM
Dual inline memory module

- **ROM chips—to store fixed start-up instructions:** Unlike RAM, to which data is constantly being added and removed, **_ROM (read-only memory)_ cannot be written on or erased by the computer user without special equipment. ROM chips contain fixed start-up instructions.** That is, ROM chips are loaded, at the factory, with programs containing special instructions for basic computer operations, such as those that start the computer or put characters on the screen. These chips are nonvolatile; their contents are not lost when power to the computer is turned off.

    In computer terminology, **_read_ means to transfer data from an input source into the computer's memory or CPU. The opposite is _write_—to transfer data from the computer's CPU or memory to an output device.** Thus, with a ROM chip, "read-only" means that the CPU can retrieve programs from the ROM chip but cannot modify or add to those programs. A variation is _PROM (programmable read-only memory)_, which is a ROM chip that allows you, the user, to load read-only programs and data. However, this can be done only once.

- **CMOS chips—to store flexible start-up instructions:** Pronounced "_see-moss_," **_CMOS (complementary metal-oxide semiconductor) chips_ are powered by a battery and thus don't lose their contents when the power is turned off. CMOS chips contain flexible start-up instructions, such as time, date, and calendar, that must be kept current even when the computer is turned off.** Unlike ROM chips, CMOS chips can be reprogrammed, as when you need to change the time for daylight savings time.

- **Flash memory chips—to store flexible programs:** Also a nonvolatile form of memory, **_flash memory chips_ can be erased and reprogrammed more than once** (unlike PROM chips, which can be programmed only once). Flash memory, which can range from 1 to 64 megabytes in capacity, is used to store programs not only in personal computers but also in pagers, cellphones, printers, and digital cameras. The 8-megabyte Memory Stick built by Sony provides more than five times the storage of a 1.44-megabyte floppy disk.

How Cache Works: Level 1 (Internal) & Level 2 (External)

Pronounced "cash," **_cache_ temporarily stores instructions and data that the processor is likely to use frequently. Thus, cache speeds up processing.**
There are two kinds of cache—Level 1 and Level 2:

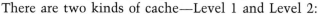

- **Level 1 (L1) cache—part of the microprocessor chip:** _Level 1 (L1) cache,_ also called _internal cache,_ is built into the processor chip. Ranging from 8 to 256 kilobytes, its capacity is less than that of Level 2 cache, although it operates faster.
- **Level 2 (L2) cache—not part of the microprocessor chip:** This is the kind of cache usually referred to in computer ads. _Level 2 (L2) cache,_ also called _external cache,_ resides outside the processor chip and consists of SRAM chips. Capacities range from 64 kilobytes to 2 megabytes. (Intel ads call L2 Advanced Transfer Cache.)

In addition, most current computer operating systems allow for the use of **_virtual memory_—that is, some free hard-disk space is used to extend the capacity of RAM.** First L1, then L2, then RAM, then hard disk (or CD-ROM)—that is the order in which the processor goes about searching for data or program instructions. In this progression, each kind of memory or storage is slower than its predecessor.

Describe the following: word size, CPU, the control unit and the machine cycle, the ALU and arithmetic and logical operations, registers, and buses.

What is RAM and what are three variants?

What is ROM and its variant CPROM?

Discuss CMOS chips and flash memory chips.

How does L1 cache differ from L2 cache?

Ports & Cables

A _port_ **is a connecting socket or jack on the outside of the system unit into which are plugged different kinds of cables.** A port allows you to plug in a cable to connect a peripheral device, such as a monitor, printer, or modem, so that it can communicate with the computer system.

Ports are of several types. _(See ● Panel 4.8, next page.)_

- Serial ports—for transmitting slow data over long distances: **A line connected to a _serial port_ will send bits one after the other,** like cars on a one-lane highway. Because bits must follow each other one at a time, a serial port is usually used to connect devices that do not require fast transmission of data, such as keyboard, mouse, monitors, and modems. It is also useful for sending data over a long distance. The standard for PC serial ports is the 9-pin or 25-pin RS-232C connector.

- Parallel ports—for transmitting fast data over close distances: **A line connected to a _parallel port_ allows 8 bits (1 byte) to be transmitted simultaneously,** like cars on an eight-lane highway. Parallel lines move information faster than serial lines do, but they can transmit information efficiently only up to 15 feet. Thus, parallel ports are used principally for connecting printers or external disk or magnetic-tape backup storage devices.

- SCSI ports—for transmitting fast data to up to seven devices in a daisy chain: Pronounced "scuzzy," a **_SCSI (small computer system interface) port_ allows data to be transmitted in a "daisy chain" for up to 7 devices at speeds (32 bits at a time) higher than those possible with serial and parallel ports.** Among the devices that may be connected are external hard-disk drives, CD-ROM drives, scanners, and magnetic-tape backup units. The term _daisy chain_ means that several devices are connected in series to each other, so that data for the seventh device, for example, has to go through the other six devices first. Sometimes the equipment on the chain is inside the computer, an internal daisy chain; sometimes it is outside the computer, an external daisy chain.

2 USB Ports

- USB ports—for transmitting data to up to 127 devices in a daisy chain: A **_USB (universal serial bus) port_ can theoretically connect up to 127 peripheral devices daisy-chained to one general-purpose port.** USB ports are useful for peripherals such as digital cameras, digital speakers, scanners, high-speed modems, and joysticks. The so-called USB _hot plug_ or _hot swappable_ allows such devices to be connected or disconnected even while the PC is running. In addition, USB permits **_Plug and Play_, which allows peripheral devices and expansion cards to be automatically configured while they are being installed.** This avoids the hassle of setting switching and creating special files that plagued earlier users. In 2000, computer makers were planning to introduce USB 2.0, expected to run 40 times faster than the earlier version.[12]

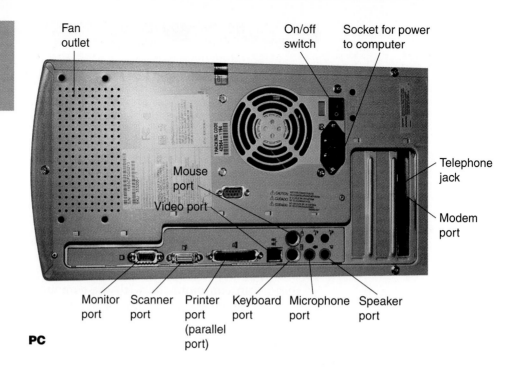

Fan outlet · On/off switch · Socket for power to computer · Telephone jack · Modem port · Mouse port · Video port · Monitor port · Scanner port · Printer port (parallel port) · Keyboard port · Microphone port · Speaker port

PC

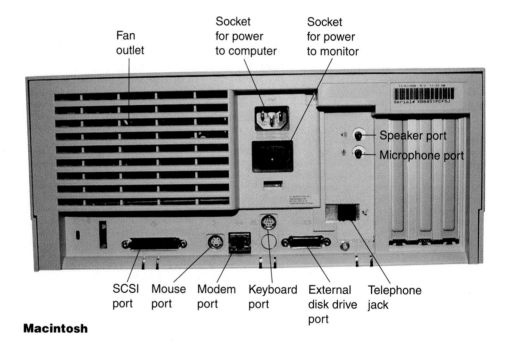

Fan outlet · Socket for power to computer · Socket for power to monitor · Speaker port · Microphone port · SCSI port · Mouse port · Modem port · Keyboard port · External disk drive port · Telephone jack

Macintosh

Can you really connect up to 127 devices on a single chain? An Intel engineer did set a world record at an industry trade show before a live audience by connecting 111 peripheral devices to a single USB port on a PC. But respected technology writer Walter Mossberg says that "almost none of the USB peripherals I've seen support this [daisy-chaining] feature."[13] Thus, though most PCs contain only two USB ports, it's worth shopping around to find a model with extra USB connectors.

- **Dedicated ports—for keyboard, mouse, phone, and so on:** So far, we have been considering general-purpose ports, but the back of a computer also has other, *dedicated ports*—ports for special purposes. Among these are the round ports for connecting the keyboard and the mouse. There are also jacks for speakers and microphones and modem-to-telephone jacks. Finally, there is one connector that is not a port at all—the power plug socket, into which you insert the power cord that brings electricity from a wall plug.

- **Infrared ports—for cableless connections over a few feet:** When you use a handheld remote unit to change channels on a TV set, you're using invisible radio waves of the type known as infrared waves. An **_infrared port_ allows a computer to make a cableless connection with infrared-capable devices,** such as some printers. This type of connection requires an unobstructed line of sight between transmitting and receiving ports, and they can only be a few feet apart.

Expandability: Buses & Cards

Today many new microcomputer systems can be expanded. As mentioned earlier, *expansion* is a way of increasing a computer's capabilities by adding hardware to perform tasks that are not part of the basic system. *Upgrading* means changing to newer, usually more powerful or sophisticated versions. (Computer ads often make no distinction between "expansion" and "upgrading." Their main interest is simply to sell you more hardware or software.)

Whether or not a computer can be expanded depends on its "architecture"—closed or open. *Closed architecture* means a computer has no expansion slots; *open architecture* means it does have expansion slots. **_Expansion slots_ are sockets on the motherboard into which you can plug expansion cards. _Expansion cards_—also known as expansion boards, adapter cards, interface cards, plug-in boards, controller cards, add-ins, or add-ons—are circuit boards that provide more memory or that control peripheral devices.** *(See ● Panel 4.9.)* Common expansion cards connect to

● **PANEL 4.9**
Expandability
How an expansion card fits into an expansion slot

Expansion card   Expansion slot

the monitor (graphics card), speakers and microphones (sound card), and network (network card), as we'll discuss. Most computers have four to eight expansion slots, some of which may already contain expansion cards included in your initial PC purchase.

Expansion cards are made to connect with different types of buses on the motherboard. (As we mentioned, *buses* are electrical data roadways through which bits are transmitted.) The bus that connects the CPU within itself and to main memory is the *system bus*. The bus that connects the CPU with expansion slots on the motherboard and thus with peripheral devices is the *expansion bus*. We already alluded to the universal serial bus (USB), whose purpose, in fact, is to *eliminate* the need for expansion slots and expansion cards, since you can just connect USB devices in a daisy chain outside the system unit. Three expansion buses to be aware of are *ISA*, *PCI*, and *AGP*.

- ISA bus—for ordinary low-speed uses: The **ISA *(industry standard architecture) bus* is the most widely used expansion bus.** It is also the oldest and, at 8 or 16 bits, the slowest at transmitting data, though it is still used for mouses, modem cards, and low-speed network cards.
- PCI bus—for higher-speed uses: The **PCI *(peripheral component interconnect) bus* is a higher-speed bus,** and at 32 or 64 bits wide it is over four times faster than ISA buses. PCI is widely used to connect graphics cards, sound cards, modems, and high-speed network cards.
- AGP bus—for even higher speeds and 3D graphics: The **AGP *(accelerated graphics port) bus* transmits data at even higher speeds and was designed to support video and three-dimensional (3D) graphics.** An AGP bus is twice as fast as a PCI bus.

Among the types of expansion cards are graphics, sound, modem, and network interface cards. A special kind of card is the PC card.

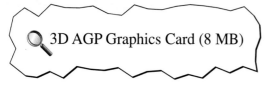
3D AGP Graphics Card (8 MB)

PCI Wavetable Sound Card

- Graphics cards—for monitors: Graphics cards are included in all PCs. **Also called a video card or video adapter, a *graphics card* converts signals from the computer into video signals that can be displayed as images on a monitor.** A three-dimensional AGP card is an example of a graphics card. The power of an AGP graphics card is often expressed in megabytes, as in 8, 16, or 32 MB.
- Sound cards—for speakers and audio output: A **sound card is used to transmit digital sounds through speakers, microphones, and headsets.** Sound cards come installed on most new PCs. Cards such as PCI wavetable sound cards are used to add music and sound effects to computer video games. *Wavetable synthesis* is a method of creating music based on a wave table, which is a collection of digitized sound samples taken from recordings of actual instruments. The sound samples are then stored on a sound card and are edited and mixed together to produce music. Wavetable synthesis produces higher quality audio output than other sound techniques.
- Modem cards—for remote communication via phone lines: Very occasionally you may see a modem that is outside the computer. Most new PCs, however, come with internal modems—modems installed inside as circuit cards. The modem not only sends and receives digital data over telephone lines to and from other computers but can also transmit voice and fax signals.
- Network interface cards—for remote communication via cable: A **network interface card allows the transmission of data over a cable network,** which connects various computers and other devices such as printers.

PANEL 4.10
PC card
An example of a PC card
used in a laptop

- PC cards—for laptop computers: Originally called *PCMCIA cards* (for the Personal Computer Memory Card International Association), __PC cards__ **are thin, credit-card size (2.1 by 3.4 inches) devices used principally on laptop computers to expand capabilities.** *(See ● Panel 4.10.)* Examples are extra memory (flash RAM), sound, modem, hard disks, and even pagers and cellular communicators. At present there are three sizes for PC cards—I (thin), II (thick), and III (thickest). Type I is used primarily for flash memory cards. Type II, the kind you'll find most often, is used for fax modems and network-interface cards. Type III is for rotating disk devices, such as hard-disk drives, and for wireless communication devices.

CONCEPT CHECK

Distinguish among the following ports: serial, parallel, SCSI, USB, dedicated, and infrared.

Distinguish closed architecture from open architecture.

Define the following three expansion buses: ISA, PCI, AGP.

Why would you need the following cards: graphics, sound, modem, network interface, and PC?

4.2 Secondary Storage

KEY QUESTIONS
What are the features of floppy disks, hard disks, optical disks, magnetic tape, smart cards, and online secondary storage?

You're on a trip with your laptop, or maybe just a cell phone or a personal digital assistant, and you don't have a crucial file of data. Or maybe you need to look up a phone number that you can't get through the phone company's directory assistance. Fortunately, you backed up your data online, using any one of several services (Desktop.com, Driveway, FreeDesk.com, i-drive.com, MagicalDesk.com, MyInternetDesktop.com, MyWebOS.com, Visto.com, or X:Drive, for example), and are able to access it through your modem.[14]

Here is yet another example of how the World Wide Web is offering alternatives to traditional computer functions that once resided within stand-alone machines. We are not, however, fully into the all-online era just yet. Let us consider more traditional forms of __secondary storage hardware__, **devices that permanently hold data and information as well as programs.** We will look at floppy disks, hard disks, optical disks, magnetic tape, smart cards, and online storage.

Hardware

161

Floppy Disks

A _floppy disk_, **often called a diskette or simply a disk, is a removable flat piece of mylar plastic packaged in a 3.5-inch plastic case.** Data and programs are stored on the disk's coating by means of magnetized spots, following standard on/off patterns of data representation (such as ASCII). The plastic case protects the mylar disk from being touched by human hands. Originally, when most disks were larger (5.25 inches), the disks actually were "floppy," not rigid; now only the plastic disk inside is flexible or floppy.

Floppy disks are inserted into a floppy-disk drive, a device that holds, spins, reads data from, and writes data to a floppy disk. _Read_ means that the data in secondary storage is converted to electronic signals and a copy of that data is transmitted to the computer's memory (RAM). _Write_ means that a copy of the electronic information processed by the computer is transferred to secondary storage. Floppy disks have a **_write-protect notch_, which allows you to prevent a diskette from being written to.** In other words, it allows you to protect the data already on the disk. To write-protect, use your thumbnail or the tip of a pen to move the small sliding tab on the lower right side of the disk (viewed from the back), thereby uncovering a square hole. _(See ● Panel 4.11.)_

On the diskette, **data is recorded in concentric circles called _tracks_.** Unlike on a vinyl phonograph record, these tracks are neither visible grooves nor a single spiral. Rather, they are closed concentric rings. On a formatted disk each track is divided into **_sectors_, invisible wedge-shaped sections used for storage reference purposes.** When you save data from your computer to a diskette, the data is distributed by tracks and sectors on the disk. That is, the system software uses the point at which a sector intersects a track to reference the data location.

When you insert a floppy disk into the slot (the _drive gate_ or _drive door_) in the front of the disk drive, the disk is fixed in place over the spindle of the drive mechanism. The **_read/write head_ is used to transfer data between the computer and the disk.** When the disk spins inside its case, the read/write head moves back and forth over the _data access area_ on the disk. When the disk is not in the drive, a metal clip covers this access area. An access light goes on when the disk is in use. After using the disk, you can retrieve it by pressing an eject button beside the drive. _(See ● Panel 4.12.)_ Note: _Do not remove the disk when the access light is on._

Let's compare the 3.5-inch floppy disk with some 3.5-inch **_floppy-disk cartridges_, or higher-capacity removable disks**—Zip disks, SuperDisks, and HiFD disks. Let's consider these:

Front

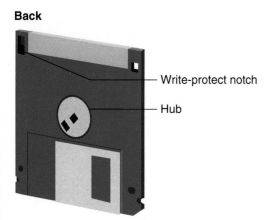

- Label
- Hard plastic jacket
- Data access area
- Metal protective plate (shutter) that moves aside (in disk drive) to expose data access area on disk

Back

- Write-protect notch
- Hub

Tracks and sectors

1 sector

track

Bits on 1 track

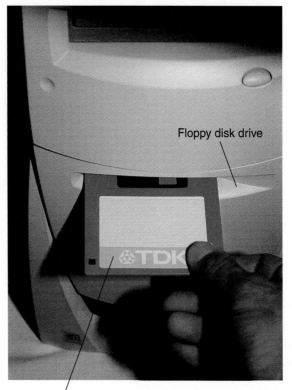

Floppy disk drive

Floppy disk

- **3.5-inch floppy disks—1.44 megabytes:** The present-day standard for traditional floppy disks is 1.44 megabytes, the equivalent of 400 type-written pages. Today's floppy carries the label *2HD*, in which the *2* stands for "double-sided" (it holds data on both sides) and the *HD* stands for "high density" (which means it stores more data than the previous standard—*DD*, for "double density").

  When you buy a box of floppies, be sure to check whether they are "IBM formatted" (for PCs only) or "Macintosh formatted" (for Apple machines). You can also buy "unformatted" disks, which means you have to *format* or *initialize* them yourself—that is, prepare the disks for use so that the operating system can write information on them. The software commands for formatting are described in your computer's user manual.

- **Zip disks—100 megabytes:** Produced by Iomega Corp., **_Zip disks_** **are special disks with a capacity of 100 or even 250 megabytes.** At 100 megabytes, this is 70 times the storage capacity of the standard floppy. Among other uses, Zip disks are used to store large spreadsheet files, database files, image files, multimedia presentation files, and Web sites. Zip disks require their own Zip disk drives, which may come installed on new computers, although external Zip drives are also available.

- **SuperDisks—120 megabytes:** Produced by Imation, **_SuperDisks_** **are disks with a capacity of 120 megabytes; the SuperDisk drive can also read standard 1.44-megabyte floppy disks,** which Zip drives cannot do.

🔍 Iomega 100 MB Zip Drive

Zip drive
External Zip drive and disks

Hardware

163

● PANEL 4.13

Hard disk

In a microcomputer, the hard disk is enclosed within the system unit. Unlike a floppy disk, it is not accessible. This hard-disk drive is for a notebook computer.

- **HiFD disks—200 megabytes:** Made by Sony Corp., **_HiFD disks_ have a capacity of 200 megabytes; the disk drive can also read standard 1.44-megabyte floppies.** HiFD disks have 140 times the capacity of today's standard floppy disks.

Hard Disks

Floppy disks use flexible plastic, but hard disks use metal. **_Hard disks_ are thin but rigid metal platters covered with a substance that allows data to be held in the form of magnetized spots.** Hard disks are tightly sealed within an enclosed hard-disk-drive unit to prevent any foreign matter from getting inside. Data may be recorded on both sides of the disk platters. *(See ● Panel 4.13.)*

Hard disks are quite sensitive devices. The read/write head does not actually touch the disk but rather rides on a cushion of air about 0.000001 inch thick. *(See ● Panel 4.14.)* The disk is sealed from impurities within a container, and the whole apparatus is manufactured under sterile conditions. Otherwise, all it would take is a human hair, a dust particle, a fingerprint smudge, or a smoke particle to cause what is called a head crash. A *head crash* happens when the surface of the read/write head or particles on its surface come into contact with the surface of the hard-disk platter, causing the loss of some or all of the data on the disk. A head crash can also happen when you bump a computer too hard or drop something heavy on the system cabinet. An incident of this sort could, of course, be a disaster if the data has not been backed up. There are firms that specialize in trying to retrieve (for a hefty price) data from crashed hard disks, though this cannot always be done.

There are two types of hard disks—nonremovable and removable.

- Nonremovable hard disks: A **_nonremovable hard disk_, also known as a fixed disk, is housed in a microcomputer system unit and is used to store nearly all programs and most data files.** Usually it consists of four 3.5-inch metallic platters sealed inside a drive case the size of a small sandwich, which contains disk platters on a drive spindle, read/write

● PANEL 4.14

Gap between hard disk read/write head and platter

Were the apparatus not sealed, all it would take is a human hair, dust particle, fingerprint, or smoke particle to cause a head crash.

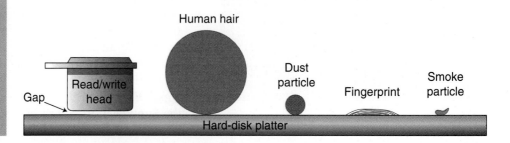

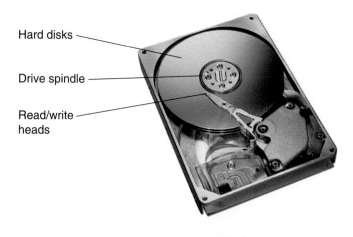

Hard disks

Drive spindle

Read/write heads

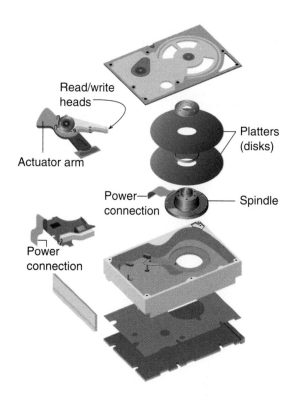

Read/write heads

Actuator arm

Platters (disks)

Power—connection

Spindle

Power connection

● PANEL 4.15

Inside a microcomputer's nonremovable hard disk

These platters are installed inside a drive case to prevent contaminants from affecting the read/write heads.

heads mounted on an access arm that moves back and forth, and power connections and circuitry. (*See ● Panel 4.15.*) Operation is much the same as for a diskette drive: The read/write heads locate specific instructions or data files according to track or sector.

Microcomputer hard drives with capacities measured in tens of gigabytes—up to 40 gigabytes, according to current ads—are becoming essential because today's programs are so huge. Microsoft Office alone is 500 megabytes. As for speed, hard disks allow faster access to data than floppy disks do, because a hard disk spins many times faster. Computer ads frequently specify speeds in revolutions per minute. A floppy disk drive rotates at only 360 rpm; a 7200-rpm hard drive is going about 300 miles per hour.

🔍 40 GB Ultra ATA 7200 RPM Hard Drive

● PANEL 4.16

Removable hard disk

The Jaz disk from Iomega is a hard-disk cartridge system with a capacity of 2 gigabytes.

In addition, ads may specify the type of **_hard-disk controller_, a special-purpose circuit board that positions the disk and read/write heads and manages the flow of data and instructions to and from the disk.** Popular hard-disk controllers are *Ultra ATA* (or *EIDE*) and *SCSI*. Commonly found on new PCs, *Ultra ATA (advanced technology attachment)* allows fast data transfer and high storage capacity; it is also known as *EIDE (enhanced integrated drive electronics)*. Ultra ATA can support only one or two hard disks. By contrast, *SCSI (small computer system interface)*, pronounced "scuzzy," supports several disk drives as well as other peripheral devices by linking them in a daisy chain of up to seven devices. SCSI controllers are faster and have more storage capacity than EIDE controllers; they are typically found in servers and workstations.

● Removable hard disks: **_Removable hard disks_, or hard-disk cartridges, consist of one or two platters enclosed along with read/write heads in a hard plastic case, which is inserted into a microcomputer's cartridge drive.** (*See ● Panel 4.16.*) Typical capacity is 2 gigabytes. Two popular systems are the Iomega's Jaz and

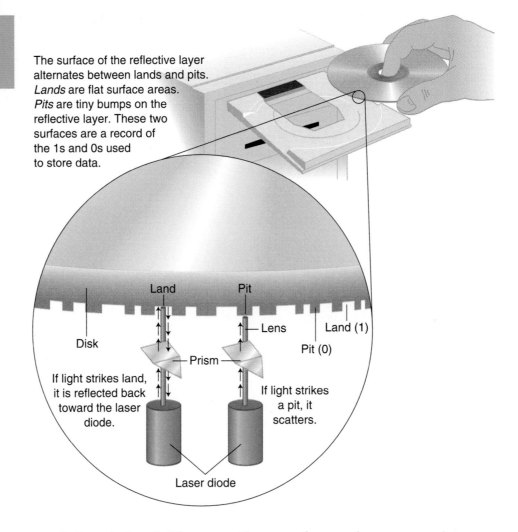

The surface of the reflective layer alternates between lands and pits. *Lands* are flat surface areas. *Pits* are tiny bumps on the reflective layer. These two surfaces are a record of the 1s and 0s used to store data.

Land

Pit

Disk

Lens

Land (1)

Pit (0)

Prism

If light strikes land, it is reflected back toward the laser diode.

If light strikes a pit, it scatters.

Laser diode

SyQuest's SparQ. These cartridges are often used to transport huge files, such as desktop-publishing files with color and graphics and large spreadsheets. They are also frequently used to back up data.

Optical Disks: CDs & DVDs

Everyone who has ever played an audio CD is familiar with optical disks. An ***optical disk* is a removable disk, usually 4.75 inches in diameter and less than one-twentieth of an inch thick, on which data is written and read through the use of laser beams.** An audio CD holds up to 74 minutes (2 billion bits' worth) of high-fidelity stereo sound. Some optical disks are used strictly for digital data storage, but many are used to distribute multimedia programs that combine text, visuals, and sound.

With an optical disk, there is no mechanical arm, as with floppy disks and hard disks. Instead, a high-power laser beam is used to write data by burning tiny pits or indentations into the surface of a hard plastic disk. To read the data, a low-power laser light scans the disk surface: Pitted areas are not reflected and are interpreted as 0 bits; smooth areas are reflected and are interpreted as 1 bits. (*See* ● *Panel 4.17.*) Because the pits are so tiny, a great deal more data can be represented than is possible in the same amount of space on a diskette and many hard disks. An optical disk can hold over 4.7 gigabytes of data, the equivalent of 1 million typewritten pages.

Nearly every PC marketed today contains a CD-ROM or DVD-ROM drive, which can also read audio CDs. (*See* ● *Panel 4.18.*) These, along with their recordable and rewritable variations, are the two principal types of optical-disk technology used with computers.

CD-ROM,
label side up

Slide-out
tray for
CD-ROM
disk drive

Types of Optical Disks	Storage Capacity	Typical Uses
CD-ROM (read-only)	650 megabytes	To distribute software, databases, reference data
CD-R (for recording on once)	650 megabytes	To store enormous amounts of data
CD-RW (reusable)	650 megabytes	To develop multimedia presentations
DVD-ROM (read-only)	4.7 gigabytes	To distribute movies and sound presentations
DVD-R (for recording on once)	4.7 gigabytes	To store enormous amounts of data
DVD-RW/DVD-RAM (reusable)	2.6–5.2 gigabytes	To develop large multimedia presentations

44X Max CD-ROM Drive

- **CD-ROM—for reading only:** For microcomputer users, the best-known type of optical disk is the CD-ROM. ***CD-ROM (compact disk-read-only memory) is an optical-disk format that is used to hold prerecorded text, graphics, and sound.*** Like music CDs, a CD-ROM is a read-only disk. *Read-only* means the disk's content is recorded at the time of manufacture and cannot be written on or erased by the user. As the user, you have access only to the data imprinted by the disk's manufacturer. A CD-ROM disk can hold up to 650 megabytes of data, equal to over 300,000 pages of text.

  A CD-ROM drive's speed is important because with slower drives images and sounds may appear choppy. In computer ads, drive speeds are indicated by the symbol "X," as in "44X," which is a high speed. "X" denotes the original data transfer rate of 150 kilobytes per second. (The data transfer rate is the time the drive takes to transmit data to another device. A 44X drive runs at 44 times 150, or 6600 kilobytes (6.6 megabytes) per second. If an ad carries the word "Max," as in "44X Max," this indicates the device's maximum speed. Drives range in speed from 16X to 48X; the faster ones are more expensive.

CD-R/RW Drive

- **CD-R—for recording on once:** ***CD-R (compact disk-recordable) disks can be written to only once but can be read many times.*** This allows consumers to make their own CD disks, though it's a slow process. (Recording a full disk takes 20–60 minutes.) Also, once recorded, the information cannot be erased. CD-R is often used by companies for archiving—that is, to store vast amounts of information. A variant is the Photo CD, an optical disk developed by Kodak that can digitally store photographs taken with an ordinary 35-millimeter camera. Once you've shot a roll of color photographs, you take it for processing to a photo shop, which produces a disk containing your images.

Hardware

You can view the disk on any personal computer with a CD-ROM drive and the right software.

- **CD-RW—for rewriting many times:** A ***CD-RW (compact disk-rewritable) disk***, **also known as an erasable optical disk, allows users to record and erase data so that the disk can be used over and over again.** Special CD-RW drives and software are required. CD-RW disks are useful for archiving and backing up large amounts of data or work in multimedia production or desktop publishing. However, because they are relatively slow they are no substitute for a hard disk.

- **DVD-ROM—the versatile video disk:** A ***DVD-ROM (digital versatile disk or digital video disk, with read-only memory)*** **is a CD disk with extremely high capacity, able to store 4.7–17 gigabytes.** How is this done? Like a CD or CD-ROM, the surface of a DVD contains microscopic pits, which represent the 0s and 1s of digital code that can be read by a laser. The pits on the DVD, however, are much smaller and grouped more closely together than those on a CD, allowing far more information to be represented. Also, the laser beam used focuses on pits roughly half the size of those on current audio CDs. In addition, the DVD format allows for two layers of data-defining pits, not just one. Finally, engineers have succeeded in squeezing more data into fewer pits, principally through data compression.

Many new computer systems now come with a DVD drive as standard equipment; these drives can also take standard CD-ROM disks. DVDs have enormous potential to replace CDs for archival storage, mass distribution of software, and entertainment. They not only store far more data but are different in quality from CDs. As one writer points out, "DVDs encompass much more: multiple dialogue tracks and screen formats, and best of all, smashing sound and video."[15] The theater-quality video and sound, of course, is what makes DVD a challenger to videotape as a vehicle for movie rentals.

As with CDs, DVDs have their recordable and rewritable variants. ***DVD-R (DVD-recordable) disks*** **allow one-time recording by the consumer.** Two types of reusable disks are *DVD-RW (DVD-rewritable)* and DVD-RAM (DVD-random access memory), both of which can be recorded on and erased more than once.

Magnetic Tape

Similar to the tape used on an audio tape recorder (but of higher density), ***magnetic tape*** **is thin plastic tape coated with a substance that can be magnetized. Data is represented by magnetized spots (representing 1s) or non-magnetized spots (representing 0s).** Today, "mag tape" is used mainly for backup and archiving—that is, for maintaining historical records—where there is no need for quick access.

On large computers, tapes are used on magnetic-tape units or reels, and in cartridges. On microcomputers, tape is used in the form of ***tape cartridges***, **modules resembling audio cassettes that contain tape in rectangular, plastic housings.** There are three common types of drive tapes—QIC, DAT, and (most expensive) DLT. *(See ● Panel 4.19.)*

Smart Cards

Today in the United States most credit cards are old-fashioned magnetic-strip cards. A magnetic-strip card has a strip of magnetically encoded data on its back. The encoded data might include your name, account number, and PIN (personal identification number). Two other kinds of cards, smart cards and optical cards, which hold far more information, are already popular in Europe. Manufacturers are betting they will soon be popular in the U.S.[16]

Type and Abbreviation	Storage Capacity
Quarter-inch cartridge (QIC)	40 megabytes to 5 gigabytes
Digital audio tape (DAT)	2 to 24 gigabytes
Digital linear tape (DLT)	20 to 40 gigabytes

● **PANEL 4.19**
Magnetic-tape drive and three types of tape cartridges

Magnetic-tape cartridge drive

- **Smart cards:** A **_smart card_** **looks like a credit card but contains a microprocessor and memory chip.** When inserted into a reader, it transfers data to and from a central computer, and it can store some basic financial records. Smart cards can be used as telephone debit cards. You insert the card into a slot in the phone, wait for a tone, and dial the number. The time of your call is automatically calculated on the chip inside the card and deducted from the balance. Many colleges and universities issue student cards as smart cards.
- **Optical cards:** The conventional magnetic-stripe credit card holds the equivalent of a half page of data. The smart card with a microprocessor and memory chip holds the equivalent of 250 pages. The optical card presently holds about 2000 pages of data. Optical cards use the same type of technology as music compact disks but look like silvery credit cards. **_Optical cards_** **are plastic, laser-recordable, wallet-type cards used with an optical-card reader.** Because they can cram so much data (6.6 megabytes) into so little space, they may become popular in the future. With an optical card, for instance, there's enough room for a person's health card to hold not only his or her medical history and health-insurance information but also digital images, such as electrocardiograms.

CONCEPT CHECK

What are four types of floppy disks and what are their features?

What are characteristics of hard disks, both nonremovable and removable?

Explain the various types of optical disks—CD-ROM, CD-R, CD-RW, and DVD-ROM.

How is magnetic tape used?

What are three kinds of smart cards?

Hardware

Online Secondary Storage

If the network computer or thin-client computer actually becomes as popular as its promoters hope, the Internet itself will become, in effect, your hard disk. We described the concept of online storage at the start of this section. The services we mentioned included some applications software, such as calendar, address book, and even word processors. Other services, however, simply offer online storage for backup purposes. Examples are @Backup *(www.atbackup.com)*, Connected Online Backup *(www.connected.com)*, Network Associates Quick Backup to Personal Vault *(www.mcafee.com)*, and SafeGuard Interactive *(www.sgii.com)*. Monthly prices are generally in the $10–$15 range. When you sign up with the service, you usually download from a Web site free software that lets you upload whatever files you wish to the company's server. For security, you are given a password, and the files are supposedly encrypted to guard against anyone giving them an unwanted look.

From a practical standpoint, online backup should be used only for vital files. Removable hard-disk cartridges are the best medium for backing up entire hard disks, including files and programs.

4.3 Input Hardware

KEY QUESTIONS

What are the three categories of input hardware, what devices do they include, and what are their features?

Input hardware **consists of devices that translate data into a form the computer can process.** The people-readable form of the data may be words, but the computer-readable form consists of 0s and 1s, represented as off and on electrical signals.

Input hardware devices are categorized as three types: *keyboards, pointing devices,* and *source data-entry devices.* (See ● *Panel 4.20.)* Quite often a computer system will combine all three.

Keyboards

A *keyboard* **is a device that converts letters, numbers, and other characters into electrical signals that can be read by the computer's processor.** The keyboard may look like a typewriter keyboard to which some special keys have

● PANEL 4.20
Three types of input devices

Keyboards	Pointing Devices	Source Data-entry Devices
Traditional computer keyboards	Mice, trackballs, pointing sticks, touchpads	Scanning devices: imaging systems, bar-code readers, mark- and character-recognition devices (MICR, OMR, OCR), fax machines
Specialty keyboards and terminals: dumb terminals, intelligent terminals (ATMs, POS terminals), Internet terminals	Touch screens	Audio input devices
	Pen-based computer systems, light pens, digitizer (digitizing tablet)	Video-input devices
		Digital cameras
		Voice-recognition systems
		Sensors
		Radio-frequency identification
		Human-biology input devices

been added. Alternatively, it may look like the keys on a bank ATM or the keypad of a pocket computer. It may even be a Touch-Tone phone or cable-TV set-top box.

Let's look at *traditional computer keyboards* and various kinds of *specialty keyboards and terminals.*

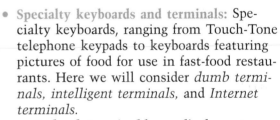

104-Key Keyboard

- **Traditional computer keyboards:** Conventional computer keyboards have all the keys on typewriter keyboards, plus other keys unique to computers. This totals 104–105 keys for desktop computers and 85 keys for laptops. Newer keyboards such as Rocket-Board and Microsoft Internet Keyboard Pro have extra buttons for fast Web access.[17] The keyboard is built into laptop computers or attached to desktop computers with a cable. Wireless keyboards (which use an infrared signal) are available for those who find portable computer keyboards too small for efficient typing. The keyboard illustration in Chapter 3 shows keyboard functions. *(See ● Panel 3.17, pp. 110–111.)*

In so-called *ergonomic keyboards,* the keyboard is divided in the middle and sloped in order to relieve repetitive strain injuries in the wrist. **The purpose of _ergonomics_ is to make working conditions and equipment safer and more efficient.** And if you're the type who tends to spill food on your keyboard there is the Flexboard, a device like a rubber bathmat, which can be washed under running water and hung out to dry.[18]

Ergonomic keyboard

- **Specialty keyboards and terminals:** Specialty keyboards, ranging from Touch-Tone telephone keypads to keyboards featuring pictures of food for use in fast-food restaurants. Here we will consider *dumb terminals, intelligent terminals,* and *Internet terminals.*

A **_dumb terminal_ has a display screen and a keyboard and can input and output but not process data.** These terminals you see being used by airline reservations clerks to access a mainframe computer containing flight information. Dumb terminals can perform no functions independent of the mainframe to which they are linked.

An **_intelligent terminal_ has its own memory and processor, as well as a display screen and keyboard.** Such a terminal can perform some functions independent of any mainframe to which it is linked. One example is the *automated teller machine (ATM),* a self-service banking machine that is connected through a telephone network to a central computer. Another example is the *point-of-sale (POS) terminal,* used to record purchases at the point in a store where the customer purchases goods or services. Recently, many intelligent terminals have been replaced by personal computers.

Palm PDA with collapsible keyboard

An **_Internet terminal_ provides access to the Internet.** There are several variants: (1) the *set-top box* or *Web terminal,* which displays Web pages on a TV set; (2) the *network computer,* a cheap (less than $500), stripped-down computer that connects people to networks; (3) the *online game player,* which not only lets you play games but also connects to the Internet; (4) the full-blown *PC/TV* (or *TV/PC*), which merges the personal computer with the television set; and (5) the *wireless pocket PC* or *personal digital assistant (PDA),* a handheld computer with a tiny keyboard that can do two-way wireless messaging.

Top

Cable

Right button

Left button

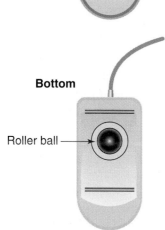

Bottom

Roller ball

Trackball

Pointing stick

Pointing Devices

One of the most natural of all human gestures, the act of pointing, is incorporated in several kinds of input devices. <u>**Pointing devices**</u> **control the position of the cursor or pointer on the screen.** Pointing devices include the *mouse* and its variants, the *touch screen*, and various forms of *pen input*.

- **The mouse and its variants—trackball, pointing stick, and touchpad:** The principal pointing tool used with microcomputers is the mouse, a device that is rolled about on a desktop and directs a pointer on the computer's display screen. The mouse pointer—an arrow, a rectangle, a pointing finger—is the symbol that indicates the position of the mouse on the display screen. The pointer may change to the shape of an I-beam to indicate that it is a cursor showing the place where text may be entered.

On the bottom side of the mouse is a ball that translates the mouse movement into digital signals. On the top side are one to five buttons. The first button is used for common functions, such as clicking and dragging. The functions of the other buttons are determined by whatever software you're using. Microsoft's IntelliMouse Optical has five programmable buttons, a scroll wheel, and no rolling ball to get gummed up, which allows the mouse to work on almost any surface.[19]

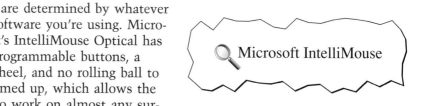

Microsoft IntelliMouse

There are three main variations on the mouse:

The <u>*trackball*</u> **is a movable ball, mounted on top of a stationary device, that can be rotated using your fingers or palm.** In fact, the trackball looks like the mouse turned upside down. Instead of moving the mouse around on the desktop, you move the trackball with the tips of your fingers.

A <u>*pointing stick*</u> **looks like a pencil eraser protruding from the keyboard between the G, H, and B keys. You move the pointing stick with your forefinger while using your thumb to press buttons located in front of the space bar.** (A forerunner of the pointing stick is the *joystick*, which consists of a vertical handle like a gearshift lever mounted on a base with one or two buttons.)

A <u>*touchpad*</u> **is a small, flat surface over which you slide your finger, using the same movements as you would with a mouse.** The cursor follows the movement of your finger. You "click" by tapping your finger on the pad's surface or by pressing buttons positioned close by the pad.

Touchpad

- **Touch screen:** A _**touch screen**_ is a video display screen that has been sensitized to receive input from the touch of a finger. *(See ● Panel 4.21.)* The screen is covered with a plastic layer, behind which are invisible beams of infrared light. You can input requests for information by pressing on buttons or menus displayed. The answers to your requests are displayed as output in words or pictures on the screen. (There may also be sound.) You find touch screens in kiosks, ATMs, airport tourist directories, hotel TV screens (for guest checkout), and campus information kiosks making available everything from lists of coming events to (with proper ID and personal code) student financial-aid records and grades.

- **Pen input:** Some input devices use variations on an electronic pen. Examples are *pen-based systems, light pens,* and *digitizers.*

    **Pen-based computer systems** allow users to enter handwriting and marks onto a computer screen by means of a penlike stylus rather than typing on a keyboard. *(See ● Panel 4.22.)* Pen computers use handwriting recognition software that translates handwritten characters made by the stylus into data that is usable by the computer. Many handheld computers and PDAs have pen input, as do digital notebooks.

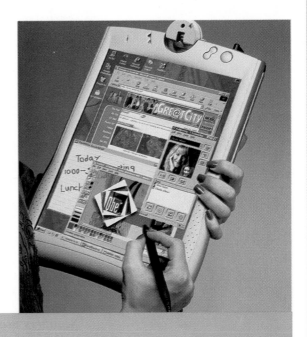

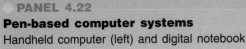

● **PANEL 4.22**
Pen-based computer systems
Handheld computer (left) and digital notebook

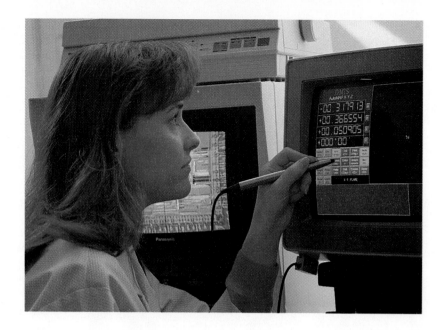

● PANEL 4.23
Light pen
This person is using a light pen to input to the computer.

The ***light pen*** **is a light-sensitive penlike device connected by a wire to the computer terminal.** The user brings the pen to a desired point on the display screen and presses the pen button, which identifies that screen location to the computer. *(See ● Panel 4.23.)* Light pens are used by engineers, graphic designers, and illustrators.

A ***digitizer*** **uses a mouselike copying device called a puck, which can convert drawings and photos to digital data.** One form of digitizer is the ***digitizing tablet*,** **an electronic plastic board on which each specific location corresponds to a location on the screen.** When you use a puck, the tablet converts your movements into digital signals that are input to the computer. Digitizing tablets are often used to make maps and engineering drawings. *(See ● Panel 4.24.)*

● PANEL 4.24
Digitizing tablet
This is often used in engineering and architectural applications.

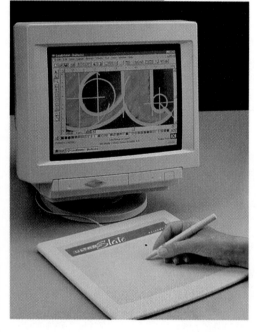

Source Data-Entry Devices

Source data-input devices do not require keystrokes in order to input data to the computer. Rather, data is entered directly from the *source*, without human intervention. ***Source data-entry devices*** **create machine-readable data on magnetic media or paper or feed it directly into the computer's processor.** In this section, we first cover *scanning devices—imaging systems, bar-code readers, mark- and character-recognition devices,* and *fax machines.* We then describe *audio-input devices* and *video* and *photographic input.* Finally, we describe *voice-recognition systems, sensors, radio-frequency identification devices,* and *human-biology input devices.*

- Scanning devices—imaging systems: Anthony J. Scalise, 80, of Utica, New York, found a 1922 picture of his father and other immigrants from the Italian city of Scandale. "It was wonderful, all those people with walrus mustaches," he said. He immediately had prints made for friends and relatives. It's easy to make such duplicates using the self-service Kodak scanners called imaging systems now found in many photo stores.[20]

    Scanners **use laser beams and reflected light to translate images of text, drawings, photos, and the like into digital form.** The images can then be processed by a computer, displayed on a monitor, stored on a

● PANEL 4.25
Image scanner
A graphics designer scans
an image into his desktop
publishing system.

storage device, or communicated to another computer. One type of scanner is the *imaging system—or image scanner, or graphics scanner—which converts text, drawings, and photographs into digital form that can be stored in a computer system and then manipulated, output, or sent via modem to another computer. (See ● Panel 4.25.)* The system scans each image—color or black and white—with light and breaks the image into light and dark dots or color dots, which are then converted to digital code. For example, the imaging system used in desktop publishing scans in artwork or photos that can then be positioned within a page of text, using desktop publishing software.

Imaging-system technology has led to a whole new art or industry called *electronic imaging*. **Electronic imaging is the software-controlled combining of separate images, using scanners, digital cameras, and advanced graphic computers.** This technology has become an important part of multimedia.

● Scanning devices—bar-code readers: Another scanning device reads **bar codes, the vertical zebra-striped marks you see on most manufactured retail products**—everything from candy to cosmetics to comic books. *(See ● Panel 4.26.)* In North America, supermarkets, food manufacturers, and others have agreed to use a bar-code system called the *Universal Product Code*. Other kinds of bar-code systems are used on everything from FedEx packages, to railroad cars, to the jerseys of long-distance

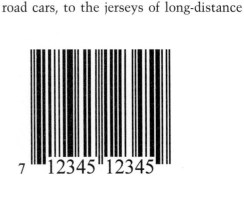

7 12345 12345

runners. Bar codes are read by ***bar-code readers***, **photoelectric scanners that translate the bar-code symbols into digital code.** The price of a particular item is set within the store's computer and appears on the salesclerk's point-of-sale terminal and on your receipt. Records of sales are input to the store's computer and used for accounting, restocking store inventory, and weeding out products that don't sell well.

- **Scanning devices—mark-recognition and character-recognition devices:** There are three types of scanning devices that sense marks or characters. They are usually referred to by their abbreviations *MICR*, *OMR*, and *OCR*.

  ***Magnetic-ink character recognition (MICR)* reads the strange-looking numbers printed at the bottom of checks.** MICR characters, which are printed with magnetized ink, are read by MICR equipment, producing a digitized signal. The bank's reader/sorter machine employs this signal to sort checks.

  ***Optical mark recognition (OMR)* uses a device that reads pencil marks and converts them into computer-usable form.** The best-known example is the OMR technology used to read the College Board Scholastic Aptitude Test (SAT) and the Graduate Record Examination (GRE).

  ***Optical character recognition (OCR)* uses a device that reads preprinted characters in a particular font (typeface design) and converts them to digital code.** OCR characters appear on utility bills and price tags on department-store merchandise; for example, the wand reader is a common OCR scanning device. *(See ● Panel 4.27.)*

- **Scanning devices—fax machines:** A ***fax machine***—or facsimile trans-**mission machine—scans an image and sends it as electronic signals over telephone lines to a receiving fax machine, which re-creates the image on paper.**

  There are two types of fax machines—*dedicated fax machines* and *fax modems*. ***Dedicated fax machines* are specialized devices that do nothing except send and receive fax documents.** These are what we usually think of as fax machines. They are found not only in offices and homes but also alongside regular phones in public places such as airports.

  A ***fax modem* is installed as a circuit board inside the computer's system cabinet. It is a modem with fax capability that enables you to send signals directly from your computer to someone else's fax machine or computer fax modem.** With this device, you don't have to print out the material from your printer and then turn around and run it through the scanner on a fax machine. The fax modem allows you to send information more quickly than if you had to feed it page by page into a machine.

● **PANEL 4.27**
Optical character recognition
Special typefaces can be read by a scanning device called a wand reader.

OCR-A
NUMERIC   0123456789
ALPHA   ABCDEFGHIJKLMNO
SYMBOLS   >$/-+-#"

OCR-B
NUMERIC   00123456789
ALPHA   ACENPSTVX
SYMBOLS   <+>-¥

The fax modem is another feature of mobile computing, and especially powerful as a receiving device. Fax modems are installed inside portable computers, including pocket PCs and PDAs. You can also link up a cellular phone to a fax modem in your portable computer and thereby send and receive wireless fax messages no matter where you are in the world.

- **Audio-input devices:** An ___audio-input device___ **records analog sound and translates it for digital storage and processing.** An analog sound signal is a continuously variable wave within a certain frequency range. For the computer to process them, these variable waves must be converted to digital 0s and 1s. The principal use of audio-input devices is to produce digital input for multimedia computers.

  An audio signal can be digitized in two ways—by an *audio board* or a *MIDI board.* Analog sound from a cassette player or a microphone goes through a special circuit board called an audio board (or card). An *audio board* is an add-on circuit board in a computer that converts analog sound to digital sound and stores it for further processing and/or plays it back, providing output directly to speakers or an external amplifier. A *MIDI board*—MIDI, pronounced "middie," stands for *Musical Instrument Digital Interface*—provides a standard for the interchange of musical information between musical instruments, synthesizers, and computers.

- **Video-input cards:** As with sound, most film and videotape is in analog form; the signal is a continuously variable wave. To be used by a computer, the signals that come from a VCR or a camcorder must be converted to digital form through a special digitizing card—a *video-capture card*—that is installed in the computer.

  Two types of video cards are frame-grabber video and full-motion video. *Frame-grabber video cards* can capture and digitize only a single frame at a time. *Full-motion video cards* can convert analog to digital signals at rates of up to 30 frames per second, giving the effect of a continuously flowing motion picture.

- **Digital cameras:** Digital cameras are particularly interesting because they foreshadow major change for the entire industry of photography.

PANEL 4.28
Digital camera

Instead of using traditional (chemical) film, a ___digital camera___ **uses a light-sensitive processor chip to capture photographic images in digital form on a small diskette inserted in the camera or on flash-memory chips.** *(See ● Panel 4.28.)* The bits of digital information can then be copied right into a computer's hard disk for manipulation and printing out.

- **Voice-recognition systems:** Can your computer tell whether you want it to "recognize speech" or "wreck a nice beach"? Voice-recognition systems have had to overcome many difficulties, such as different voices, pronunciations, and accents. Recently, however, the systems have measurably improved. Current accuracy rates are at about 90–95%. A ___voice-recognition system___, **using a microphone (or a telephone) as an input device, converts a person's speech into digital signals by comparing the electrical patterns produced by the speaker's voice with a set of prerecorded patterns stored in the computer.** *(See ● Panel 4.29, next page.)*

Voice-recognition systems are finding many uses. Warehouse workers are able to speed inventory-taking by recording inventory counts verbally. Traders on stock exchanges can communicate their trades by speaking to computers. Radiologists can dictate their interpretations of X-rays directly into transcription machines. Nurses can fill out patient

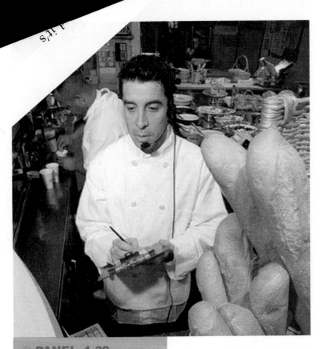

PANEL 4.29
Voice input device

charts by talking to a computer. Speakers of Chinese can speak to machines that will print out Chinese characters. Indeed, for many disabled individuals, a computer isn't so much a luxury or a simple productivity tool as a necessity. It provides freedom of expression, independence, and empowerment.

- **Sensors:** A **_sensor_ is an input device that collects specific data directly from the environment and transmits it to a computer.** Although you are unlikely to see such input devices connected to a PC in an office, they exist all around us, often in nearly invisible form. Sensors can be used to detect all kinds of things: speed, movement, weight, pressure, temperature, humidity, wind, current, fog, gas, smoke, light, shapes, images, and so on.

    Sensors are used to detect the speed and volume of traffic and adjust traffic lights. They are used on mountain highways in wintertime in the Sierra Nevada as weather-sensing devices to tell workers when to roll out snowplows. In California, sensors have been planted along major earthquake fault lines in an experiment to see whether scientists can predict major earth movements. *(See ● Panel 4.30.)* In aviation, sensors are used to detect ice buildup on airplane wings or to alert pilots to sudden changes in wind direction.

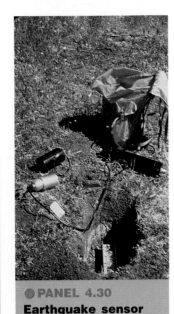

PANEL 4.30
Earthquake sensor

- **Radio-frequency identification devices:** **_Radio-frequency identification technology,_ or RF-ID tagging, consists of (1) a "tag" containing a microchip that contains code numbers that (2) can be read by the radio waves of a scanner linked to a database.** Drivers with RF-ID tags can breeze through the tollbooths without having to even roll down their windows; the toll is automatically charged to their accounts. Radio-readable ID "tags" are also used by the Postal Service to monitor the flow of mail. They are used in inventory control and warehousing. They are used in the railroad industry to keep track of rail cars. They are even injected into dogs and cats, so that veterinarians with the right scanning equipment can identify them if the pets become separated from their owners.

- **Human-biology input devices:** Characteristics and movements of the human body, when interpreted by sensors, optical scanners, voice recognition, and other technologies, can become forms of input. Two examples are *biometric systems* and *line-of-sight systems.*

    Biometrics is the science of measuring individual body characteristics. Biometric security devices identify a person through a fingerprint, voice intonation, or other biological characteristic. For example, retinal-identification devices use a ray of light to identify the distinctive network of blood vessels at the back of one's eyeball.

    Line-of-sight systems enable you to use your eyes to "point" at the screen. This technology allows some physically disabled users to direct a computer. For example, the Eyegaze System from LC Technologies allows you to operate a computer by focusing on particular areas of a display screen. A camera mounted on the computer analyzes the point of focus of the eye to determine where you are looking. You operate the computer by looking at icons on the screen and "press a key" by looking at one spot for a specified period of time.

Describe the various types of keyboards.

Discuss the mouse and its variants.

Identify the other types of pointing devices.

What are the several types of scanning devices?

Describe all the other types of source-data entry devices.

4.4 Output Hardware

Are we back to old-time radio? Almost. Except that you can call up local programs by downloading them from the Internet. Want to listen to comedian Robin Williams, with all his dopey foreign accents and cackling laughs? You can get his weekly audio show on audible.com.[21] The sound quality isn't even as good as that of AM radio, but no doubt that will improve eventually. Computer output is taking more and more innovative forms and getting better and better.

Output devices **convert machine-readable information, obtained as the result of processing, into people-readable form.** The principal kinds of output are *softcopy* and *hardcopy. (See ● Panel 4.31.)*

- Softcopy: *Softcopy* **refers to data that is shown on a display screen or is in audio or voice form.** This kind of output is not tangible; it cannot be touched.
- Hardcopy: *Hardcopy* **refers to printed output.** The principal examples are printouts, whether text or graphics, from printers. Film, including microfilm and microfiche, is also considered hardcopy output.

There are several types of output devices. In the following three sections, we discuss, first, *softcopy output—display screens;* second, *hardcopy output—printers;* and, third, *other output—sound, voice, and video.*

Softcopy Output: Display Screens

Display screens—**also variously called *monitors, CRTs,* or simply *screens*— are output devices that show programming instructions and data as they are being input and information after it is processed.**

🔍 17", .25dp Monitor (16" Display)

As for television screens, the size of a computer screen is measured diagonally from corner to corner in inches. For desktop microcomputers, the most common sizes are 13, 15, 17, 19, and 21 inches; for laptop computers, 12.1, 13.3, and 14.1 inches. Computer ads often now show the actual display area, called the *viewable image size (vis),* which may be an inch or so less. A 15-inch monitor may have a 13.8-inch vis and

● PANEL 4.31
Types of output devices

Softcopy Devices	Hardcopy Devices	Other Devices
CRT display screens	Impact printers: dot-matrix printer	Sound output
Flat-panel display screen (e.g., liquid-crystal display)	Nonimpact printers: laser, ink-jet, thermal	Voice output
		Video output

17" screen size
16" viewable image size

Monitor screen size	Viewable image area
15 inches	14 inches
17 inches	16 inches
21 inches	20 inches

a 17-inch monitor may have a 16-inch vis.

In deciding which display screen to buy, you will need to consider issues of screen clarity (dot pitch, resolution, and refresh rate), types of monitor (CRT versus flat panel, active-matrix flat panel versus passive-matrix flat panel), and color and resolution standard (SVGA and XGA).

- **Screen clarity—dot pitch, resolution, and refresh rate:** Among the factors affecting screen clarity (often mentioned in ads) are *dot pitch, resolution,* and *refresh rate.* These relate to the individual dots known as pixels, which represent the images on the screen. A **_pixel_, for "picture element," is the smallest unit on the screen that can be turned on and off or made different shades.**

  Dot pitch (dp) is the amount of space between the centers of adjacent pixels; the closer the dots, the crisper the image. For a .28dp monitor, for instance, the dots are 28/100ths of a millimeter apart. Generally, a dot pitch of .28dp will provide clear images.

  Resolution is the image sharpness of a display screen; the more pixels there are per square inch, the finer the level of detail attained. Resolution is expressed in terms of the formula *horizontal pixels × vertical pixels.* Each pixel can be assigned a color or a particular shade of gray. Standard resolutions are 640×480, 800×600, 1024×768, 1280×1024, and 1600×1200 pixels.

  Refresh rate is the number of times per second that the pixels are recharged so that their glow remains bright. In general, displays are refreshed 45–100 times per second. The higher the refresh rate, the more solid the image looks on the screen—that is, the less it flickers.

- **Two types of monitors—CRT and flat-panel:** Display screens are of two types: *CRT* and *flat-panel. (See ● Panel 4.32.)*

  A **_CRT_, for cathode-ray tube, is a vacuum tube used as a display screen in a computer or video display terminal.** This same kind of technology is found not only in the screens of desktop computers but also in television sets and flight-information monitors in airports. *Note:* Advertisements for desktop computers often *do not* include a monitor as part of the price of the system. You need to be prepared to spend a few hundred dollars extra for the monitor.

  Compared to CRTs, flat-panel displays are much thinner, weigh less, and consume less power. Thus, they are better for portable computers, although they are available for desktop computers as well. **_Flat-panel displays_ are made up of two plates of glass with a substance in between them in which light is manipulated.** One technology used is **_liquid crystal display (LCD)_, in which molecules of liquid crystal line up in a way that alters their optical properties, creating images on the screen by transmitting or blocking out light.**

  Flat-panel monitors are available for desktop computers. Because they are smaller than CRTs, they fit more easily onto a crowded desk. However, CRTs are considerably cheaper. A new 15-inch CRT costs as little as $100. A flat-panel costs 5–10 times as much. (You can get flat-panel TVs, too—at even scarier prices.)[22]

- **Active-matrix versus passive-matrix flat-panel displays:** Flat-panel screens are either active-matrix or passive-matrix displays, according to where their transistors are located.

PANEL 4.32
CRT (*top*) versus flat-panel display (*right*)

In an ***active matrix display***, **also known as TFT (thin-film transistor) display, each pixel on the screen is controlled by its own transistor.** Active-matrix screens are much brighter and sharper than passive-matrix screens, but they are more complicated and thus more expensive. They also require more power, affecting the battery life in laptop computers.

In a ***passive-matrix display***, **a transistor controls a whole row or column of pixels.** Passive matrix provides a sharp image for one-color (monochrome) screens but is more subdued for color. The advantage is that passive-matrix displays are less expensive and use less power than active-matrix displays, but they aren't as clear and bright and can leave "ghosts" when the display changes quickly. Passive-matrix displays go by the abbreviations HPA, STN, or DSTN.

- **Color and resolution standards for monitors—SVGA and XGA:** We mentioned earlier in the chapter that PCs come with *graphics cards* (also known as *video cards* or *video adapters*) that convert signals from the computer into video signals that can be displayed as images on a monitor. The monitor then separates the video signal into three colors: red, green, and blue signals. Inside the monitor, these three colors combine to make up each individual pixel.

Two common color and resolution standards for monitors are *SVGA* and *XGA*. (See ● *Panel 4.33, next page.*)

SVGA (super video graphics array) **supports a resolution of 800 × 600 pixels, or variations, producing 16 million possible simultaneous colors.** SVGA is the most common standard used today.

XGA (extended graphics array) **has a resolution of up to 1024 × 768 pixels, with 65,536 possible colors.** It is used mainly on workstation systems, such as those employed by engineering designers.

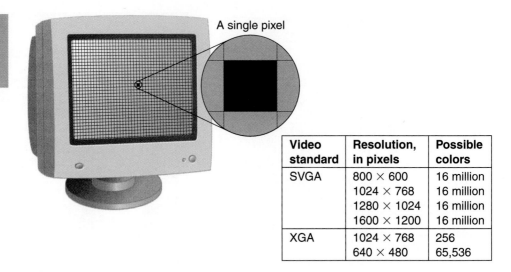

A single pixel

Video standard	Resolution, in pixels	Possible colors
SVGA	800 × 600	16 million
	1024 × 768	16 million
	1280 × 1024	16 million
	1600 × 1200	16 million
XGA	1024 × 768	256
	640 × 480	65,536

Hardcopy Output: Printers

The prices for computer systems in ads often do not include a printer. Thus, you will need to budget an additional $150 or more for a printer. A **_printer_ is an output device that prints characters, symbols, and perhaps graphics on paper or another hardcopy medium.** The resolution, or quality of sharpness of the image, is indicated by **_dpi (dots per inch),_ which is a measure of the number of dots that are printed in a linear inch.** For microcomputer printers, the resolution is in the range 60–1500 dpi.

HP DeskJet 970Cse Printer

Printers can be separated into two categories, according to whether or not the image produced is formed by physical contact of the print mechanism with the paper. *Impact printers* do have contact with paper; *nonimpact printers* do not. We will also consider *multifunction printers.*

- Impact printers: An **_impact printer_ forms characters or images by striking a mechanism such as a print hammer or wheel against an inked ribbon, leaving an image on paper. A _dot-matrix printer_ contains a print head of small pins, which strike an inked ribbon against paper, to form characters or images.** Print heads are available with 9, 18, or 24 pins; the 24-pin head offers the best quality. Dot-matrix printers can print *draft quality,* a coarser-looking 72 dpi, or *near-letter-quality (NLQ),* a crisper-looking 144 dpi. The machines print 40–300 characters per second and can handle graphics as well as text. A disadvantage is the noise they produce. Nowadays impact printers are more commonly used with mainframes than with personal computers.

- Nonimpact printers: Nonimpact printers are faster and quieter than impact printers because they have fewer moving parts. **<u>Nonimpact printers</u> form characters and images without direct physical contact between the printing mechanism and paper.** Two types of nonimpact printers often used with microcomputers are *laser printers* and *ink-jet printers.* A third kind, the *thermal printer,* is seen less frequently.

  Like a dot-matrix printer, a **_laser printer_ creates images with dots. However, as in a photocopying machine, these images are produced on a drum, treated with a magnetically charged ink-like toner (powder), and then transferred from drum to paper.** *(See ● Panel 4.34.)*

  There are good reasons that laser printers are among the most common types of nonimpact printer. They produce sharp, crisp images of both text and graphics. They are quiet and fast—able to print 4–32 text-only pages per minute for individual microcomputers and up to 200 pages per minute for mainframes. They can print in different

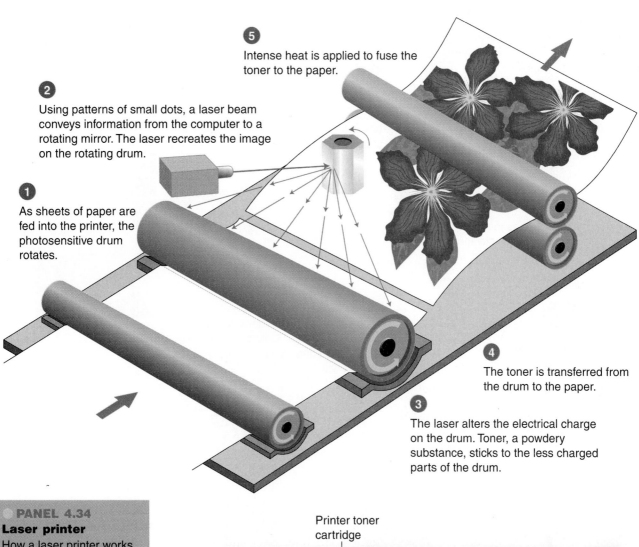

5 Intense heat is applied to fuse the toner to the paper.

2 Using patterns of small dots, a laser beam conveys information from the computer to a rotating mirror. The laser recreates the image on the rotating drum.

1 As sheets of paper are fed into the printer, the photosensitive drum rotates.

4 The toner is transferred from the drum to the paper.

3 The laser alters the electrical charge on the drum. Toner, a powdery substance, sticks to the less charged parts of the drum.

● **PANEL 4.34**
Laser printer
How a laser printer works (above). Replacing a toner cartridge (right).

Printer toner cartridge

fonts—that is, type styles and sizes. The more expensive models can print in different colors. Finally, a freshly printed page from a laser won't smear, as one from an ink-jet might.

To be able to manage graphics and complex page design, a laser printer works with a page description language. A ***page description language* is software that describes the shape and position of characters and graphics to the printer.** PostScript (from Adobe Systems) is one common type of page description language; Hewlett-Packard Graphic Language (HPGL) is another.

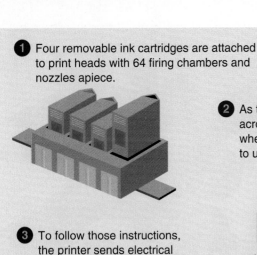

1 Four removable ink cartridges are attached to print heads with 64 firing chambers and nozzles apiece.

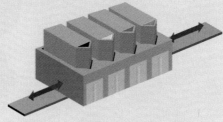

2 As the print heads move back and forth across the page, software instructs them where to apply dots of ink, what colors to use, and in what quantity.

3 To follow those instructions, the printer sends electrical pulses to thin resistors at the base of the firing chambers behind each nozzle.

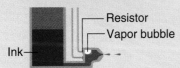

Resistor
Vapor bubble
Ink

5 A matrix of dots forms characters and pictures. Colors are created by layering multiple color dots in varying densities.

4 The resistor heats a thin layer of ink, which in turn forms a vapor bubble. That expansion forces ink through the nozzle and onto the paper at a rate of about 6000 dots per second.

● **PANEL 4.35**
Ink-jet printer

Ink-jet printers **spray small, electrically charged droplets of ink from four nozzles through holes in a matrix at high speed onto paper.** *(See* ● *Panel 4.35.)* Like laser and dot-matrix printers, ink-jet printers form images with little dots.

The advantages of ink-jet printers are that they can print in color, are quieter, and are much less expensive than color laser printers. The disadvantages are that they print in a somewhat lower resolution than laser printers and they are slower. Printing a document with high-resolution color graphics may take 10 minutes or more for a single page. Ink-jets, which spray ink onto the page a line at a time, can produce both high-quality black-and-white text and high-quality color graphics. If you print a lot of color, you'll find color ink-jets much slower and more expensive to operate than color laser printers. Still, the initial cost of a color ink-jet printer is considerably less than a color laser printer. The rock-bottom price is only about $150.

Thermal printers **use colored waxes and heat to produce images by burning dots onto special paper.** The colored wax sheets are not required for black-and-white output. However, thermal printers are expensive, and they require expensive paper. For people who want the highest-quality color printing available with a desktop printer, thermal printers are the answer.

● **Multifunction printers—printers that do more than print:** _Multifunction printers_ **combine several capabilities, such as printing, scanning, copying, and faxing, all in one device.** Both Xerox and Hewlett-Packard make machines that combine four pieces of office equipment in one—photocopier, fax machine, scanner, and laser printer. Multifunction printers take up less space and cost less than the four separate office machines that they replace. The downside, however, is

Do I need color, or will black-only do? Are you mainly printing text or will you need to produce color charts and illustrations (and, if so, how often)? If you print lots of black text, consider getting a laser printer. If you might occasionally print color, get an ink-jet that will accept cartridges for both black and color.

Do I have other special output requirements? Do you need to print envelopes or labels? special fonts (type styles)? multiple copies? transparencies or on heavy stock? Find out if the printer comes with envelope feeders, sheet feeders holding at least 100 sheets, or whatever will meet your requirements.

Is the printer easy to set up? Can you easily put the unit together, plug in the hardware, and adjust the software (the "driver" programs) to make the printer work with your computer?

Is the printer easy to operate? Can you add paper, replace ink/toner cartridges or ribbons, and otherwise operate the printer without much difficulty?

Does the printer provide the speed and quality I want? Will the machine print at least three pages a minute of black text and two pages a minute of color? Are the blacks dark enough and the colors vivid enough?

Will I get a reasonable cost per page? Special paper, ink or toner cartridges (especially color), and ribbons are all ongoing costs. Ink-jet color cartridges, for example, may last 100–500 pages and cost $25–$30 new. Laser toner cartridges are cheaper. Ribbons for dot-matrix printers are cheaper still. Ask the seller what the cost per page works out to.

Does the manufacturer offer a good warranty and good telephone technical support? Find out if the warranty lasts at least 2 years. See if the printer's manufacturer offers telephone support in case you have technical problems. The best support systems offer toll-free numbers and operate evenings and weekends as well as weekdays.

that, if one component breaks, nothing works. The accompanying box gives some questions to consider when you're buying a printer. *(See ● Panel 4.36.)*

Other Output: Sound, Voice, Animation, & Video

Most PCs are now multimedia computers, capable of outputting not only text and graphics but also sound, voice, and video, as we consider next.

- Sound output: ***Sound-output devices* produce digitized sounds, ranging from beeps and chirps to music.** To use sound output, you need appropriate software and a sound card. The sound card could be Sound Blaster or, since that brand became a de facto standard, one that is "Sound Blaster compatible." Well-known brands include Creative Labs, Diamond, and Turtle Beach. The sound card plugs into an expansion slot in your computer; on newer computers it is integrated with the motherboard.

- Voice output: ***Voice-output devices* convert digital data into speech-like sounds.** You hear such forms of voice output on telephones ("Please hang up and dial your call again"), in soft-drink machines, in cars, in toys and games, and recently in mapping software for vehicle-navigation devices. Some uses of speech output are simply frivolous or amusing. You can replace your computer start-up beep with the sound of James Brown screaming "I feel gooooooood!" But some uses are quite serious. For people with physical challenges, computers with voice output help to level the playing field.

- Video output: ***Video* consists of photographic images, which are played at 15–29 frames per second to give the appearance of full**

motion. Video is input into a multimedia system using a video camera or VCR and, after editing, is output on a computer's display screen. Because video files can require a great deal of storage—a 3-minute video may require 1 gigabyte of storage—video is often compressed (a topic we discuss in Chapter 6).

Another form of video output is *__videoconferencing__*, **in which people in different geographical locations can have a meeting—can see and hear one another—using computers and communications.** Videoconferencing systems range from videophones to group conference rooms with cameras and multimedia equipment to desktop systems with small video cameras, microphones, and speakers. (We discuss this topic further in Chapter 5.)

CONCEPT CHECK

What is the difference between softcopy and hardcopy output?

What are the different characteristics of display screens?

What is the difference between impact and nonimpact printers?

Identify the characteristics of dot-matrix, laser, ink-jet, thermal, and multifunction printers.

Describe sound, voice, and video output.

PRACTICAL ACTION BOX

How to Buy a Laptop

"Selecting a laptop computer is much more complicated than buying a desktop PC," observes *Wall Street Journal* technology writer Walter Mossberg.[a] The reason: Laptops (now often called "notebooks") vary more than desktop "generic boxes," which tend to be similar, at least within a price class.[b] Trying to choose among the many Windows-based laptops is a particularly brow-wrinkling experience. (Macintosh laptops tend to be more straightforward since there are only two models.)

Nevertheless, here are some suggestions:[c]

Purpose. What are you going to use your laptop for? You can get a laptop that's essentially a desktop replacement and won't be moved much. If you expect to use the machine a lot in class, in libraries, or on airplanes, however, weight and battery life are important.

Budget & Weight. Laptops range from $1000 to $4000, with high-end brands aimed mainly at businesspeople. You may wish to concentrate on those in the range $1000–$1500 (for an eMachines eSlate or Toshiba), or $1500–$2200 (for an Apple, Dell, or Gateway), or $2300–$2400 (for a Compaq, Hewlett-Packard, or Sony).

In the $2000 range are the light machines, 3–4 pounds. Designed for mobility, they tend to lack internal disk drives and all the standard ports. The heavy machines (7 pounds and up) in the $3000-plus range generally include all the features, including DVD drives and big screens.

Batteries: The Life–Weight Trade-off. A rechargeable lithium-ion battery lasts longer than the nickel–metal hydride battery. Even so, a battery in the less expensive machines will usually run continuously for only about 2¼ hours. (DVD players are particularly voracious consumers of battery power, so that it's the rare laptop that will allow you to finish watching a 2-hour movie.)

The trade-off is that heavier machines usually have longer battery life. (The Apple iBook, which weighs 6.6 pounds, is the champ at 4.5–6 hours.) The lightweight machines tend to get less than two hours, and toting extra batteries offsets the weight savings. (A battery can weigh a pound or so.)

Software. Many laptops come with less software than you would get with a typical desktop, though what you get will probably be adequate for most student purposes. On laptop PCs, count on getting Microsoft Works rather than the more powerful Microsoft Office 2000 to handle word processing, spreadsheets, databases, and the like.

Keyboards & Pointing Devices. The keys on a laptop keyboard are usually the same size as those on a desktop machine, although they may be smaller. However, the up-and-down action feels different, and the keys may feel wobbly. In addition, some keys may be omitted altogether or keys may do double duty or appear in unaccustomed

arrangements.

Most laptops have a small touch-sensitive pad in lieu of a mouse—you drag your finger across the touchpad to move the cursor. Others use a pencil-eraser-size pointing stick in the middle of the keyboard.

Screens. If you're not going to carry the laptop around much, go for a big, bright screen. Most people find they are comfortable with a 12- to 14-inch display, measured diagonally, though screens can be as small as 10.4 inches and as large as 15 inches.

The best screens are active-matrix display (TFT). However, some low-priced models have the cheaper passive-matrix screens (HPA, STN, or DSTN), which are harder to read, though you may find you can live with them. XGA screens (1024 × 768 pixels) have a higher resolution than SVGA screens (800 × 600 pixels), but fine detail may not be important to you.

Memory, Speed, & Storage Capacity. If you're buying a laptop to complement your desktop, you may be able to get along with reduced memory, slow processor, small hard disk, and no CD-ROM. Otherwise, all these matters become important.

Memory (RAM) is the most important factor in computer performance, even though processor speed is more heavily hyped. Most laptops have at least 64 megabytes (MB) of memory. Cheaper models have only 32 megabytes. In that case, the hard drive has to compensate, with corresponding loss of speed and battery life. A microprocessor running 350–500 megahertz (MHz) or higher is adequate, and more recent models are faster than this.

Sometimes laptops are referred to as "three-spindle" or "two-spindle" machines. In a three-spindle machine, a hard drive, a floppy-disk drive, and a CD-ROM drive all reside internally (not as external peripherals). A two-spindle machine has a hard drive and space for either a floppy-disk drive or a CD-ROM drive (or a second battery). A hard drive of 6 gigabytes (GB) or more is sufficient for most people.

For more information about buying computers, go to *www.mhhe.com/cit/uit4e, www.zdnet.com/computershopper,* and *micro.uoregon/buyersguide.*

Hardware

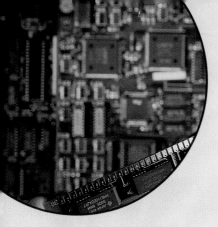

Visual Summary

active-matrix display (p. 181, KQ 4.4) Also known as *TFT (thin-film transistor) display*; each pixel on the screen is controlled by its own transistor. Why it's important: *Active-matrix screens are much brighter and sharper than passive-matrix screens, but they are more complicated and thus more expensive. They also require more power, affecting the battery life in laptop computers.*

AGP (accelerated graphics port) bus (p. 160, KQ 4.1) Bus that transmits data at high speeds; designed to support video and three-dimensional (3D) graphics. Why it's important: *An AGP bus is twice as fast as a PCI bus.*

arithmetic/logic unit (ALU) (p. 153, KQ 4.1) Part of the CPU that performs arithmetic operations and logical operations and controls the speed of those operations. Why it's important: *Arithmetic operations are the fundamental math operations: addition, subtraction, multiplication, and division. Logical operations are comparisons such as is "equal to," "greater than," or "less than."*

ASCII (American Standard Code for Information Interchange) (p. 148, KQ 4.1) Binary code used with microcomputers. Besides more conventional characters, ASCII includes such characters as math symbols and Greek letters. Why it's important: *ASCII is the binary code most widely used in microcomputers.*

audio-input device (p. 177, KQ 4.3) Hardware that records analog sound and translates it for digital storage and processing. Why it's important: *Analog sound signals are continuously variable waves within a certain frequency range. For the computer to process them, these variable waves must be converted to digital 0s and 1s. The principal use of audio-input devices is to produce digital input for multimedia computers. An audio signal can be digitized in two ways—by an audio board or a MIDI board.*

bar-code reader (p. 176, KQ 4.3) Photoelectric scanner that translates bar codes into digital code. Why it's important: *With bar-code readers and the appropriate software system, store clerks can total purchases and produce invoices with increased speed and accuracy; and stores and other businesses can monitor inventory and services with increased efficiency.*

bar codes (p. 175, KQ 4.3) Vertical zebra-striped marks imprinted on most manufactured retail products. Why it's important: *In North America, supermarkets, food manufacturers, and others have agreed to use a bar-code system called the Universal Product Code. Other kinds of bar-code systems are used on everything from FedEx packages, to railroad cars, to the jerseys of long-distance runners.*

bay (p. 150, KQ 4.1) Shelf or opening in the computer case used for the installation of electronic equipment, generally storage devices such as a hard drive or DVD drive. Why it's important: *A computer may come equipped with four or seven bays, which permit the expansion of system capabilities.*

binary system (p. 147, KQ 4.1) A two-state system used for data representation in computers; has only two digits— 0 and 1. Why it's important: *In the computer, 0 can be represented by electrical current being off and 1 by the current being on. All data and program instructions that go into the computer are represented in terms of these binary numbers.*

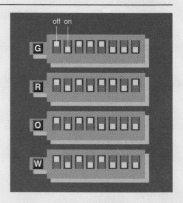

biometrics (p. 178, KQ 4.3) Science of measuring individual body characteristics. Why it's important: *Biometric security devices identify a person through a fingerprint, voice intonation, or other biological characteristic. For example, retinal-identification devices use a ray of light to identify the distinctive network of blood vessels at the back of one's eyeball.*

bit (p. 148, KQ 4.1) Short for "binary digit," which is either a 0 or a 1 in the binary system of data representation in computer systems. Why it's important: *The bit is the fundamental element of all data and information processed and stored in a computer system.*

bus (p. 154, KQ 4.1) Also called *bus line;* electrical data roadway through which bits are transmitted within the CPU and between the CPU and other components of the motherboard. Why it's important: *A bus resembles a multilane highway: The more lanes it has, the faster the bits can be transferred.*

byte (p. 148, KQ 4.1) Group of 8 bits. Why it's important: *A byte represents one character, digit, or other value. It is the basic unit used to measure the storage capacity of main memory and secondary storage devices (kilobytes and megabytes).*

cache (p. 156, KQ 4.1) Special high-speed memory area on a chip that the CPU can access quickly. It temporarily stores instructions and data that the processor is likely to use frequently. Why it's important: *Cache speeds up processing.*

CD-R (compact disk-recordable) disks (p. 167, KQ 4.2) Optical-disk form of secondary storage that can be written to only once but can be read many times. Why it's important: *This format allows consumers to make their own CD disks, though it's a slow process. Once recorded, the information cannot be erased. CD-R is often used by companies for archiving—that is, to store vast amounts of information. A variant is the Photo CD, an optical disk developed by Kodak that can digitally store photographs taken with an ordinary 35-millimeter camera.*

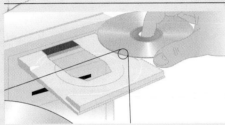

CD-ROM (compact disk-read-only memory) (p. 167, KQ 4.2) Optical-disk form of secondary storage that is used to hold prerecorded text, graphics, and sound. Why it's important: *Like music CDs, a CD-ROM is a read-only disk. Read-only means the disk's content is recorded at the time of manufacture and cannot be written on or erased by the user. As the user, you have access only to the data imprinted by the disk's manufacturer. A CD-ROM disk can hold up to 650 megabytes of data, equal to over 300,000 pages of text.*

CD-RW (compact disk-rewritable) disk (p. 168, KQ 4.2) Optical-disk form of secondary storage; also known as *erasable optical disk.* Allows users to record and erase data so that the disk can be used over and over again. Special CD-RW drives and software are required. Why it's important: *CD-RW disks are useful for archiving and backing up large amounts of data or work in multimedia production or desktop publishing; however, they are relatively slow.*

chip (p. 151, KQ 4.1) Also called a *semiconductor,* or *integrated circuit;* consists of millions of electronic circuits printed on a tiny piece of silicon. Silicon is an element widely found in sand that has desirable electrical (or "semiconducting") properties. Why it's important: *Chips have made possible the development of small computers.*

CISC (complex instruction set computing) chips (p. 150, KQ 4.1) Design that allows a microprocessor to support a large number of instructions. Why it's important: *CISC chips are used mostly in PCs and in conventional mainframes. CISC chips are generally slower than RISC chips.*

CMOS (complementary metal-oxide semiconductor) chip (p. 156, KQ 4.1) Chip powered by a battery and that doesn't lose its contents when the power is turned off. Why it's important: *CMOS chips contain flexible start-up instructions, such as time, date, and calendar, that must be kept current even when the computer is turned off. Unlike ROM chips, CMOS chips can be reprogrammed—for example, when you need to change the time for daylight savings time.*

control unit (p. 153, KQ 4.1) Part of the CPU that deciphers each instruction stored in it and then carries out the instruction. *Why it's important:* *The control unit directs the movement of electronic signals between main memory and the arithmetic/logic unit. It also directs these electronic signals between main memory and the input and output devices.*

CPU (central processing unit) (p. 153, KQ 4.1) The processor; it follows the instructions of the software (program) to manipulate data into information. The CPU consists of two parts—(1) the control unit and (2) the arithmetic/logic unit (ALU), which both contain registers, or high-speed storage areas. All are linked by a kind of electronic "roadway" called a bus. *Why it's important:* *The CPU is the "brain" of the computer.*

CRT (cathode-ray tube) (p. 180, KQ 4.4) Vacuum tube used as a display screen in a computer or video display terminal. *Why it's important:* *This same kind of technology is found not only in the screens of desktop computers but also in television sets and flight-information monitors in airports.*

dedicated fax machine (p. 176, KQ 4.3) Specialized device that does nothing except send and receive fax documents. *Why it's important:* *This is what we usually think of as a fax machine. Fax machines permit the transmission of text and graphic data over telephone lines quickly and inexpensively. They are found not only in offices and homes but also alongside regular phones in public places such as airports.*

digital camera (p. 177, KQ 4.3) Electronic camera that uses a light-sensitive processor chip to capture photographic images in digital form on a small diskette inserted in the camera or on flash-memory chips. *Why it's important:* *The bits of digital information can then be copied right into a computer's hard disk for manipulation and printing out.*

digitizer (p. 174, KQ 4.3) Input device that uses a mouselike copying device called a *puck* that can convert drawings and photos to digital data. *Why it's important:* *See digitizing tablet.*

digitizing tablet (p. 174, KQ 4.3) One form of digitizer; an electronic plastic board on which each specific location corresponds to a location on the screen. When you use a puck, the tablet converts your movements into digital signals that are input to the computer. *Why it's important:* *Digitizing tablets are often used to make maps and engineering drawings, as well as to "trace" drawings.*

display screen (p. 179, KQ 4.4) Also called *monitors, CRTs,* or simply *screens;* output devices that show programming instructions and data as they are being input and information after it is processed. *Why it's important:* *Screens are needed to display softcopy output.*

dot-matrix printer (p. 182, KQ 4.4) Impact printer with a print head of small pins that strike an inked ribbon against paper to form characters or images. Print heads are available with 9, 18, or 24 pins; the 24-pin head offers the best quality. *Why it's important:* *Dot-matrix printers can print draft-quality, a coarser-looking 72 dpi, or near-letter-quality (NLQ), a crisper-looking 144 dpi. The machines print 40–300 characters per second and can handle graphics as well as text. Dot-matrix printers are used much less frequently than laser printers and ink-jet printers.*

dot pitch (dp) (p. 180, KQ 4.4) Amount of space between the centers of adjacent pixels; the closer the dots, the crisper the image. *Why it's important:* *Dot pitch is one of the measures of display-screen crispness. For a .28dp monitor, for instance, the dots are 28/100ths of a millimeter apart. Generally, a dot pitch of .28dp will provide clear images.*

dpi (dots per inch) (p. 182, KQ 4.4) Measure of the number of dots that are printed in a linear inch. For microcomputer printers, the resolution is in the range 60–1500 dpi. *Why it's important:* *The higher the dpi, the better the resolution.*

dumb terminal (p. 171, KQ 4.3) Display screen and a keyboard hooked up to a computer system; it can input and output but not process data. *Why it's important:* *Dumb terminals are used, for example, by airline reservations clerks to access a mainframe computer containing flight information.*

DVD-R (DVD-recordable) disks (p. 168, KQ 4.2) DVD disks that allow one-time recording by the consumer. Two types of reusable disks are DVD-RW (DVD–rewritable) and DVD-RAM (DVD-random access memory), both of which can be recorded on and erased more than once. Why it's important: *Recordable DVD disks offer the user yet another option for storing large amounts of data.*

DVD-ROM (digital versatile disk or digital video disk, with read-only memory) (p. 168, KQ 4.2) CD-type disk with extremely high capacity, able to store 4.7–17 gigabytes. Why it's important: *Like a CD or CD-ROM, the surface of a DVD contains microscopic pits, which represent the 0s and 1s of digital code that can be read by a laser. The pits on the DVD, however, are much smaller and grouped more closely together than those on a CD, allowing far more information to be recorded. Also, the laser beam focuses on pits roughly half the size of those on current audio CDs. In addition, the DVD format allows for two layers of data-defining pits, not just one.*

EBCDIC (Extended Binary Coded Decimal Interchange Code) (p. 148, KQ 4.1) Binary code used with large computers. Why it's important: *EBCDIC is commonly used in mainframes.*

electronic imaging (p. 175, KQ 4.3) Software-controlled combining of separate images, using scanners, digital cameras, and advanced graphic computers. Why it's important: *This technology has become an important part of multimedia.*

ergonomics (p. 171, KQ 4.3) Study, or science, of working conditions and equipment with the goal of making them safer and more efficient. Why it's important: *Ergonomic principles are used in designing ways to use computers to further productivity while avoiding stress, illness, and injuries.*

expansion (p. 150, KQ 4.1) Way of increasing a computer's capabilities by adding hardware that performs tasks that are not part of the basic system. Why it's important: *Expansion allows users to customize and/or upgrade their computer systems.*

expansion card (p. 159, KQ 4.1) Also known as *expansion board, adapter card, interface card, plug-in board, controller card, add-in,* or *add-on;* circuit boards that provide more memory or that control peripheral devices. Why it's important: *Common expansion cards connect to the monitor (graphics card), speakers and microphones (sound card), and network (network card). Most computers have four to eight expansion slots, some of which may already contain expansion cards included in your initial PC purchase.*

expansion slot (p. 159, KQ 4.1) Socket on the motherboard into which you can plug an expansion card. Why it's important: *See* expansion card.

fax machine (p. 176, KQ 4.3) Facsimile transmission machine; scans an image and sends it as electronic signals over telephone lines to a receiving fax machine, which re-creates the image on paper. Why it's important: *See* dedicated fax machine and fax modem.

fax modem (p. 176, KQ 4.3) Installed as a circuit board inside the computer's system cabinet; a modem with fax capability that enables you to send signals directly from your computer to someone else's fax machine or computer fax modem. Why it's important: *With this device, you don't have to print out the material from your printer and then turn around and run it through the scanner on a fax machine. The fax modem allows you to send information more quickly than if you had to feed it page by page into a machine. Fax modems are installed inside portable computers, including pocket PCs and PDAs. You can also link up a cellular phone to a fax modem in your portable computer and thereby send and receive wireless fax messages no matter where you are in the world.*

flash memory chips (p. 156, KQ 4.1) Chips that can be erased and reprogrammed more than once (unlike PROM chips, which can be programmed only once). Why it's important: *Flash memory, which can range from 1 to 64 megabytes in capacity, is used to store programs not only in personal computers but also in pagers, cellphones, printers, and digital cameras. Unlike standard RAM chips, flash memory is nonvolatile—data is retained when the power is turned off.*

flat-panel display (p. 180, KQ 4.4) Display screens that are much thinner, weigh less, and consume less power than CRTs. Flat-panel displays are made up of two plates of glass with a substance in between them in which light is manipulated. Why it's important: *Flat-panel displays are used in portable computers, although they are available for desktop computers as well.*

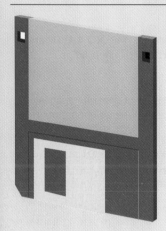

floppy disk (p. 162, KQ 4.2) Often called a *diskette* or simply a *disk;* removable flat piece of mylar plastic packaged in a 3.5-inch plastic case. Data and programs are stored on the disk's coating by means of magnetized spots, following standard on/off patterns of data representation (such as ASCII). The plastic case protects the mylar disk from being touched by human hands. Why it's important: *Floppy disks are used on all microcomputers.*

floppy-disk cartridges (p. 162, KQ 4.2) High-capacity removable 3.5-inch disks—Zip disks, SuperDisks, and HiFD disks. Why it's important: *These cartridges store more data than regular floppy disks and are just as portable.*

gigabyte (G, GB) (p. 148, KQ 4.1) Approximately 1 billion bytes (1,073,741,824 bytes); a measure of storage capacity. Why it's important: *This measure was formerly used mainly with "big iron" (mainframe) types of computers but is typical of secondary storage (hard disk) capacity of today's microcomputers.*

gigahertz (GHz) (p. 152, KQ 4.1) Measure of speed used for the latest generation of processors: a billion cycles per second. Why it's important: *Since a new high-speed processor can cost many hundred dollars more than the previous generation of chip, experts often recommend that buyers fret less about the speed of the processor (since the work most people do on their PCs doesn't even tax the limits of the current hardware) and more about spending money on extra memory.*

graphics card (p. 160, KQ 4.1) Also called a *video card* or *video adapter*, converts signals from the computer into video signals that can be displayed as images on a monitor. Why it's important: *Graphics cards are necessary for the display output.*

hardcopy (p. 179, KQ 4.4) Refers to printed output. Why it's important: *The principal examples are printouts, whether text or graphics, from printers. Film, including microfilm and microfiche, is also considered hardcopy output.*

hard disk (p. 164, KQ 4.2) Secondary storage medium; thin but rigid metal platter covered with a substance that allows data to be stored in the form of magnetized spots. Hard disks are tightly sealed within an enclosed hard-disk-drive unit to prevent any foreign matter from getting inside. Data may be recorded on both sides of the disk platters. Why it's important: *Hard disks hold much more data than do floppy disks. All microcomputers use hard disks as their principal storage medium.*

hard-disk controller (p. 165, KQ 4.2) Special-purpose circuit board that positions the disk and read/write heads and manages the flow of data and instructions to and from the disk. Why it's important: *Common PC hard-disk controllers are Ultra ATA (or EIDE) and SCSI.*

HiFD disk (p. 164, KQ 4.2) Made by Sony Corp.; has a capacity of 200 megabytes. The disk drive can also read standard 1.44-megabyte floppies. Why it's important: *HiFD disks have 140 times the capacity of today's standard floppy disks.*

impact printer (p. 182, KQ 4.4) Printer that forms characters or images by striking a mechanism such as a print hammer or wheel against an inked ribbon, leaving an image on paper. Why it's important: *Nonimpact printers are more commonly used than impact printers, but dot-matrix printers are still used in some businesses.*

infrared port (p. 159, KQ 4.1) Port that allows a computer to make a cableless connection with infrared-capable devices, such as some printers. Why it's important: *This type of connection requires an unobstructed line of sight between transmitting and receiving ports, and they can be only a few feet apart. Remote-control units use infrared waves.*

ink-jet printer (p. 184, KQ 4.4) Printer that sprays small, electrically charged droplets of ink from four nozzles through holes in a matrix at high speed onto paper. Like laser and dot-matrix printers, ink-jet printers form images with little dots. Why it's important: *Because they produce high-quality images on special paper, ink-jet printers are often used in graphic design and desktop publishing. However, ink-jet printers are slower than laser printers and print at a lower resolution on regular paper.*

input hardware (p. 170, KQ 4.3) Devices that translate data into a form the computer can process. Why it's important: *The people-readable form of the data may be words but the computer-readable form consists of 0s and 1s, represented as off and on electrical signals. Input hardware devices are categorized as three types: keyboards, pointing devices, and source data-entry devices.*

intelligent terminal (p. 171, KQ 4.3) Hardware unit with its own memory and processor, as well as a display screen and keyboard, hooked to a larger computer system. Why it's important: *Such a terminal can perform some functions independent of any mainframe to which it is linked. One example is the automated teller machine (ATM), a self-service banking machine that is connected through a telephone network to a central computer. Another example is the point-of-sale (POS) terminal, used to record purchases at the point in a store where the customer purchases goods or services. Recently, many intelligent terminals have been replaced by personal computers.*

Intel-type chip (p. 151, KQ 4.1) Processor chip for PCs; made principally by Intel Corp., but also by Advanced Micro Devices (AMD), Cyrix, DEC, and others. Why it's important: *These chips are used by manufacturers such as Compaq, Dell, Gateway 2000, Hewlett-Packard, and IBM. Since 1993, Intel has marketed its chips under the names Pentium, Pentium Pro, Pentium MMX, Pentium II, Pentium III, and Celeron. Many ads for PCs contain the logo "Intel inside" to show that the systems run an Intel microprocessor.*

Internet terminal (p. 171, KQ 4.3) Terminal that provides access to the Internet. Why it's important: *There are several variants of Internet terminal: (1) the set-top box or Web terminal, which displays Web pages on a TV set; (2) the network computer, a cheap, stripped-down computer that connects people to networks; (3) the online game player, which not only lets you play games but also connects to the Internet; (4) the full-blown PC/TV (or TV/PC), which merges the personal computer with the television set; and (5) the wireless pocket PC or personal digital assistant (PDA), a handheld computer with a tiny keyboard that can do two-way wireless messaging.*

ISA (industry standard architecture) bus (p. 160, KQ 4.1) The most widely used expansion bus. Why it's important: *ISA is also the oldest and, at 8 or 16 bits, the slowest at transmitting data, though it is still used for mouses, modem cards, and low-speed network cards.*

keyboard (p. 170, KQ 4.3) Input device that converts letters, numbers, and other characters into electrical signals that can be read by the computer's processor. Why it's important: *Keyboards are the most popular kind of input device.*

kilobyte (K, KB) (p. 148, KQ 4.1) Approximately 1000 bytes (1024 bytes); a measure of storage capacity. Why it's important: *The kilobyte was a common unit of measure for memory or secondary-storage capacity on older computers.*

laser printer (p. 182, KQ 4.4) Nonimpact printer that creates images with dots, similar to a photocopying machine. Images are produced on a drum, treated with a magnetically charged ink-like toner (powder), and then transferred from drum to paper. Why it's important: *Laser printers produce much better image quality than do dot-matrix printers and can print in many more colors; they are also quieter. Laser printers, along with page description languages, enabled the development of desktop publishing.*

light pen (p. 174, KQ 4.3) Light-sensitive penlike device connected by a wire to the computer terminal. The user brings the pen to a desired point on the display screen and presses the pen button, which identifies that screen location to the computer. Why it's important: *Light pens are used by engineers, graphic designers, and illustrators.*

liquid crystal display (LCD) (p. 180, KQ 4.4) Flat-panel display in which molecules of liquid crystal line up in a way that alters their optical properties, creating images on the screen by transmitting or blocking out light. Why it's important: *LCD is useful not only for portable computers but also as a display for various electronic devices, such as watches and radios.*

machine cycle (p. 153, KQ 4.1) Series of operations performed by the control unit to execute a single program instruction. It (1) fetches an instruction, (2) decodes the instruction, (3) executes the instruction, and (4) stores the result. Why it's important: *The machine cycle is the essence of computer-based processing.*

magnetic-ink character recognition (MICR) (p. 176, KQ 4.3) Type of scanning technology that reads magnetized-ink characters printed at the bottom of checks and converts them to digital form. Why it's important: *MICR technology is used by banks to sort checks.*

magnetic tape (p. 168, KQ 4.2) Thin plastic tape coated with a substance that can be magnetized. Data is represented by magnetized spots (representing 1s) or nonmagnetized spots (representing 0s). Why it's important: *Today, "mag tape" is used mainly for backup and archiving—that is, for maintaining historical records—where there is no need for quick access.*

megabyte (M, MB) (p. 148, KQ 4.1) Approximately 1 million bytes (1,048,576 bytes); measure of storage capacity. Why it's important: *Microcomputer primary-storage capacity is expressed in megabytes.*

megahertz (MHz) (p. 152, KQ 4.1) Measure of microcomputer processing speed, controlled by the system clock. Why it's important: *Generally, the higher the megahertz rate, the faster the computer can process data. Today a 550-MHz Pentium III–based microcomputer processes 550 million cycles per second.*

microprocessor (p. 150, KQ 4.1) Miniaturized circuitry of a computer processor. It stores program instructions that process, or manipulate, data into information. The key parts of the microprocessor are transistors. Why it's important: *Microprocessors enabled the development of microcomputers.*

Motorola-type chips (p. 152, KQ 4.1) Microprocessors made by Motorola for Apple Macintosh computers. Why it's important: *Since 1993, Motorola has joined forces with IBM and Apple to produce the PowerPC family of chips.*

multifunction printer (p. 184, KQ 4.4) Hardware device that combines several capabilities, such as printing, scanning, copying, and faxing. Why it's important: *Multifunction printers take up less space and cost less than the four separate office machines that they replace. The downside, however, is that, if one component breaks, nothing works.*

network interface card (p. 160, KQ 4.1) Electronic circuitry card that allows the transmission of data over a cable network, which connects various computers and other devices such as printers. Why it's important: *Network interface cards are necessary for all network functions.*

nonimpact printer (p. 182, KQ 4.4) Printer that forms characters and images without direct physical contact between the printing mechanism and paper. Why it's important: *Two types of nonimpact printers often used with microcomputers are laser printers and ink-jet printers. A third kind, the thermal printer, is seen less frequently.*

nonremovable hard disk (p. 164, KQ 4.2) Also known as a *fixed disk;* hard disk housed in a microcomputer system unit and used to store nearly all programs and most data files. Usually it consists of four 3½-inch metallic platters sealed inside a drive case the size of a small sandwich, which contains disk platters on a drive spindle, read/write heads mounted on an access arm that moves back and forth, and power connections and circuitry. Operation is much the same as for a diskette drive: The read/write heads locate specific instructions or data files according to track or sector. Hard disks can also come in removable cartridges. Why it's important: *See hard disk.*

optical card (p. 169, KQ 4.2) Plastic, laser-recordable, wallet-type cards used with an optical-card reader. Why it's important: *Because they can cram so much data (6.6 megabytes) into so little space, they may become popular in the future. With an optical card, for instance, there's enough room for a person's health card to hold not only his or her medical history and health-insurance information but also digital images, such as electrocardiograms.*

optical disk (p. 166, KQ 4.2) Removable disk, usually 4.75 inches in diameter and less than one-twentieth of an inch thick, on which data is written and read through the use of laser beams. Why it's important: *An audio CD holds up to 74 minutes (2 billion bits' worth) of high-fidelity stereo sound. Some optical disks are used strictly for digital data storage, but many are used to distribute multimedia programs that combine text, visuals, and sound.*

NUMERIC	0123456789
ALPHA	ABCDEFGHIJKLMNOPQRSTUVWXYZ
SYMBOLS	>$/-+-#"

optical character recognition (OCR) (p. 176, KQ 4.3) Type of scanning technology that reads special preprinted characters in a particular font (typeface design) and converts them to digital code. Why it's important: *OCR characters appear on utility bills and price tags on department-store merchandise.*

optical mark recognition (OMR) (p. 176, KQ 4.3) Type of scanning technology that reads pencil marks and converts them into computer-usable form. Why it's important: *OMR technology is used to read the College Board Scholastic Aptitude Test (SAT) and the Graduate Record Examination (GRE).*

output devices (p. 179, KQ 4.4) Hardware devices that convert machine-readable information, obtained as the result of processing, into people-readable form. The principal kinds of output are softcopy and hardcopy. Why it's important: *Without output devices people would have no access to processed data and information.*

page description language (p. 183, KQ 4.4) Software that describes the shape and position of characters and graphics to the printer. PostScript (from Adobe Systems) is one common type of page description language; Hewlett-Packard Graphic Language (HPGL) is another. Why it's important: *Page description languages enabled the development of desktop publishing.*

parallel port (p. 157, KQ 4.1) Part of the computer where you can plug in a device that allows 8 bits (1 byte) to be transmitted simultaneously, like cars on an eight-lane highway. Why it's important: *Parallel lines move information faster than serial lines do, but they can transmit information efficiently only up to 15 feet. Thus, parallel ports are used principally for connecting printers or external disk or magnetic-tape backup storage devices.*

passive-matrix display (p. 181, KQ 4.4) Flat-panel display in which a transistor controls a whole row or column of pixels. Passive matrix provides a sharp image for one-color (monochrome) screens but is more subdued for color. Why it's important: *Passive-matrix displays are less expensive and use less power than active-matrix displays, but they aren't as clear and bright and can leave "ghosts" when the display changes quickly. Passive-matrix displays go by the abbreviations HPA, STN, or DSTN.*

PC card (p. 161, KQ 4.1) Thin, credit-card size (2.1 by 3.4 inches) hardware device. Why it's important: *PC cards are used principally on laptop computers to expand capabilities.*

PCI (peripheral component interconnect) bus (p. 160, KQ 4.1) High-speed bus; at 32 or 64 bits wide, it is more than four times faster than ISA buses. Why it's important: *PCI is widely used in microcomputers to connect graphics cards, sound cards, modems, and high-speed network cards.*

pen-based computer system (p. 173, KQ 4.3) Input system that allows users to enter handwriting and marks onto a computer screen by means of a penlike stylus rather than by typing on a keyboard. Pen computers use handwriting recognition software that translates handwritten characters made by the stylus into data that is usable by the computer. Why it's important: *Many handheld computers and PDAs have pen input, as do digital notebooks.*

petabyte (P, PB) (p. 148, KQ 4.1) Approximately 1 quadrillion bytes (1,048,576 gigabytes); measure of storage capacity. Why it's important: *The huge storage capacities of modern databases are now expressed in petabytes.*

pixel (p. 180, KQ 4.4) Short for "picture element"; the smallest unit on the screen that can be turned on and off or made different shades. Why it's important: *Pixels are the building blocks that allow text and graphical images to be displayed on a screen.*

Plug and Play (p. 157, KQ 4.1) USB peripheral connection standard that allows peripheral devices and expansion cards to be automatically configured while they are being installed. Why it's important: *Plug and Play avoids the hassle of setting switches and creating special files that plagued earlier users.*

pointing devices (p. 172, KQ 4.3) Hardware that controls the position of the cursor or pointer on the screen. Why it's important: *Pointing devices include the mouse and its variants, the touch screen, and various forms of pen input.*

pointing stick (p. 172, KQ 4.3) Pointing device that looks like a pencil eraser protruding from the keyboard between the G, H, and B keys. You move the pointing stick with your forefinger while using your thumb to press buttons located in front of the space bar. Why it's important: *Pointing sticks are used principally in videogames, computer-aided design systems, and in robots.*

port (p. 157, KQ 4.1) Connecting socket or jack on the outside of the system unit into which are plugged different kinds of cables. Why it's important: *A port allows you to plug in a cable to connect a peripheral device, such as a monitor, printer, or modem, so that it can communicate with the computer system.*

printer (p. 182, KQ 4.4) Output device that prints characters, symbols, and perhaps graphics on paper or another hardcopy medium. Why it's important: *Printers provide one of the principal forms of computer output.*

radio-frequency identification technology (p. 178, KQ 4.3) RF-ID tagging consists of (1) a "tag" containing a microchip that contains code numbers that can be read by the radio waves of a scanner linked to a database. Why it's important: *Drivers with RF-ID tags can breeze through the tollbooths without having to even roll down their windows; the toll is automatically charged to their accounts. Radio-readable ID "tags" are also used by the Postal Service to monitor the flow of mail. They are used in inventory control and warehousing. They are used in the railroad industry to keep track of rail cars.*

RAM (random access memory) chips (p. 155, KQ 4.1) Also called *primary storage* and *main memory;* these chips temporarily hold software instructions and data before and after it is processed by the CPU. RAM is a volatile form of storage. Why it's important: *RAM is the working memory of the computer. Having enough RAM is critical; without sufficient RAM you will not be able to run many software programs.*

read (p. 156, KQ 4.1) To transfer data from an input source into the computer's memory or CPU. Why it's important: *Reading, along with writing, is an essential computer activity.*

read/write head (p. 162, KQ 4.2) Mechanism used to transfer data between the computer and the disk. When the disk spins inside its case, the read/write head moves back and forth over the data access area on the disk. Why it's important: *The read/write head enables the essential activities of reading and writing data.*

refresh rate (p. 180, KQ 4.4) Number of times per second that screen pixels are recharged so that their glow remains bright. In general, displays are refreshed 45–100 times per second. Why it's important: *The higher the refresh rate, the more solid the image looks on the screen—that is, the less it flickers.*

registers (p. 153, KQ 4.1) High-speed storage areas that temporarily store data during processing. Why it's important: *Registers may store a program instruction while it is being decoded, store data while it is being processed by the ALU, or store the results of a calculation.*

removable hard disks (p. 165, KQ 4.2) Also called *hard-disk cartridges;* one or two platters enclosed along with read/write heads in a hard plastic case, which is inserted into a microcomputer's cartridge drive. Typical capacity is 2 gigabytes. Two popular systems are Iomega's Jaz and SyQuest's SparQ. Why it's important: *These cartridges offer users greater storage capacity than do floppy disks but with the same portability.*

resolution (p. 180, KQ 4.4) Clarity or sharpness of display-screen images; the more pixels there are per square inch, the finer the level of detail attained. Resolution is expressed in terms of the formula horizontal pixels × vertical pixels. Each pixel can be assigned a color or a particular shade of gray. Standard resolutions are 640 × 480, 800 × 600, 1024 × 768, 1280 × 1024, and 1600 × 1200 pixels. Why it's important: *Users need to know what screen resolution is appropriate for their purposes.*

RISC (reduced instruction set computing) chips (p. 151, KQ 4.1) Type of chip in which the complexity of the microprocessor is reduced by reducing the amount of seldom-used instructions. Why it's important: *RISC chips are used mostly in workstations. As a result, workstations can work up to 10 times faster then most PCs. RISC chips have been used in many Macintosh computers since 1993.*

ROM (read-only memory) (p. 156, KQ 4.1) Memory chip that cannot be written on or erased by the computer user without special equipment. Why it's important: *ROM chips contain fixed start-up instructions. They are loaded, at the factory, with programs containing special instructions for basic computer operations, such as those that start the computer or put characters on the screen. These chips are nonvolatile; their contents are not lost when power to the computer is turned off.*

scanner (p. 174, KQ 4.3) Source-data input device that uses laser beams and reflected light to translate images of text, drawings, photos, and the like into digital form. Why it's important: *Scanners simplify the input of complex data. The images can be processed by a computer, displayed on a monitor, stored on a storage device, or communicated to another computer.*

SCSI (small computer system interface) port (p. 157, KQ 4.1) Interface that allows data to be transmitted in a "daisy chain" for up to 7 devices at speeds (32 bits at a time) higher than those possible with serial and parallel ports. Why it's important: *Among the devices that may be connected are external hard-disk drives, CD-ROM drives, scanners, and magnetic-tape backup units.*

secondary storage hardware (p. 161, KQ 4.2) Devices that permanently hold data and information as well as programs. Why it's important: *Secondary storage—as opposed to primary storage— is nonvolatile; that is, saved data and programs are permanent, or remain intact, when the power is turned off.*

sectors (p. 162, KQ 4.2) Wedge-shaped sections on a formatted diskette used for storage reference purposes. Why it's important: *When you save data from your computer to a diskette, the data is distributed by tracks and sectors on the disk. That is, the system software uses the point at which a sector intersects a track to reference the data location.*

sensor (p. 178, KQ 4.3) Input device that collects specific data directly from the environment and transmits it to a computer. Why it's important: *Although you are unlikely to see such input devices connected to a PC in an office, they exist all around us, often in nearly invisible form. Sensors can be used to detect all kinds of things: speed, movement, weight, pressure, temperature, humidity, wind, current, fog, gas, smoke, light, shapes, images, and so on.*

serial port (p. 157, KQ 4.1) Port for connecting a cable that will send bits one after the other, like cars on a one-lane highway. Why it's important: *Because bits must follow each other one at a time, a serial port is usually used to connect devices that do not require fast transmission of data, such as keyboard, mouse, monitors, and modems. It is also useful for sending data over a long distance.*

smart card (p. 169, KQ 4.2) Card similar to a credit card that contains a microprocessor and memory chip. Why it's important: *When inserted into a reader, a smart card transfers data to and from a central computer, and it can store some basic financial records. Smart cards can be used as telephone debit cards. You insert the card into a slot in the phone, wait for a tone, and dial the number. The time of your call is automatically calculated on the chip inside the card and deducted from the balance. Many colleges and universities issue student cards as smart cards.*

softcopy (p. 179, KQ 4.4) Refers to data that is shown on a display screen or is in audio or voice form. This kind of output is not tangible; it cannot be touched. Why it's important: *This term is used to distinguish nonprinted output from printed (hardcopy) output.*

sound card (p. 160, KQ 4.1) Circuit board used to transmit digital sounds through speakers, microphones, and headsets. Why it's important: *Sound cards come installed on most new PCs. Cards such as PCI wavetable sound cards are used to add music and sound effects to computer video games.*

sound-output devices (p. 185, KQ 4.4) Hardware that produces digitized sounds, ranging from beeps and chirps to music. Why it's important: *To use sound output, you need appropriate software and a sound card.*

source data-entry devices (p. 174, KQ 4.3) Non-keyboard data entry devices that create machine-readable data on magnetic media or paper or feed it directly into the computer's processor. Categories include scanning devices—imaging systems, bar-code readers, mark- and character-recognition devices, fax machines, audio-input devices, video and photographic input, voice-recognition systems, sensors, radio-frequency identification devices, and human-biology input devices. Why it's important: *Source-data entry devices lessen reliance on keyboards for data entry and can make data entry more accurate.*

SuperDisk (p. 163, KQ 4.2) Produced by Imation; disks with a capacity of 120 megabytes. The SuperDisk drive can also read standard 1.44-megabyte floppy disks, which Zip drives cannot do. Why it's important: *See floppy-disk cartridges.*

SVGA (super video graphics array) (p. 181, KQ 4.4) Graphics board standard that supports a resolution of 800 × 600 pixels, or variations, producing 16 million possible simultaneous colors. Why it's important: *SVGA is the most common standard used today.*

system clock (p. 152, KQ 4.1) Internal timing device that uses fixed vibrations from a quartz crystal to deliver a steady stream of digital pulses or "ticks" to the CPU. These ticks are called *cycles*. Why it's important: *Faster clock speeds will result in faster processing of data and execution of program instructions, as long as the computer's internal circuits can handle the increased speed.*

tape cartridge (p. 168, KQ 4.2) Modules resembling audio cassettes that contain tape in rectangular, plastic housings. There are three common types of drive tapes—QIC, DAT, and (most expensive) DLT. Why it's important: *On large computers, tapes are used on magnetic-tape units or reels, and in cartridges. On microcomputers, tape is used in the form of cassettes. Tape is used mainly for archiving purposes and backup.*

terabyte (T, TB) (p. 148, KQ 4.1) Approximately 1 trillion bytes (1,009,511,627,776 bytes); measure of storage capacity. Why it's important: *The storage capacities of some mainframes and supercomputers are expressed in terabytes.*

thermal printer (p. 184, KQ 4.4) Printer that uses colored waxes and heat to produce images by burning dots onto special paper. The colored wax sheets are not required for black-and-white output. Thermal printers are expensive, and they require expensive paper. Why it's important: *For people who want the highest-quality color printing available with a desktop printer, thermal printers are the answer.*

touchpad (p. 172, KQ 4.3) Input device; a small, flat surface over which you slide your finger, using the same movements as you would with a mouse. The cursor follows the movement of your finger. You "click" by tapping your finger on the pad's surface or by pressing buttons positioned close by the pad. Why it's important: *Touchpads let you control the cursor/pointer with your finger, and they require very little space to use. Most laptops have touchpads.*

touch screen (p. 173, KQ 4.3) Video display screen that has been sensitized to receive input from the touch of a finger. The screen is covered with a plastic layer, behind which are invisible beams of infrared light. Why it's important: *You can input requests for information by pressing on buttons or menus displayed. The answers to your requests are displayed as output in words or pictures on the screen. (There may also be sound.) You find touch screens in kiosks, ATMs, airport tourist directories, hotel TV screens (for guest checkout), and campus information kiosks making available everything from lists of coming events to (with proper ID and personal code) student financial-aid records and grades.*

trackball (p. 172, KQ 4.3) Movable ball, mounted on top of a stationary device, that can be rotated using your fingers or palm (it looks like the mouse turned upside down). Instead of moving the mouse around on the desktop, you move the trackball with the tips of your fingers. Why it's important: *Trackballs require less space to use than does a mouse.*

tracks (p. 162, KQ 4.2) The rings on a diskette along which data is recorded. Why it's important: *See sectors.*

transistor (p. 150, KQ 4.1) Tiny electronic device that acts as an on/off switch, switching between on and off millions of times per second. Why it's important: *Transistors are part of the microprocessor.*

Unicode (p. 148, KQ 4.1) Binary coding scheme that uses two bytes (16 bits) for each character, rather than one byte (8 bits). Why it's important: *Instead of the 256 character combinations of ASCII, Unicode can handle 65,536 character combinations. Thus it allows almost all the written languages of the world to be represented using a single character set.*

upgrading (p. 150, KQ 4.1) Refers to changing to newer, usually more powerful or sophisticated versions, such as a more powerful microprocessor or more memory chips. Why it's important: *Through upgrading, users can improve their computer systems without buying completely new ones.*

USB (universal serial bus) port (p. 157, KQ 4.1) Port that can theoretically connect up to 127 peripheral devices daisy-chained to one general-purpose port. Why it's important: *USB ports are useful for peripherals such as digital cameras, digital speakers, scanners, high-speed modems, and joysticks. The so-called USB hot plug or hot swappable allows such devices to be connected or disconnected even while the PC is running.*

video (p. 185, KQ 4.3) Photographic images played at 15–29 frames per second to give the appearance of full motion. Why it's important: *Video is input into a multimedia system using a video camera or VCR and, after editing, is output on a computer's display screen. Because video files can require a great deal of storage—a 3-minute video may require 1 gigabyte of storage—video is often compressed.*

videoconferencing (p. 186, KQ 4.4) Form of video output in which people in different geographical locations can have a meeting—can see and hear one another—using computers and communications. Why it's important: *Videoconferencing systems range from videophones to group conference rooms with cameras and multimedia equipment to desktop systems with small video cameras, microphones, and speakers. Many organizations use videoconferencing to take the place of face-to-face meetings.*

virtual memory (p. 156, KQ 4.1) Type of hard disk space that mimics primary storage (RAM). Why it's important: *When RAM space is limited, virtual memory allows users to run more software at once, provided the computer's CPU and operating system are equipped to use it. The system allocates some free disk space as an extension of RAM; that is, the computer swaps parts of the software program between the hard disk and RAM as needed.*

voice-output device (p. 185, KQ 4.4) Hardware that converts digital data into speech-like sounds. Why it's important: *You hear such forms of voice output on telephones ("Please hang up and dial your call again"), in soft-drink machines, in cars, in toys and games, and recently in mapping software for vehicle-navigation devices.*

voice-recognition system (p. 177, KQ 4.3) Input system that uses a microphone (or a telephone) as an input device and converts a person's speech into digital signals by comparing the electrical patterns produced by the speaker's voice with a set of prerecorded patterns stored in the computer. *Why it's important: Voice-recognition technology is useful for inputting situations in which people are unable to use their hands or need their hands free for other purposes.*

volatile (p. 155, KQ 4.1) Temporary storage, as in RAM; the contents are lost when the power is turned off. *Why it's important: Save your work to a secondary-storage medium, such as a hard disk, in case the electricity goes off while you're working.*

word size (p. 153, KQ 4.1) Number of bits that the processor may process at any one time. *Why it's important: The more bits in a word, the faster the computer. A 32-bit computer—that is, one with a 32-bit-word processor—will transfer data within each microprocessor chip in 32-bit chunks, or 4 bytes at a time.*

write (p. 156, KQ 4.1) To transfer data from the computer's CPU or memory to an output device. *Why it's important: See read.*

write-protect notch (p. 162, KQ 4.2) Floppy disk feature that prevents a diskette from being written to. *Why it's important: This feature allows you to protect the data already on the disk. To write-protect, use your thumbnail or the tip of a pen to move the small sliding tab on the lower right side of the disk (viewed from the back), thereby uncovering the square hole.*

XGA (extended graphics array) (p. 181, KQ 4.4) Graphics board display standard; has a resolution of up to 1024 × 768 pixels, with 65,536 possible colors. It is used mainly on workstation systems, such as those employed by engineering designers. *Why it's important: XGA offers the most sophisticated standard for color and resolution.*

Zip disk (p. 163, KQ 4.2) Floppy disk cartridge with a capacity of 100 or 250 megabytes. At 100 megabytes, this is 70 times the storage capacity of the standard floppy. *Why it's important: Among other uses, Zip disks are used to store large spreadsheet files, database files, image files, multimedia presentation files, and Web sites. Zip disks require their own Zip disk drives, which may come installed on new computers, although external Zip drives are also available. See also floppy-disk cartridges.*

Chapter Review

Self-Test Questions

1. A(n) _kilobyte_ is about 1000 bytes; a(n) _megabyte_ is about 1 million bytes.

2. The _control unit_ is the part of the microprocessor that tells the rest of the computer how to carry out a program's instructions.

3. A(n) _dumb_ terminal is entirely dependent for all its processing activities on the computer system to which it is connected.

4. The two main categories of printer are _impact_ and _nonimpact_.

5. The process of retrieving data from a storage device is referred to as _read_; the process of copying data to a storage device is called _writing_. ? 6

6. To avoid losing data, users should always _____ their files.

7. Formatted diskettes have _tracks_ and _sectors_ that the system software uses to reference data locations.

8. The _CPU_ is often referred to as the "brain" of a computer.

9. _ergonomics_ is the study of the physical relationships between people and their work environment.

10. A(n) _mouse_ is an input device that is rolled about on a desktop and directs a pointer on the computer's display screen.

Multiple-Choice Questions

1. Which of the following is another term for *primary storage*?
   a. ROM
   b. ALU
   c. CPU
   d. RAM ⟵
   e. CD-R

2. Which of the following is not included on a computer's motherboard?
   a. RAM chips
   b. ROM chips
   c. keyboard ⟵
   d. microprocessor
   e. expansion slots

3. Which of the following is not a pointing device?
   a. mouse
   b. touchpad
   c. keyboard ⟵

4. Which of the following is used to hold data and instructions that will be used shortly by the CPU?
   a. ROM chips
   b. peripheral devices
   c. RAM chips ⟵
   d. CD-R
   e. hard disk

5. Which of the following coding schemes is widely used on microcomputers?
   a. EBCDIC
   b. Unicode
   c. ASCII ⟵
   d. Microcode
   e. Unix

6. Which of the following is used to measure processing speed in microcomputers?
   a. MIPS
   b. flops
   c. picoseconds
   d. megahertz ⟵
   e. millihertz

True/False Questions

T **F** ⟵ 1. A bus connects a computer's control unit and ALU.

T F ⟵ 2. The machine cycle comprises the instruction cycle and the execution cycle.

T F ⟵ 3. On a computer screen, the more pixels that appear per square inch, the higher the resolution.

T F ⟵ 4. Magnetic tape is the most common secondary storage medium used with microcomputers.

T **F** ⟵ 5. Main memory is nonvolatile.

T F ⟵ 6. Photos taken with a digital camera can be downloaded to a computer's hard disk.

Short-Answer Questions

1. What is ASCII, and what do the letters stand for?
2. Why should measures of capacity matter to computer users?
3. What's the difference between RAM and ROM?
4. What is the significance of the term *megahertz*?
5. What is a motherboard? Name at least four components of a motherboard.
6. What determines how a keyboard's function keys work?
7. What characteristics determine the clarity of a computer screen?
8. Describe two situations in which scanning is useful.
9. What is source-data entry?
10. What advantage does a floppy-disk cartridge have over a regular floppy disk?
11. Why is it important for your computer to be expandable?
12. What is *pixel* short for? What is a pixel?

Concept Mapping

On a separate sheet of paper, draw a concept map, or visual diagram, linking concepts. Show how the following terms are related.

audio-input device	megahertz
bay	microprocessor
bus	mouse
CD-ROM	optical disk
CPU	output hardware
display screen	page description
dpi	language
expansion card	pixel
expansion slot	port
fax modem	RAM
flat-panel display	read/write head
floppy disk	resolution
gigabyte	SCSI
hardcopy	sector
hard disk	softcopy
ink-jet printer	source-data entry
input hardware	touchpad
keyboard	track
kilobyte	USB
laser printer	XGA
LCD	Zip disk
megabyte	

Knowledge in Action

1. If you're using Windows 95 or 98, you can easily determine what microprocessor is in your computer and how much RAM it has. To begin, click the Start button in the Windows desktop pull-up menu bar and then choose Settings, Control Panel. Then locate the System icon in the Control Panel window and double-click on the icon.

The System Properties dialog box will open, which contains four tabs: General, Device Manager, Hardware Profiles, and Performance. The name of your computer's microprocessor will display on the General tab. To see how much RAM is in your computer, click the Performance tab.

2. The Blue Mountain supercomputer is one of the most powerful computers in the world. Use an Internet search engine such as *www.yahoo.com* or *www.dogpile.com* to locate information on this machine. Determine why the computer was developed, how much RAM and how many processors the computer has, and what kind of secondary storage it uses.

3. The objective of this project is to introduce you to an online encyclopedia that's dedicated to computer technology. The *www.webopaedia* Web site is a good resource for deciphering computer ads and clearing up difficult concepts. For practice, visit the site and type "main memory" into the Search text box and then press the enter key. Print out the page that displays. Then locate information on other topics of interest to you.

4. Visit a local computer store and note the system requirements listed on five software packages. What are the requirements for: processor? RAM? operating system? available hard disk space? CD-ROM speed? Are there any output hardware requirements?

5. Visit a local electronics store or an online shopping site such as *www.beyond.com/hardware/printers.htm* and investigate five different types of printer for sale. Note the (a) type of printer, (b) price, and (c) resolution, as well as whether the printer is PC- or Mac-compatible. Click on "Specs" for additional information on each printer. Which printer would you choose? Why?

6. Cut out an advertisement from a newspaper or a magazine that features a new microcomputer system. Circle all the terms that are familiar to you now that you have read the first four chapters of this text. Define these terms on a separate sheet of paper. Is this computer expandable? How much does it cost? Is the monitor included in the price? A printer?

7. If you're using Windows 95 or 98, you can easily determine the storage capacity of your computer. Double-click the My Computer icon on the Windows desktop. Then right-click the hard disk icon ("C:"). From the mouse's right-click menu, choose Properties. Note your hard disk's capacity, amount of used space, and amount of free space.

8. Develop a binary system of your own. Use any two objects, states, or conditions, and encode the following statement: I am a rocket scientist.

9. *Paperless office* is a term that has been around for some time. However, the paperless office has not yet been achieved. Do you think the paperless office is a good idea? Do you think it's possible? Why do you think it has not yet been achieved?

Telecommunications

Networks & Communications— The "New Story" in Computing

Key Questions

You should be able to answer the following questions.

5.1   **From the Analog to the Digital Age** How do digital and analog data differ, and what does a modem do?

5.2   **The Practical Uses of Communications** What are some offerings of new telecommunications technology?

5.3   **Some Factors Affecting Data Communications** What important factors determine how data is transmitted?

5.4   **Networks** What are the benefits of networks, and what are their types, components, and variations?

5.5   **Cyberethics: Controversial Material, Censorship, & Privacy Issues** What are considerations to be aware of regarding cyberethics?

A Visual Overview of This Chapter

1 **From the Analog to the Digital Age.** Computers use **digital** signals, which presents information in a binary way. Most other phenomena, such as telephones and TV, use **analog** signals, which continuously vary in strength or quality. A **modem** converts digital signals into analog signals, so that computer signals can be sent over phone lines.

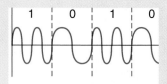

2 **The Practical Uses of Communications.** Connectivity, or communications connections, takes many forms. (1) The Internet offers research and news, games and other entertainment, software, travel services, shopping, and financial management, among other features. (2) The **Global Positioning System (GPS),** which consists of satellites that transmit timed radio signals that can be used to identify earth location, is useful for car, boat, and plane navigation. (3) **Videoconferencing,** the linking of people through TV video and sound plus computers, is useful for long-distance meetings. (4) **Workgroup computing,** the use of microcomputer networks to enable workers to cooperate on projects, allows people to work on the same information at the same time. (5) **Telecommuting,** working at home with telecommunications, can increase productivity. (6) **Virtual offices,** nonpermanent, mobile offices run with telecommunications and computers, add workplace flexibility. (7) Home networks enable households to link and share all kinds of peripheral devices. (8) Information appliances deliver all types of data anywhere at any time.

3 **Some Factors Affecting Data Communications.** The following affect how data is transmitted.
The **radio frequency spectrum,** fields of electrical and magnetic energy, carries communications signals, which vary according to frequency, or repeating waves. A range of frequencies is called a **band** or **bandwidth.** The wider the band, the faster data can be transmitted. **Broadband** connections are very high speed.

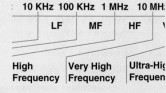

A **communications channel** is the path over which information travels in a telecommunications system. Channels may be wired or wireless.

Three types of *wired channels* are the following. **Twisted-pair wire,** or standard telephone wire, consists of two strands of insulated copper wire twisted around each other; it is used for both voice and data transmission. **Coaxial cable** consists of insulated copper wire wrapped in other materials; it is better than twisted pair for resisting noise. **Fiber-optic cable** consists of thin strands of glass or plastic that transmit beams of light rather than electricity; it is very fast and noise-resistant.

Four types of *wireless channels* are the following. **Infrared transmission** sends data via infrared-light waves, as is done with some wireless mouses. **Broadcast radio** sends data over long distances, as between states. **Microwave radio** transmits voice and data via super-high-frequency radio waves, as between hilltops. **Communications satellites** are microwave relay stations that orbit the earth, occupying low, medium, or high (geostationary) earth orbits.

Compression is a method of removing repetitive elements from a file so that it requires less time to transmit; at the receiving end, the file is *decompressed*—the repeated patterns are restored. Two methods of compression are lossless and lossy. *Lossless compression* uses mathematical techniques to replace repetitive patterns of bits with a kind of coded summary; on decompression, the bits are restored, so that the data is the same as what went in—important in database and similar computer data. *Lossy compression* permanently discards some data during compression; it is often used for graphics files and sound files. Two compression standards are **JPEG,** used for still images, and **MPEG,** used for moving images.

4 **Networks.** A communications **network** is a system of interconnected computers, phones, or other communications devices that can share applications and data. Among the benefits: Networks enable sharing of peripheral devices, programs, and data; better communications; improved security of information; and access to numerous databases.

Types of networks are as follows. A **wide area network (WAN)** covers a wide geographical area, such as a country. A **metropolitan area network (MAN)** covers a city or suburb. A **local area network (LAN)** covers a limited area such as an office or a building. Most large networks have a **host computer,** a mainframe or midsize central computer to control the network. Any device attached to a network is called a **node.** MANs and LANs may be connected to the Internet by a high-speed network called a **backbone.**

Two types of LANs are client/server and peer-to-peer. A **client/server LAN** consists of microcomputers requesting data *(clients)* and powerful computers supplying data *(servers).* A **file server**, for example, stores programs and data files; other servers are database server, printer server, Web server, and mail server. In a **peer-to-peer LAN,** there is no server; microcomputers on a network communicate with each other directly. Several standard components of a LAN are the connection or cabling system, microcomputers with network interface cards, network operating system (Novell NetWare, Microsoft Windows NT/2000, Unix, or Linux), and other shared devices (printers, storage devices). Other components are a **router,** a special computer that directs communicating messages when networks are tied together; a **bridge,** an interface used to connect the same types of networks; and a **gateway,** an interface permitting communication between dissimilar networks.

Organizations now use two variant networks that use the Internet's infrastructure and standards. One is an **intranet,** an organization's internal private network for employee use. The other is an **extranet,** for selected suppliers and other strategic parties as well as employees. Security for such networks is maintained through a **firewall,** a system of hardware and software that blocks unauthorized users inside and outside the organization.

5 **Cyberethics: Controversial Material, Censorship, & Privacy Issues.** Two important issues of cyberethics are as follows.

To protect children against access to controversial material, parents may employ blocking software that screens objectionable material based on keywords; browsers that contain built-in ratings for Internet and World Wide Web use; and the V-chip, which allows the screening of TV programs high in violence, sex, and the like.

Privacy, the right of people not to reveal information about themselves, is under pressure from information technology. Web **cookies,** files stored on a user's hard drive when he or she visits a Web site, allows Web site operators to track user movements online. Medical records and office e-mail are other areas in which there may be privacy intrusions.

The essence of all revolution, stated philosopher Hannah Arendt, is the start of a *new story* in human experience.

Before the 1950s, computing devices processed data into information, and communications devices communicated information over distances. The two streams of technology developed pretty much independently, like rails on a railroad track that never merge. Now we have a new story, a revolution.

For us, the new story has been *digital convergence*—the gradual merger of computing and communications into a new information environment, in which *the same information is exchanged among many kinds of equipment, using the language of computers. (See ● Panel 5.1.)* At the same time, there has been a convergence of several important industries—computers, telecommunications, consumer electronics, entertainment, mass media—producing new electronic products that perform multiple functions.

An example is what's happening in television. *WebTV* consists of a set-top box powered by WebTV Networks, a subsidiary of Microsoft. The box makes a conventional television set function like three different devices: a television with satellite service, an Internet-linked computer, and an enhanced digital video recorder. The result of this convergence of technologies is that, besides watching TV programs, you can also interact with them in real time; for instance, you might participate in *Jeopardy* along with the contestants you're viewing on the screen.[1] Or you can order a Domino's pizza by pointing and clicking with a remote control.[2] Still to come, though delayed by birth pangs (expense and lack of common standards), is *digital television (DTV)* and the variant called *high-definition television (HDTV)*, which are supposed to deliver not only crisper images but also more channels and computer connections.[3]

● PANEL 5.1
Digital convergence—the fusion of computer and communications technologies
Today's new information environment came about gradually from the merger of two separate streams of technological development—computers and communications.

Computer Technology

1621 AD	1642	1833	1843
Slide rule invented (Edmund Gunther)	First mechanical adding machine (Blaise Pascal)	Babbage's difference engine (automatic calculator)	World's first computer programmer, Ada Lovelace, publishes her notes

Communications Technology

1562	1594	1639	1827	1835	1846	1866	1876	1888
First monthly newspaper (Italy)	First magazine (Germany)	First printing press in North America	Photographs on metal plates	Telegraph (first long-distance digital communication system)	High-speed printing	Trans-atlantic telegraph cable laid	Telephone invented	Radio waves identified

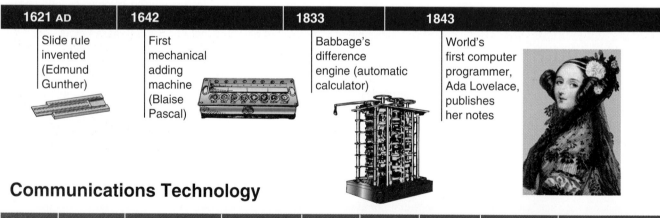

Why have the worlds of computers and of telecommunications been so long in coming together? Because *computers are digital, but most of the world has been analog.* Let's take a look at what this means.

The Digital Basis of Computers: Electrical Signals as Discontinuous Bursts

Computers may seem like incredibly complicated devices, but, as we saw in Chapter 4, their underlying principle is simple. Because they are based on on/off electrical states, they use the *binary system,* which consists of only two digits—0 and 1. Today **_digital_ specifically refers to communications signals or information represented in a two-state (binary) way.** More generally, **digital** is usually synonymous with "computer-based."

Digital data consists of data (expressed as 0s and 1s) represented by on/off electrical pulses. These pulses are transmitted in discontinuous bursts rather than (as with analog devices) in continuous waves.

The Analog Basis of Life: Electrical Signals as Continuous Waves

"The shades of a sunset, the flight of a bird, or the voice of a singer would seem to defy the black or white simplicity of binary representation," points out one writer.[4] Indeed, these and most other phenomena of the world are **_analog_, continuously varying in strength and/or quantity.** Sound, light, temperature, and pressure values, for instance, can fall anywhere on a continuum or range. The highs, lows, and in-between states have historically been represented with analog devices rather than in digital form. Examples of analog devices are a speedometer, a thermometer, and a tire-pressure gauge, all

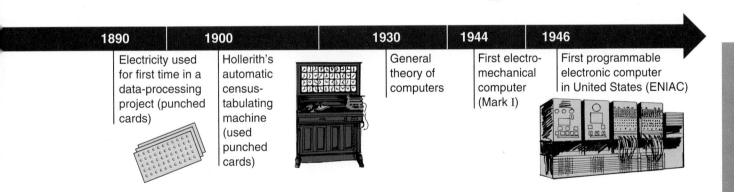

1890	1900		1930	1944	1946
Electricity used for first time in a data-processing project (punched cards)	Hollerith's automatic census-tabulating machine (used punched cards)		General theory of computers	First electro-mechanical computer (Mark I)	First programmable electronic computer in United States (ENIAC)

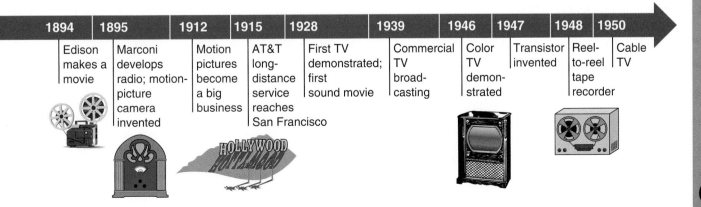

1894	1895	1912	1915	1928	1939	1946	1947	1948	1950
Edison makes a movie	Marconi develops radio; motion-picture camera invented	Motion pictures become a big business	AT&T long-distance service reaches San Francisco	First TV demonstrated; first sound movie	Commercial TV broad-casting	Color TV demon-strated	Transistor invented	Reel-to-reel tape recorder	Cable TV

of which can measure continuous fluctuations. The electrical signals on a telephone line, for instance, have traditionally been analog-data representations of the original voices.

Thus, *analog data* is transmitted in a continuous form—a continuous electrical signal in the shape of a wave (called a *carrier wave*). Telephone, radio, television, and cable-TV technologies have long been based on analog data.

Purpose of the Modem: Converting Digital Signals to Analog Signals & Back

To understand the differences between digital and analog transmission, look at a graphic representation of an on/off digital signal emitted from a computer. Like a regular light switch, this signal has only two states—on and off. Compare this with a graphic representation of a wavy analog signal emitted as a signal. The changes in this signal are gradual, as in a dimmer switch, which gradually increases or decreases brightness.

Because telephone lines have traditionally been analog, you need to have a *modem* if your computer is to send communications signals over a telephone line. The modem translates the computer's digital signals into the telephone line's analog signals. The receiving computer also needs a modem to translate the analog signals back into digital signals.

<u>Modem</u> is short for "*modulate/demodulate*"; a sending modem modulates digital signals into analog signals for transmission over phone lines. A receiving modem demodulates the analog signals back into digital signals. The modem provides a means for computers to communicate with one another using the standard copper-wire telephone network, an analog system that was built to transmit the human voice but not computer signals. *(See ● Panel 5.2.)*

How, in fact, does a modem convert the continuous analog wave to a dis-

1952	1963	1964	1967	1969	1970	1971	1975	1977	1978
UNIVAC computer correctly predicts election of Eisenhower as U.S. President	BASIC developed at Dartmouth	IBM introduces 360 line of computers	Hand-held calculator	ARPA-Net established, led to Internet	Micro-processor chips come into use; floppy disk introduced for storing data	First pocket calculator	First micro-computer (MITs Altair 8800)	Apple II computer (first personal computer sold in assembled form)	5¼" floppy disk; Atari home videogame

Fusing of computer and communications lines of development

1952	1957	1961	1968	1975	1976	1977	1979	1982
Direct-distance dialing (no need to go through operator); transistor radio introduced	First satellite launched (Russia's Sputnik)	Push-button telephones	Portable video recorders; video cassettes	Flat-screen TV	First wide-scale marketing of TV computer games (Atari)	First inter-active cable TV	3-D TV demonstrated	Compact disks; European consortium launches multiple communications satellites

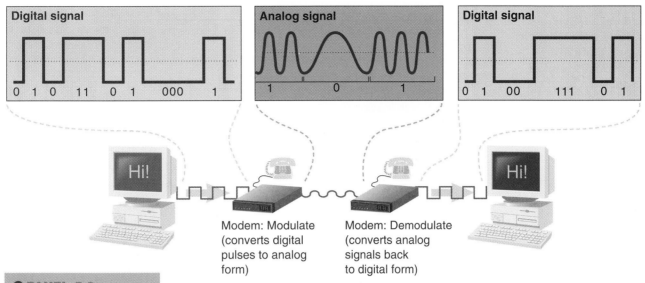

Digital signal	Analog signal	Digital signal
0 1 0 11 0 1 000 1	1 0 1	0 1 00 111 0 1

Modem: Modulate
(converts digital
pulses to analog
form)

Modem: Demodulate
(converts analog
signals back
to digital form)

● PANEL 5.2

Analog versus digital signals, and the modem

Note that an analog signal represents a continuous electrical signal in the form of a wave. A digital signal is discontinuous, expressed as discrete bursts in on/off electrical pulses.

continuous digital pulse that can represent 0s and 1s? The modem can make adjustments to either the *frequency*—the number of cycles per second, or the number of times a wave repeats during a specific time interval (the fastness/slowness)—or it can make adjustments to the analog signal's *amplitude*—the height of the wave (the loudness/softness). Thus, in frequency, a slow wave might represent a 0 and a quick wave might represent a 1. In amplitude, a low wave might represent a 0 and a high wave might represent a 1. *(See ● Panel 5.3, next page.)*

Our concern goes far beyond telephone transmission. How can the analog realities of the world be expressed in digital form? How can light, sounds,

Computer Technology

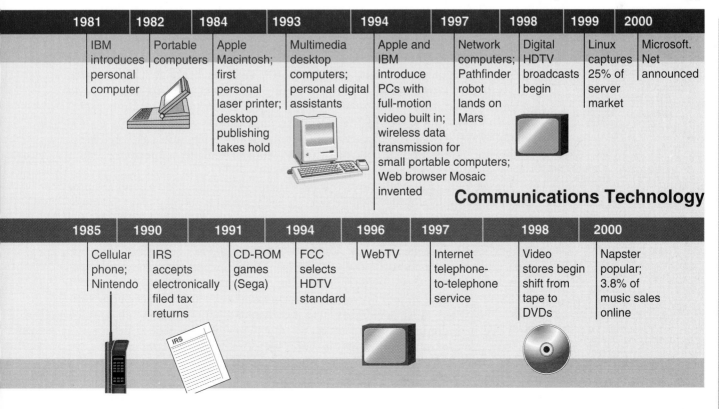

1981	1982	1984	1993	1994	1997	1998	1999	2000
IBM introduces personal computer	Portable computers	Apple Macintosh; first personal laser printer; desktop publishing takes hold	Multimedia desktop computers; personal digital assistants	Apple and IBM introduce PCs with full-motion video built in; wireless data transmission for small portable computers; Web browser Mosaic invented	Network computers; Pathfinder robot lands on Mars	Digital HDTV broadcasts begin	Linux captures 25% of server market	Microsoft. Net announced

Communications Technology

1985	1990	1991	1994	1996	1997	1998	2000
Cellular phone; Nintendo	IRS accepts electronically filed tax returns	CD-ROM games (Sega)	FCC selects HDTV standard	WebTV	Internet telephone-to-telephone service	Video stores begin shift from tape to DVDs	Napster popular; 3.8% of music sales online

The continuous, even cycle of an analog wave...

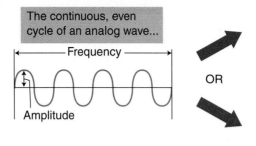

Frequency

Amplitude

OR

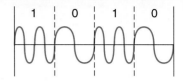

... is converted to digital form through *frequency modulation*—the frequency of the cycle increases to represent a 1 and stays the same to represent a 0.

1 0 1 0

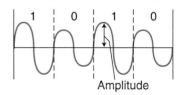

... or is converted to digital form through *amplitude modulation*—the height of the wave is increased to represent a 1 and stays the same to represent a 0.

1 0 1 0

Amplitude

colors, temperatures, and other dynamic values be represented so that they can be manipulated by a computer? Let us consider this.

Converting Reality to Digital Form

Suppose you are using an analog tape recorder to record a singer during a performance. The analog process will produce a near duplicate of the sounds. This will include distortions, such as buzzings and clicks, or electronic hums if an amplified guitar is used.

The digital recording process is different. The way in which music is captured for audio CDs does not provide a duplicate of a musical performance. Rather, the digital process uses *representative selections (samples)* to record the sounds. The copy obtained is virtually exact and free from distortion and noise. Computer-based equipment takes samples of sounds at regular intervals—nearly 44,100 times a second. The samples are converted to numbers that the computer then uses to express the sounds. The sample rate of 44,100 times per second and the high precision fool our ears into hearing a smooth, continuous sound. Similarly, for visual material, a computer can take samples of values such as brightness and color. The same is true of other aspects of real-life experience, such as pressure, temperature, and motion.

Are we being cheated out of our experience of "reality" by allowing computers to sample sounds, images, and so on? Actually, people willingly made this compromise years ago, before computers were invented. Movies, for instance, carve up reality into 24 frames a second. Television frames are drawn at 30 lines per second. These processes happen so quickly that our eyes and brains easily jump the visual gaps. Digital processing of analog experience is just one more way of expressing or translating reality.

Turning analog reality into digital form provides tremendous opportunities. One of the most important is that all kinds of multimedia can now be changed into digital form and transmitted as data to all kinds of devices.

Now let us examine the digital world of telecommunications.

Distinguish between digital and analog signals.

Explain what a modem does.

5.2 The Practical Uses of Communications

Guyana is the lone English-speaking country in South America. In the village of Lethem (population 2000), a collective of women from two native tribes has revived the ancient art of hand-weaving large hammocks. Thanks to donated computer equipment and Internet training for one member, plus free satellite-assisted online access, the collective sold 17 hammocks around the world in 1999 through its Web site (*www.gol.net.gy/rweavers*). Though it might take 600 hours to make, a hammock could bring as much as $1000—a gigantic sum in this area.[5]

Such is the power of communications connections, or *connectivity*, which gives us all instant, around-the-clock information from all over the globe. Let us consider some of the forms this connectivity takes: Internet offerings, the global positioning system (GPS), videoconferencing and videophones, workgroup computing and groupware, telecommuting and virtual offices, home networks, and information appliances.

Internet Offerings

Earlier (Chapter 2) we discussed such Internet features as e-mail, telephony, chat rooms, and bulletin boards. We also discussed e-commerce and will do so again in Chapter 6. But other offerings abound. Examples:

Electronic auction

In Alsmeer, Netherlands, an early-morning flower auction takes place. Buyers bid via electronic network. Within three hours, 17 million flowers will have been snapped up. By noon, the flowers will be on jets, bound for shops around the world.

- **Research and news:** The information resources available online are mind-boggling. Indeed, the only restriction on how much research you can do online is the limit on your credit card, if you're using fee (rather than free) databases. You can also avail yourself of encyclopedias and other reference works. Many databases offer access, for a fee, to unabridged text from newspapers, journals, and magazines. As a result, a third of the public now goes for news at least once a week, although the credibility of Internet news sources varies widely.[6]

- **Games and other entertainment:** Online computer games are extremely popular. In single-player games, you play against the computer. In multiplayer games, you play against others, whether someone in your household or someone overseas. Other entertainments include cartoons, sound clips, pictures of show-business celebrities, and reviews of movies and CDs. You can also join online clubs with others who share your interests, whether science fiction, popular music, or cooking. Two software programs, Napster and Gnutella, which allow file sharing among thousands of computers around the world, have made it easy to get music over the Internet at no charge.[7]

- **Software:** Many users download freeware, shareware, and commercial demonstration programs from online sources. They can also download software updates (called *patches*).

- **Travel services:** Services such as Eaasy Sabre or Travelshopper are streamlined versions of the reservations systems travel agents use. You can search for flights and book reservations through the computer and have tickets sent to you by Federal Express. You can also view hotel and restaurant guides, such as the Zagat Restaurant Directory. If you're going to visit a friend during semester break, you may find it beneficial to print out maps and driving directions from a free source such as MapQuest (*www.Mapquest.com*), Map-It (*www.map-it.com*), or Yahoo! (*www.yahoo.com*).

- **Shopping:** If you can't stand parking hassles, limited store hours, and checkout lines, online services may provide a shopping alternative for everything from prescription drugs to furniture. In some cities it's even possible to order groceries through online services. Peapod Inc. is an online grocery service serving 10,000 households in the Chicago and San Francisco areas. Peapod offers more than 18,000 items, from laundry detergent to lettuce. The orders are delivered in temperature-controlled containers to a designated location, such as a lockbox or garage.

- **Financial management:** The Internet offers access to investment brokerages so that you can invest money and keep tabs on your portfolio and on the stock market. Banks and credit unions also offer online services. Many organizations allow bill paying online.

Overnight delivery
Online shopping has been a boon to package delivery services.

The Global Positioning System

A $10 billion infrastructure developed by the military in the mid-1980s, the ***Global Positioning System (GPS)*** **consists of a series of earth-orbiting satellites continuously transmitting timed radio signals that can be used to identify earth locations.** A GPS receiver—handheld or mounted in a vehicle, plane, or boat—can pick up transmissions from any of four satellites, interpret the information from each, and calculate to within a few hundred feet or less the receiver's longitude, latitude, and altitude. Some GPS receivers include map software for finding your way around, as with the Guidestar system available with some rental cars.

GPS
This car-navigation system is guided by GPS.

Videoconferencing & Videophones: Video/Voice Communication

"I was a little nervous about going in front of the camera," said job applicant Mark Dillard, "but I calmed down pretty quickly after we got going, and it went well."[8]

Interviewing for a job can be uncomfortable for many people. However, Dillard had just undergone a high-tech interview. He had talked to a job recruiter in New York while sitting in front of a video camera in a booth at a local Kinko's store in Atlanta.

Videoconferencing, **also called teleconferencing, is the use of television video and sound technology as well as computers to enable people in different locations to see, hear, and talk with one another.** For a videoconference, people may go to conference rooms or booths with specially equipped television cameras. Alternatively, videoconferencing equipment can be set up on people's desks, with a camera and microphone to capture the person speaking and a monitor and speakers for the person being spoken to. The *videophone* is a telephone with a TV-like screen and a built-in camera that allows you to see the person you're calling, and vice versa.

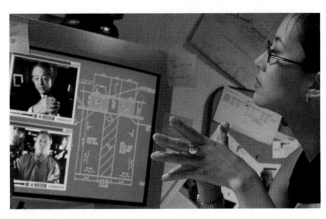

Videoconferencing
Three people hold a virtual meeting to discuss a diagram.

The main difficulty with videoconferencing and videophones is POTS ("plain old telephone service") equipment based on standard copper wire, which cannot transmit or receive images very rapidly. Thus, unless you can afford expensive high-speed communications lines, present-day screens will convey a series of jerky, stop-action images of the participants' faces.

Workgroup Computing & Groupware

When microcomputers were first brought into the workplace, they were used simply as another personal-productivity tool, like typewriters or calculators. Gradually, however, companies began to link microcomputers together on a network, usually to share an expensive piece of hardware, such as a laser printer. Then employees found that networks allowed them to share files and databases. Networking using common software also allowed users to buy equipment from different manufacturers—a mix of computers from both Sun Microsystems and Hewlett-Packard, for example. Sharing resources has led to workgroup computing.

Workgroup computing, **also called collaborative computing, enables teams of co-workers to use networks of microcomputers to share information and to cooperate on projects.** Workgroup computing is made possible not only by networks and microcomputers but also by *groupware*. As we stated in Chapter 3, *groupware* is software that allows two or more people on a network to work on the same information at the same time.

In general, groupware, such as Lotus Notes, permits office workers to collaborate with colleagues and to tap into company information through computer networks. It also enables them to link up with crucial contacts outside their organization.

Telecommuter
Work-at-home employment is becoming more commonplace.

Telecommuting & Virtual Offices

Computers and communications tools have led to telecommuting and virtual offices.

- Telecommuting: **Working at home while in telecommunication with the office is called** *telecommuting*. The number of U.S. households in which someone telecommutes—works at home at least three days a month—rose to 10.7 million in 1999 and is expected to hit 13.9 million in 2002.[9] In one study, 41% of workers surveyed said they believe they could perform part of their work as telecommuters and would like to do so. Only 9% are doing so now.[10]

Telecommuting has many benefits. The advantages to society are reduced traffic congestion, energy consumption, and air pollution. What are the advantages to employers? Productivity increases, because telecommuters experience fewer distractions at home and can work flexible hours. Teamwork improves, and the labor pool is expanded, because hard-to-get employees don't have to uproot themselves from where they want to live. A disadvantage is that people may feel isolated.

A related term is *telework,* which includes not only those who work at least part time from home but also those who work at remote or satellite offices, removed from organizations' main offices. Such satellite offices are sometimes called *telework centers.*

- **Virtual offices:** The ***virtual office*** **is an often nonpermanent and mobile office run with computer and communications technology.** Employees work from their homes, cars, and other new work sites. They use pocket pagers, portable computers, fax machines, and various phone and network services to conduct business.

  Could you stand not having a permanent office at all? Here's how one variant called "hoteling" works: You call ahead to book a room and speak to the concierge. However, the "hotel" isn't a Hilton but an organization such as advertising agency Chiat/Day. The concierge is an administrator who handles the scheduling of office cubicles. There are fewer cubicles than there are employees, who are often out in the field. When you check in, you pick up your personal effects and files from your locker and take them to the cubicle you will use for the next few days or weeks. This system depends on computers, which facilitate cubicle scheduling and the reprogramming of phones. Laptops allow employees to carry their work around with them, stored on their hard drives. Cellular phones, fax machines, and e-mail permit employees to stay in touch with supervisors and coworkers.

The rise in telecommuting and virtual offices is only part of a larger trend. "Powerful economic forces are turning the whole labor force into an army of freelancers—temps, contingents, and independent contractors and consultants," says business strategy consultant David Kline. The result, he believes, is that "computers, the Net, and telecommuting systems will become as central to the conduct of 21st-century business as the automobile, freeways, and corporate parking lots were to the conduct of mid-20th-century business."[11] The transformation will really gather momentum, some observers feel, when homes have broadband Internet access (as explained shortly).[12]

Home Networks

As we shall see, computers linked by telephone lines, cable, or wireless systems are an established component of information technology. Today, however, many new buildings and even homes are built as "network enabled." The new superconnected home or small office is equipped with a *local area network (LAN),* which allows all the personal computers under the same roof to share peripherals (such as a printer or a fax machine) and a single modem and Internet service.[13] The next development is supposed to be the networking of home appliances, linking stereos, lights, heating systems, phones, and TV sets. Once that has been accomplished, you could walk into your house and give a voice command to turn on the lights or bring up music, for example. Even kitchen appliances would be linked, so that your refrigerator, for instance, could alert a grocery store that you need more milk.[14]

The Information/Internet Appliance

An *information appliance* is a device merging computing capabilities with communications gadgets. Examples include TV set-top boxes, Internet phones, and personal digital assistants. Especially as cable and wireless channels become speedier, these devices will offer the ability to deliver all types of data—text, audio, video, film, still pictures—anywhere at any time. The various types of information appliance are discussed in the box on the next page.

CONCEPT CHECK

What are some of the offerings of the Internet?

What is a Global Positioning System?

Explain videoconferencing, workgroup computing, telecommuting and virtual offices, and home networks.

Describe the information appliance.

5.3 Some Factors Affecting Data Communications

KEY QUESTION
What important factors determine how data is transmitted?

Several factors affect how data is transmitted. Here we discuss the radio spectrum and bandwidth, wired and wireless communications channels, and also data compression and decompression.

The Radio Spectrum & Bandwidth

Telephone signals, radar waves, and the invisible commands from a garage-door opener all represent different waves on what is called the electromagnetic spectrum. In the middle of this spectrum is the radio frequency spectrum. The **radio frequency spectrum consists of fields of electrical energy and magnetic energy, which travel in waves. These fields carry communications signals.** *(See ● Panel 5.4.)* The waves vary according to frequency—

● **PANEL 5.4**
The radio spectrum

Frequency	0	10 Hz	100 Hz	1 KHz	10 KHz	100 KHz	1 MHz	10 MHz	100 MHz	1 GHz	10 GHz	100 GHz
Band Designation			VLF			LF	MF	HF	VHF	UHF	SHF	EHF

| Very Low Frequency | Low Frequency 30–300 KHz: Marine and aeronautical navigation equipment | Medium Frequency 300–3000 KHz (3 MHz): AM radio broadcast; LORAN maritime navigation; long-distance aeronautical and maritime navigation | High Frequency 3 MHz– 30 MHz: Shortwave broadcast; amateur radio; CB (citizen band) radio | Very High Frequency 30 MHz– 300 MHz: Private radio land mobile services such as police, fire, and taxi dispatch; TV channels (2–13); FM broadcasting; cordless phones; baby monitors | Ultra-High Frequency 300 MHz– 3000 MHz (3 GHz): UHF TV channels; cellular phones; common carrier point-to-point microwave transmission used by long-distance phone companies | Super High Frequency 3 GHz and above: Radar, microwave, and satellite transmission | Extreme High Frequency |

PRACTICAL ACTION BOX

Information Appliances: All Kinds of Smart Gadgets for All Kinds of Digital Data

The information appliance is a specialized computer that does just a few things, such as traversing the Internet or exchanging e-mail. Here are some examples.

- **Internet appliances:** Oracle Corp. unveiled the concept of an *Internet appliance* in 1996. This device—also called a "network computer," or NC—is really just an Internet terminal, consisting of a monitor and a keyboard. The newest versions are the i-Opener (from Netpliance), the NetVista Internet Appliance (IBM), the MSN Web Companion (Compaq), the Qubit (Qubit Technology), and the i-Station (Acer America).[a] Some of these come with factory-set function keys that put weather, news, e-mail, and e-commerce services at your fingertips. For instance, there may be a button that allows you to order pizza from a local pizza parlor.

  Experts say Internet appliances represent an important transition for computers. After two decades of "faster and more powerful" PCs, Net appliances signal a major shift to "simpler and cheaper." "While it's gotten simpler, the typical PC is still complex for a lot of people," says the director of marketing for IBM's personal systems group. "The Internet appliance is all about connectivity. You plug it in, turn it on, and you're connected."[b]

- **Internet phones:** *Internet phones,* also called *Web phones,* offer wireless access to the Internet. Their primary drawback at present is the limited amount of information they can access and the limited number of Web sites they are able to log on to. On most of these devices, the screens display only 4–11 lines of information. Moreover, they can't show graphics. However, a number of companies (Yahoo!, ABCNews.com, and others) have adapted a format called *Wireless Application Protocol,* which requires sites to strip out graphics and shorten stories. Other drawbacks are the small area of the keypad and the slow procedure for typing in letters and numbers.[c]

  Among the portable Internet phones available are PDQ 800 Smart Phone (Qualcomm), Timeport (Motorola), and Touchpoint (Denso). A desktop phone, iPhone (Infogear), combines e-mail, voice mail, and Internet access in a touch-screen device.

- **Set-top boxes:** The *set-top box* is a keypad that allows TV viewers to change channels or, in the case of interactive systems, to display Web pages on a TV set. For example, WebTV, offered by Microsoft, allows you to interact with the Internet and World Wide Web through your television set.

  Incidentally, experts differentiate between *Internet TV,* such as the services offered by WebTV and Liberate, which allow Internet access through your TV set, and *interactive TV,* such as that offered by OpenTV, which allows you to interact with the show you're watching (for example, *Jeopardy*). These are also distinct from *personalized TV,* or *personal video recorders (PVRs),* such as the stand-alone boxes offered by TiVo and Replay TV, which allow viewers to pause, rewind, and replay live TV and to record shows.[d]

the number of times a wave repeats, or makes a cycle, in a second. The radio spectrum ranges from low-frequency waves, such as those used for aeronautical and marine navigation equipment, through the medium frequencies for CB radios, cordless phones, and baby monitors, to ultrahigh frequency bands for cell phones and also microwave bands for communications satellites.

A range of frequencies is called a <u>band</u> or <u>bandwidth</u>. Bandwidth is a measure of the amount of information that can be delivered within a given period of time. For analog signals, bandwidth is expressed in hertz (Hz), or cycles per second. For digital signals, bandwidth is expressed in bits per second (bps). In the United States, certain bands are assigned by the Federal Communications Commission (FCC) for certain purposes—cell phones

- **PC/TVs:** The full-blown *PC/TV* (or TV/PC) merges the personal computer with the television set. A circuit board in your computer allows you to receive TV programs. Thus, for instance, you can watch TV shows in a window in one corner of the screen while you're working on your computer. A different kind of device is the iCE-BOX CounterTOP (from CMI), a cable-ready TV with Internet access that also features a built-in CD drive.

- **Online game players:** The *online game player* not only lets you play games but also connects to the Internet. The principal contenders on this battlefield are Sega, with its 200-megahertz Dreamcast, and Sony, with its 300-megahertz PlayStation 2. Microsoft is planning to enter the fray in late 2001 with its X-Box. Nintendo will offer Dolphin. The new game players far outstrip older models. The PlayStation 2, for instance, provides unparalleled animation details in video games, plays DVD movies, and offers high-speed Internet access and a hard drive to store data.[e]

- **Personal digital assistants:** A variety of devices known as electronic organizers, personal communicators, and handheld computers may be classified together as *personal digital assistants (PDAs)*. Examples are the Palm (from Palm Inc.), the Visor (Handspring), and the Pocket PC (Hewlett-Packard). Electronic organizers, such as the Palm, have a calendar, an address book, a calculator, and a stylus with which to write notes onscreen, along with a modem for e-mail connections. Handheld computers, such as the Pocket PC, have all these plus consumer versions of desktop-computer software. Some, such as the Visor, have a slot into which you can insert specially designed cartridges, turning the device into an e-book, MP3 player, camera, and game machine.[f]

- **Two-way pagers:** Motorola offers the PageWriter 2000X, a *two-way pager* with a miniature screen and a very small keyboard.[g] Besides two-way paging, the device can be used to send and receive e-mail.

- **E-mail appliances:** Various miniature *e-mail appliances* that allow you to access your e-mail from anywhere, with or without a PDA. One such device is the BackFlip (from PocketScience), a 5.5-ounce device that you connect to your Palm. If you hold the Palm and the BackFlip against a telephone's handset, you can dial a toll-free service that will forward your email. The 5.4-ounce Minstrel (Novatel Wireless) connects to a Palm device and lets you browse the Internet and send and receive e-mail.[h] Using the TelMail E-Mail Organizer (Sharp), which has a built-in acoustic modem, you can send and receive e-mail without a PDA.[i]

- **Short-range wireless toys:** Some playthings marketed to children may well be harbingers of new technologies for adults.[j] These *wireless toys* are low-cost devices capable of beaming voice and text messages to fellow students carrying similar devices, if they are within close range. Examples are Lightning Mail (Tiger Electronics, operates up to 50 feet); Quik Writer (Tiger Electronics, up to 90 feet); V-Mail (ToyBiz, up to 100 feet); and Cybikio (up to 300 feet). Some of these come with a miniature keyboard, some with a stylus for writing on a screen.

within one range, automated teller machines within another, broadcast television within yet another, and so on.

The bandwidth is the difference between the lowest and the highest frequencies transmitted. For example, cellular phones operate within the range 800–900 megahertz—that is, their bandwidth is 100 megahertz. *The wider the bandwidth, the faster data can be transmitted. The narrower the bandwidth, the greater the loss of transmission power.* This loss of power must be overcome by using relays or repeaters that rebroadcast the original signal. **<u>Broadband connections</u> are characterized by very high speed.** For instance, the connections that carry broadcast video range in bandwidth from 10 megabits to 30 gigabits per second.

Wired Communications Channels: Transmitting Data by Physical Means

A _communications channel_ is the path over which information travels in a telecommunications system from its source to its destination. Channels may be wired or wireless.

Three types of wired channels are twisted-pair wire (conventional telephone lines), coaxial cable, and fiber-optic cable.

Twisted-pair wire

Coaxial cable

Fiber-optic cable

- Twisted-pair wire (1–128 Mbps): The telephone line that runs from your house to the pole outside, or underground, is probably twisted-pair wire. _Twisted-pair wire_ **consists of two strands of insulated copper wire, twisted around each other. This twisted-pair somewhat reduces interference from electrical fields.** Twisted-pair is relatively slow. Moreover, it does not protect well against electrical interference. However, because so much of the world is already served by twisted-pair wire, it will no doubt be used for years to come, both for voice messages and for modem-transmitted computer data.

  The prevalence of twisted-pair wire gives rise to what experts call the "final mile problem." That is, it is relatively easy for telecommunications companies to upgrade the physical connections between cities and even between neighborhoods. But it is expensive for them to replace the "final mile" of twisted-pair wire that connects to individual houses.

- Coaxial cable (up to 200 Mbps): _Coaxial cable_, **commonly called "co-ax," consists of insulated copper wire wrapped in a solid or braided metal shield, then in an external cover.** Co-ax is widely used for cable television. Thanks to the extra insulation, coaxial cable is much better than twisted-pair wiring at resisting noise. Moreover, it can carry voice and data at a faster rate (up to 200 megabits per second). Often many coaxial cables will be bundled together.

- Fiber-optic cable (100 Mbps to 2 Gbps): A _fiber-optic cable_ **consists of dozens or hundreds of thin strands of glass or plastic that transmit pulsating beams of light rather than electricity.** These strands, each as thin as a human hair, can transmit up to 2 billion pulses per second (2 Gbps), each "on" pulse representing one bit. When bundled together, fiber-optic strands in a cable 0.12 inches thick can support a quarter- to a half-million voice conversations at the same time. Moreover, unlike electrical signals, light pulses are not affected by random electromagnetic interference in the environment. Thus, they have much lower error rates than normal telephone wire and cable. In addition, fiber-optic cable is lighter and more durable than twisted-pair and co-ax cable. A final advantage is that it cannot easily be wiretapped, so transmissions are more secure.

Wireless Communications Channels: Transmitting Data through the Air

Four types of wireless channels are infrared transmission, broadcast radio, microwave radio, and communications satellite.

- Infrared transmission (1–4 Mbps): _Infrared wireless transmission_ **sends data signals using infrared-light waves.** Infrared ports can be found on some laptop computers and printers, as well as wireless mouses. The drawbacks are that _line-of-sight_ communication is

Infrared access device

This is on a laser printer.

required—there must be an unobstructed view between transmitter and receiver—and transmission is confined to short range.

- **Broadcast radio (up to 2 Mbps):** When you tune in to an AM or FM radio station, you are using ___broadcast radio,___ **a wireless transmission medium that sends data over long distances—between regions, states, or countries.** A transmitter is required to send messages and a receiver to receive them; sometimes both sending and receiving functions are combined in a *transceiver.*

  In the lower frequencies of the radio spectrum, several broadcast radio bands are reserved not only for conventional AM/FM radio but also for broadcast television, CB (citizens band) radio, ham (amateur) radio, cellular phones, and private radio land mobile services (such as police, fire, and taxi dispatch). Some organizations use specific radio frequencies and networks to support wireless communications. For example, UPC (Universal Product Code) readers are used by grocery-store clerks restocking store shelves to communicate with a main computer so that the store can control inventory levels. We discuss this form of wireless transmission (including cellular phones) in more detail in a few pages.

- **Microwave radio (45 Mbps):** ___Microwave radio___ **transmits voice and data through the atmosphere as super-high-frequency radio waves called microwaves,** which vibrate at 1 gigahertz (1 billion hertz) per second or higher. These frequencies are used not only to operate microwave ovens but also to transmit messages between ground-based Earth stations and satellite communications systems.

  Nowadays dish- or horn-shaped microwave reflective dishes, which contain transceivers and antennas, are nearly everywhere—on towers, buildings, and hilltops. Why, you might wonder, do we have to interfere with nature by putting a microwave dish on top of a mountain? As with infrared waves, microwaves are line-of-sight; they cannot bend around corners or around the earth's curvature, so there must be an unobstructed view between transmitter and receiver. Thus, microwave stations need to be placed within 25–30 miles of each other, with no obstructions in between. The size of the dish varies with the distance (perhaps 2–4 feet in diameter for short distances, 10 feet or more for long distances). A string of microwave relay stations will each receive incoming messages, boost the signal strength, and relay the signal to the next station.

  More than half of today's telephone system uses dish microwave transmission. However, the airwaves are becoming so saturated with microwave signals that future needs will have to be satisfied by other channels, such as satellite systems.

Microwave radio

These dishes are on Midway Island, 1100 miles from Hawaii.

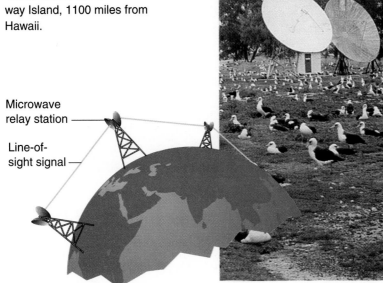

Microwave relay station

Line-of-sight signal

Communications satellite

• **Communications satellites:** To avoid some of the limitations of microwave earth stations, communications companies have added microwave "sky stations"—communications satellites. <u>*Communications satellites*</u> **are microwave relay stations in orbit around the earth.** Transmitting a signal from a ground station to a satellite is called *uplinking*; the reverse is called *downlinking*. The delivery process will be slowed if, as is often the case, more than one satellite is required to get the message delivered.

Satellite systems may occupy one of three zones in space: *GEO, MEO,* and *LEO.*

The highest level, known as *geostationary earth orbit (GEO),* is 22,300 miles up and is always directly above the equator. Because the satellites in this orbit travel at the same speed as the earth, they appear to an observer on the ground to be stationary in space—that is, they are *geostationary.* Consequently, microwave earth stations are always able to beam signals to a fixed location above. The orbiting satellite has solar-powered transceivers to receive the signals, amplify them, and retransmit them to another earth station. At this high orbit, fewer satellites are required for global coverage; however, their quarter-second delay makes two-way conversations difficult.

The *medium-earth orbit (MEO)* is 5000–10,000 miles up. It requires more satellites for global coverage than GEO.

The *low-earth orbit (LEO)* is 400–1000 miles up and has no signal delay. LEO satellites may be smaller and are cheaper to launch.

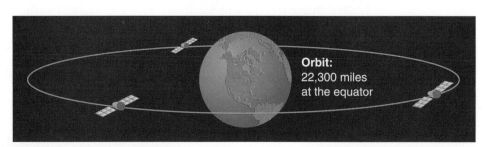

GEO

Three satellite orbits: GEO, MEO, LEO

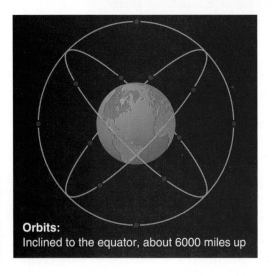

Orbits:
Inclined to the equator, about 6000 miles up

MEO   LEO

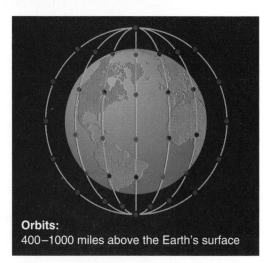

Orbits:
400–1000 miles above the Earth's surface

Compression & Decompression: Putting More Data in Less Space

The vast streams of text, audio, and visual information threaten to overwhelm us. The file of a 2-hour movie, for instance, contains so much sound and visual information that, if stored without modification on a standard

CD-ROM, it would require 360 disk changes during a single showing. A broadcast of *Oprah* that presently fits into one conventional, or analog, television channel would require 45 channels if sent in digital language.

To fit more data into less space, we use the mathematical process called compression. **_Compression,_ or digital-data compression, is a method of removing repetitive elements from a file so that the file requires less storage space and therefore less time to transmit.** Before we use the data, it is decompressed—the repeated patterns are restored. These methods are sometimes referred to as *codec* (for *compression/decompression*) techniques.

- Lossless versus lossy compression: There are two principal methods of compressing data—lossless and lossy. In any situation, which of these two techniques is more appropriate will depend on whether data quality or storage space is more critical.

  Lossless compression uses mathematical techniques to replace repetitive patterns of bits with a kind of coded summary. During decompression, the coded summaries are replaced with the original patterns of bits. In this method, the data that comes out is exactly the same as what went in; it has merely been repackaged for purposes of storage or transmission. Lossless techniques are used when it's important that nothing be lost—for instance, for computer data, database records, spreadsheets, and word processing files.

  Lossy compression techniques permanently discard some data during compression. Lossy data compression involves a certain loss of accuracy in exchange for a high degree of compression (to as little as 5% of the original file size). This method of compression is often used for graphics files and sound files. Thus, a lossy codec might discard subtle shades of color or very soft sounds. Most users wouldn't notice the absence of these details.

- Compression standards—JPEG and MPEG: Several standards exist for compression, particularly of visual data. Data recorded and compressed in one standard cannot be played back in another. The main reason for the lack of agreement is that different industries have different priorities. What will satisfy the users of still photographs, for instance, will not work for the users of movies.

  As we have seen, lossless compression schemes are used for text and numeric data files, whereas lossy compression schemes are used with graphics and video files. The principal lossy compression schemes are *JPEG* and *MPEG.*

  The leading compression standard for still images is _JPEG_ (pronounced "jay-peg"), which stands for the Joint Photographic Experts Group of the International Standards Organization. The file extension that identifies graphic images files in the JPEG format is *.jpeg* or *.jpg*. In storing and transmitting still photographs, the data must remain of high quality. The JPEG codec looks for a way to squeeze a single image, mainly by eliminating repetitive pixels (picture-element dots) within the image. Higher or lower degrees of JPEG compression may be chosen; greater compression corresponds to greater image loss.

  The leading compression standard for moving images is _MPEG_ ("em-peg"), for Motion Picture Experts Group. The file extension that identifies video (and sound) files compressed in this format is *.mpeg* or *.mpg*. People who work with videos are mainly interested in storing or transmitting an enormous amount of visual information in economical form; preserving details is a secondary consideration. The Motion Picture Experts Group sets standards for weeding out redundancies between neighboring images in a stream of video. Three MPEG standards have been developed for compressing visual information—MPEG-1, MPEG-2, and MPEG-4.

What is the radio frequency spectrum and what is bandwidth?

Describe the three types of wired channels.

Distinguish among four types of wireless channels.

Discuss compression and decompression.

5.4 Networks

Whether wired, wireless, or both, all the channels we've described can be used singly or in mix-and-match fashion to form networks. A **_network_, or communications network, is a system of interconnected computers, telephones, or other communications devices that can communicate with one another and share applications and data.** The tying together of so many communications devices in so many ways is changing the world we live in.

The Benefits of Networks

People and organizations use computers in networks for several reasons. These include the following:

- Sharing of peripheral devices: Because peripheral devices such as laser printers, disk drives, and scanners are often quite expensive, management wants to maximize their use, to justify their purchase. Usually the best way to do this is to connect the peripheral to a network serving several computer users.
- Sharing of programs and data: In most organizations, people use the same software and need access to the same information. It is less expensive for a company to buy a separate word processing program that will serve many employees than to buy a separate word processing program for each employee.

  Moreover, if all employees have access to the same data on a shared storage device, the organization can save money and avoid serious problems. If each employee has a separate machine, some employees may update customer addresses, while others remain ignorant of the changes. Updating information on a shared server is much easier than updating every user's individual system.

  Finally, network-linked employees can more easily work together online on shared projects.
- Better communications: One of the greatest features of networks is electronic mail. With e-mail, everyone on a network can easily keep others posted about important information.
- Security of information: Before networks became commonplace, an individual employee might be the only one with a particular piece of information, stored in his or her desktop computer. If the employee was dismissed—or if a fire or flood demolished the office—the company would lose that information. Today such data would be backed up or duplicated on a networked storage device shared by others.
- Access to databases: Networks enable users to tap into numerous databases, whether private company databases or public databases available online through the Internet.

Types of Networks: WANs, MANs, & LANs

Networks, which consist of various combinations of computers, storage devices, and communications devices, may be divided into three main categories, differing primarily in their geographical range.

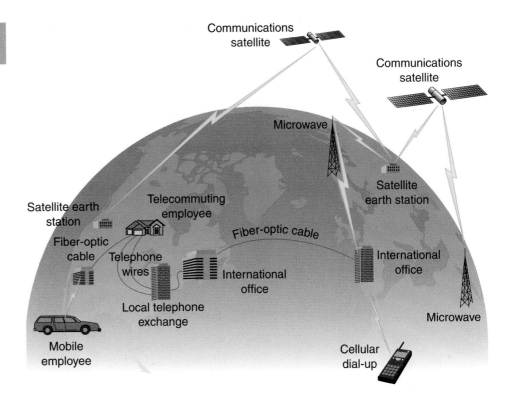

- **Wide area network:** A ***wide area network (WAN)* is a communications network that covers a wide geographical area, such as a country or the world.** Most long-distance and regional Bell telephone companies are WANs. A WAN may use a combination of satellites, fiber-optic cable, microwave, and copper wire connections and link a variety of computers, from mainframes to terminals. *(See ● Panel 5.5.)*

- **Metropolitan area network:** A ***metropolitan area network (MAN)* is a communications network covering a city or a suburb.** The purpose of a MAN is often to bypass local telephone companies when accessing long-distance services. Many cellular phone systems are MANs.

- **Local area network:** A ***local area network (LAN)* connects computers and devices in a limited geographical area,** such as one office, one building, or a group of buildings close together (for instance, a college campus). A small LAN in a modest office, or even in a home, might link a file server with a few terminals or PCs and a printer or two. Such small LANs have been called *TANs,* for "tiny area networks."

Most large computer networks have at least one ***host computer*, a mainframe or midsize central computer that controls the network.** The other devices within the network are called nodes. A ***node* is any device that is attached to a network**—for example, a microcomputer, terminal, storage device, or printer.

Networks may be connected together—LANs to MANs and MANs to WANs. A ***backbone* is a high-speed network that connects LANs and MANs to the Internet.**

Types of LANs: Client-Server & Peer-to-Peer

Local area networks consist of two principal types: client/server and peer-to-peer. *(See ● Panel 5.6, next page.)*

- **Client/server LANs:** A ***client/server LAN* consists of clients, which are microcomputers that request data, and servers, which are computers used to supply data.** The server is a powerful microcomputer that

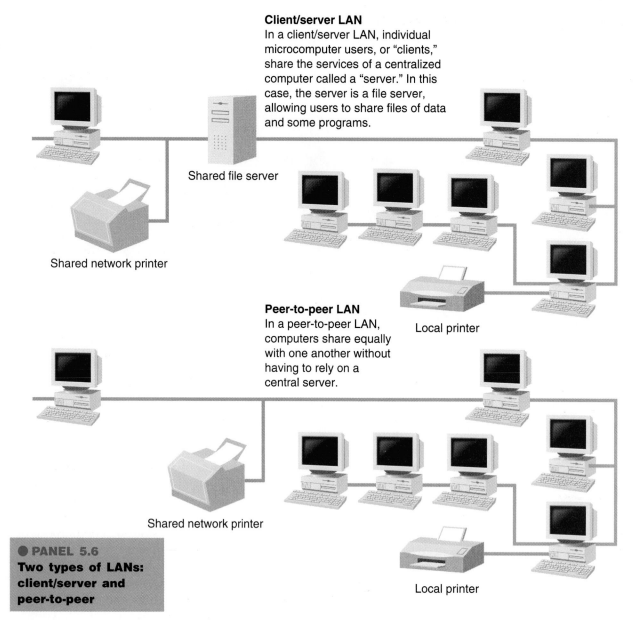

Client/server LAN
In a client/server LAN, individual microcomputer users, or "clients," share the services of a centralized computer called a "server." In this case, the server is a file server, allowing users to share files of data and some programs.

Shared file server

Shared network printer

Peer-to-peer LAN
In a peer-to-peer LAN, computers share equally with one another without having to rely on a central server.

Local printer

Shared network printer

Local printer

● PANEL 5.6
Two types of LANs: client/server and peer-to-peer

manages shared devices, such as laser printers. It runs server software for applications such as e-mail and Web browsing.

Different servers may be used to manage different tasks. A ***file server*** **is a computer that acts like a disk drive, storing the programs and data files shared by users on a LAN.** A *database server* is a computer in a LAN that stores data but doesn't store programs. A *print server* controls one or more printers and stores the print-image output from all the microcomputers on the system. *Web servers* contain Web pages that can be viewed using a browser. *Mail servers* manage e-mail.

● Peer-to-peer LANs: The word *peer* denotes one who is equal in standing with another (as in the phrases "peer pressure" and "jury of one's peers"). In a ***peer-to-peer LAN*,** **all microcomputers on the network communicate directly with one another without relying on a server.** Peer-to-peer networks are less expensive than client-server networks and work effectively for up to 25 computers. Beyond that, they slow down under heavy use. They are appropriate for small networks.

Many LANs mix elements from both client-server and peer-to-peer models.

Components of a LAN

Local area networks are made up of several standard components.

- **Connection or cabling system:** LANs may use a wired or wireless connection system. Wired connections may be twisted-pair wiring, coaxial cable, or fiber-optic cable. Wireless connections may be infrared or radio-wave transmission. Wireless networks are especially useful for mobile computers. However, they are subject to interference.

- **Microcomputers with network interface cards:** Two or more microcomputers are required, along with network interface cards. As we mentioned in Chapter 4, a *network interface card* enables the computer to send and receive messages over a cable network. The network card can be inserted into an expansion slot in a PC. Alternatively, a network card in a stand-alone box may serve a number of devices. Many new computers come with network cards already installed.

- **Network operating system:** The *network operating system (NOS)* is the system software that manages the activity of a network. The NOS supports access by multiple users and provides for recognition of users based on passwords and terminal identifications. Depending on whether the LAN is client/server or peer-to-peer, the operating system may be stored on the file server, on each microcomputer on the network, or a combination of both.

  Examples of popular NOS software are Novell NetWare, Microsoft Windows NT/2000, Unix, and Linux. Peer-to-peer networking can also be accomplished with Microsoft Windows 95/98/Me and Microsoft Windows for Workgroups.

- **Other shared devices:** Printers, scanners, storage devices, and other peripherals may be added to the network as necessary and shared by all users.

- **Routers, bridges, and gateways:** In principle, a LAN may stand alone. Today, however, it invariably connects to other networks, especially the Internet. Network designers determine the types of hardware and software necessary as interfaces to make these connections. Routers, bridges, and gateways are used for this purpose.

  A **_router_ is a special computer that directs communicating messages when several networks are connected together.** High-speed routers can serve as part of the Internet backbone, or transmission path, handling the major data traffic.

  A **_bridge_ is an interface used to connect the same types of networks.** An example is the Ethernet.

  A **_gateway_ is an interface permitting communication between dissimilar networks**—for instance, between a LAN and a WAN or between two LANs based on different network operating systems or different layouts.

These are illustrated on the next page. *(See ● Panel 5.7.)*

Intranets, Extranets, & Firewalls: Private Internet Networks

Early in the Online Age, businesses discovered the benefits of using the World Wide Web to get information to customers, suppliers, or investors. For example, in the mid-1990s Federal Express found it could save millions by allowing customers to click through Web pages to trace their parcels, instead of having FedEx customer-service agents do it. From there, it was a short step to the application of the same technology inside companies—in internal Internet networks called *intranets*.

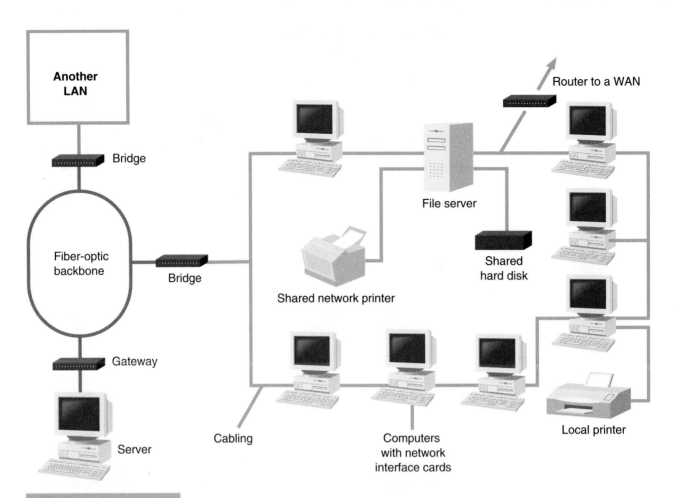

Another LAN

Bridge

Fiber-optic backbone

Bridge

Gateway

Server

Router to a WAN

File server

Shared network printer

Shared hard disk

Cabling

Computers with network interface cards

Local printer

● PANEL 5.7

Components of a typical LAN

- Intranets—for internal use only: An *__intranet__* **is an organization's internal private network that uses the infrastructure and standards of the Internet and the World Wide Web.** When a corporation develops a public Web site, it is making selected information available to consumers and other interested parties. When it creates an intranet, it enables employees to have quicker access to internal information and to share knowledge so that they can do their jobs better. Information exchanged on intranets may include employee e-mail addresses and telephone numbers, product information, sales data, employee benefit information, and lists of jobs available within the organization.

- Extranets—for certain outsiders: Taking intranet technology a few steps further, extranets offer security and controlled access. As we have seen, intranets are internal systems, designed to connect the members of a specific group or a single company. By contrast, *__extranets__* **are private intranets that connect not only internal personnel but also selected suppliers and other strategic parties.** Extranets have become popular for standard transactions such as purchasing. Ford Motor Company, for instance, has an extranet that connects more than 15,000 Ford dealers worldwide. Called FocalPt, the extranet supports sales and servicing of cars, with the aim of improving service to Ford customers.

- Firewalls: Security is essential to an intranet (or even an extranet). Sensitive company data, such as payroll information, must be kept private, by means of a *firewall*. A *__firewall__* **is a system of hardware and software that blocks unauthorized users inside and outside the organization from entering the intranet.**

  A firewall consists of two parts, a choke and a gate. The *choke* forces all data packets flowing between the Internet and the intranet to pass through a gate. The *gate* regulates the flow between the two networks. It identifies authorized users, searches for viruses, and implements other security measures. Thus, intranet users can gain access to the Internet (including key sites connected by hyperlinks), but outside Internet users cannot enter the intranet.

CONCEPT CHECK

What are the benefits of networks?

Distinguish WANs, MANs, and LANs and client-server versus peer-to-peer LANs.

What are the components of a LAN?

Discuss intranets, extranets, and firewalls.

5.5 Cyberethics: Controversial Material, Censorship, & Privacy Issues

KEY QUESTION
What are considerations to be aware of regarding cyberethics?

Communications technology gives us more choices of nearly every sort. It provides us with different ways of working, thinking, and playing. It also presents us with some different moral choices—determining right actions in the digital and online universe. Let's consider some important aspects of "cyberethics"—controversial material and censorship, and matters of privacy.

Controversial Material & Censorship

Since computers are simply another way of communicating, there should be no surprise that many people use them to communicate about sex. Yahoo!, the Internet portal company, says that the word "sex" is the most popular search word on the Net.[15] All kinds of online X-rated message boards, chat rooms, and Usenet newsgroups exist. These raise serious issues for parents. Do we want children to have access to sexual conversations, to download hard-core pictures, or to encounter criminals who might try to meet them offline? "Parents should never use [a computer] as an electronic baby sitter," says computer columnist Lawrence Magid. People online are not always what they seem to be, he points out, and a message seemingly from a 12-year-old girl could really be from a 30-year-old man. "Children should be warned never to give out personal information," says Magid, "and to tell their parents if they encounter mail or messages that make them uncomfortable."[16]

What can be done about X-rated materials? Some possibilities:

Protecting children
Blocking software can help restrict objectionable material from children.

- **Blocking software:** Some software developers have discovered a golden opportunity in making programs like SurfWatch, Net Nanny, and CYBERsitter. These "blocking" programs screen out objectionable material, typically by identifying certain unapproved keywords in a user's request or comparing the user's request for information against a list of prohibited sites.

- **Browsers with ratings:** Another proposal in the works is browser software that contains built-in ratings for Internet, Usenet, and World Wide Web files. Parents could, for example, choose a browser that has been endorsed by the local school board or the online service provider.

- **The V-chip:** The 1996 Telecommunications Law officially launched the era of the V-chip, a device that will be required equipment in most new television sets. The *V-chip* allows parents to automatically block out programs that have been labeled as high in violence, sex, or other objectionable material.

However, any attempts at restricting the flow of information are hindered by the basic design of the Internet itself, with its strategy of offering different roads to the same place. "If access to information on a computer is blocked by one route," writes the *New York Times*'s Peter Lewis, "a moderately skilled computer user can simply tap into another computer by an alternative route." Lewis cites an Internet axiom attributed to an engineer named John Gilmore: "The Internet interprets censorship as damage and routes around it."[17]

Privacy

Privacy is the right of people not to reveal information about themselves. Technology, however, puts constant pressure on this right.

Consider Web cookies, little pieces of data left in your computer by some sites you visit.[18] A *cookie* is a file that the Web server stores on your hard-disk drive when you visit a Web site. Thus, unknown to you, a Web site operator or companies advertising on the site can log your movements within the site. These records provide information that marketers can use to target customers for their products. Other Web sites can also get access to the cookies and acquire information about you.

There are other intrusions on your privacy. Think your medical records are inviolable? Actually, private medical information is bought and sold freely by various companies since there is no federal law prohibiting it. (And they simply ignore the patchwork of state laws.)

Think the boss can't snoop on your e-mail at work? The law allows employers to "intercept" employee communications if one of the parties involved agrees to the "interception." The party who agrees in this case is the employer. Indeed, employer snooping seems to be widespread.

A great many people are concerned about the loss of their right to privacy. Indeed, one survey found that 80% of the people contacted worried that they had lost "all control" of the personal information being collected and tracked by computers.[19] Although several laws restrain the government's ability to acquire and disseminate information and to listen in on private conversations, there are reasons to be alarmed.

CONCEPT CHECK

Discuss some ways to protect children from X-rated material.

What is privacy?

Visual Summary

analog (p. 207, KQ 5.1) Continuous electrical signal in the form of a wave varying in strength and/or quantity. *Why it's important: Sound, light, temperature, and pressure values, for instance, can fall anywhere on a continuum or range. The highs, lows, and in-between states have historically been represented with analog devices rather than in digital form. Examples of analog devices are a speedometer, a thermometer, and a tire-pressure gauge, all of which can measure continuous fluctuations. The electrical signals on a telephone line have traditionally been analog-data representations of the original voices. Telephone, radio, television, and cable-TV technologies have long been based on analog data.*

backbone (p. 223, KQ 5.4) High-speed network that connects LANs and MANs to the Internet. *Why it's important: The backbone is an essential part of the Internet.*

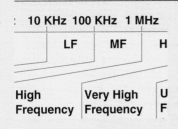

band (p. 216, KQ 5.3) Also called *bandwidth;* range of frequencies that is a measure of the amount of information that can be delivered within a given period of time. The bandwidth is the difference between the lowest and the highest frequencies transmitted. *Why it's important: For analog signals, bandwidth is expressed in hertz (Hz), or cycles per second. For digital signals, bandwidth is expressed in bits per second (bps). In the United States, certain bands are assigned by the Federal Communications Commission (FCC) for certain purposes. The wider the bandwidth, the faster data can be transmitted. The narrower the band, the greater the loss of transmission power. This loss of power must be overcome by using relays or repeaters that rebroadcast the original signal.*

bridge (p. 225, KQ 5.4) Interface used to connect the same types of networks. *Why it's important: Similar networks (local area networks) can be joined together to create larger area networks.*

broadband connections (p. 217, KQ 5.3) Connections characterized by very high speed (wide bandwidth). *Why it's important: Broadband connections are necessary for reliable, high-speed Internet hook-ups.*

broadcast radio (p. 219, KQ 5.3) Wireless transmission medium that sends data over long distances—between regions, states, or countries. A transmitter is required to send messages and a receiver to receive them; sometimes both sending and receiving functions are combined in a transceiver. *Why it's important: In the lower frequencies of the radio spectrum, several broadcast radio bands are reserved not only for conventional AM/FM radio but also for broadcast television, CB (citizens band) radio, ham (amateur) radio, cellular phones, and private radio land mobile services (such as police, fire, and taxi dispatch). Some organizations use specific radio frequencies and networks to support wireless communications.*

client/server LAN (p. 223, KQ 5.4) Type of local area network; consists of clients, which are microcomputers that request data, and servers, which are computers used to supply data. The server is a powerful microcomputer that manages shared devices, such as laser printers. It runs server software for applications such as e-mail and Web browsing. *Why it's important: Client/server networks are a common type of LAN.*

coaxial cable (p. 218, KQ 5.3) Commonly called "co-ax"; consists of insulated copper wire wrapped in a solid or braided metal shield, then in an external cover. *Why it's important: Co-ax is widely used for cable television. Because of the extra insulation, coaxial cable is much better than twisted-pair wiring at resisting noise. Moreover, it can carry voice and data at a faster rate.*

communications channel (p. 218, KQ 5.3) Path over which information travels in a telecommunications system from its source to its destination. *Why it's important: Channels may be wired or wireless. Three types of wired channels are twisted-pair wire (conventional telephone lines), coaxial cable, and fiber-optic cable.*

communications satellite (p. 220, KQ 5.3) Microwave relay stations in orbit around the earth. Why it's important: *Transmitting a signal from a ground station to a satellite is called uplinking; the reverse is called downlinking. The delivery process will be slowed if, as is often the case, more than one satellite is required to get the message delivered.*

compression (p. 221, KQ 5.3) Also called *digital-data compression;* method of removing repetitive elements from a file so that the file requires less storage space and therefore less time to transmit. Before we use the data, it is decompressed—the repeated patterns are restored. These methods are sometimes referred to as *codec* (for *compression/decompression*) techniques. Why it's important: *Many of today's files, with graphics, sound, and video, require huge amounts of storage space; data compression allows users to reduce the space required to store and transmit these files.*

digital (p. 207, KQ 5.1) Refers to communications signals or information represented in a two-state (binary) way. Why it's important: *Digital signals are the basis of computer-based communications. "Digital" is usually synonymous with "computer-based."*

extranet (p. 226, KQ 5.4) Private intranet that connects not only internal personnel but also selected suppliers and other strategic parties. Why it's important: *Extranets have become popular for standard transactions such as purchasing.*

fiber-optic cable (p. 218, KQ 5.3) Cable that consists of dozens or hundreds of thin strands of glass or plastic that transmit pulsating beams of light rather than electricity. Why it's important: *These strands, each as thin as a human hair, can transmit up to 2 billion pulses per second (2 Gbps), each "on" pulse representing one bit. When bundled together, fiber-optic strands in a cable 0.12 inches thick can support a quarter-million to a half-million voice conversations at the same time. Moreover, unlike electrical signals, light pulses are not affected by random electromagnetic interference in the environment. Thus, they have much lower error rates than normal telephone wire and cable. In addition, fiber-optic cable is lighter and more durable than twisted-pair and co-ax cable. A final advantage is that it cannot easily be wiretapped, so transmissions are more secure.*

file server (p. 224, KQ 5.4) Computer in a client/server network that acts like a disk drive, storing the programs and data files shared by users. Why it's important: *A file server enables users of a LAN to all have access to the same programs and data.*

firewall (p. 226, KQ 5.4) System of hardware and software that blocks unauthorized users inside and outside the organization from entering the intranet. Why it's important: *A firewall consists of two parts, a choke and a gate. The choke forces all data packets flowing between the Internet and the intranet to pass through a gate. The gate regulates the flow between the two networks. It identifies authorized users, searches for viruses, and implements other security measures. Thus, intranet users can gain access to the Internet (including key sites connected by hyperlinks), but outside Internet users cannot enter the intranet.*

gateway (p. 225, KQ 5.4) Interface permitting communication between dissimilar networks. Why it's important: *Gateways permit communication between a LAN and a WAN or between two LANs based on different network operating systems or different layouts.*

Global Positioning System (GPS) (p. 212, KQ 5.2) System of a series of earth-orbiting satellites continuously transmitting timed radio signals that can be used to identify earth locations. Why it's important: *A GPS receiver—handheld or mounted in a vehicle, plane, or boat—can pick up transmissions from any four satellites, interpret the information from each, and calculate to within a few hundred feet or less the receiver's longitude, latitude, and altitude. Some GPS receivers include map software for finding your way around, as with the Guidestar system available with some rental cars.*

host computer (p. 223, KQ 5.4) Mainframe or midsize central computer that controls a network. Why it's important: *The host is responsible for managing the entire network.*

infrared wireless transmission (p. 218, KQ 5.3) Sends data signals using infrared-light waves. Why it's important: *Infrared ports can be found on some laptop computers and printers, as well as wireless mouses. The advantage is that no physical connection is required among devices. The drawbacks are that line-of-sight communication is required—there must be an unobstructed view between transmitter and receiver—and transmission is confined to short range.*

intranet (p. 226, KQ 5.4) Organization's internal private network that uses the infrastructure and standards of the Internet and the World Wide Web. *Why it's important: When an organization creates an intranet, it enables employees to have quicker access to internal information and to share knowledge so that they can do their jobs better. Information exchanged on intranets may include employee e-mail addresses and telephone numbers, product information, sales data, employee benefit information, and lists of jobs available within the organization.*

JPEG (p. 221, KQ 5.3) Stands for Joint Photographic Experts Group of the International Standards Organization; JPEG is the leading compression standard for still images. *Why it's important: The file extension that identifies graphic images files in the JPEG format is .jpeg or .jpg. JPEG compression is commonly used to store and transmit still graphic images. Higher or lower degrees of JPEG compression may be chosen; greater compression corresponds to greater image loss.*

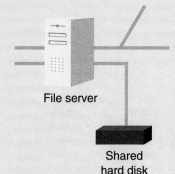

File server

Shared
hard disk

local area network (LAN) (p. 223, KQ 5.4) Network that connects computers and devices in a limited geographical area, such as one office, one building, or a group of buildings close together (for instance, a college campus). *Why it's important: LANs have replaced large computers for many functions and are considerably less expensive.*

metropolitan area network (MAN) (p. 223, KQ 5.4) Communications network covering a city or a suburb. *Why it's important: The purpose of a MAN is often to bypass local telephone companies when accessing long-distance services. Many cellular phone systems are MANs.*

microwave radio (p. 219, KQ 5.3) Transmission of voice and data through the atmosphere as super-high-frequency radio waves called *microwaves*. *Why it's important: These frequencies are used not only to operate microwave ovens but also to transmit messages between ground-based Earth stations and satellite communications systems. Nowadays dish- or horn-shaped microwave reflective dishes, which contain transceivers and antennas, are nearly everywhere—on towers, buildings, and hilltops. Like infrared waves, microwaves are line-of-sight; they cannot bend around corners or around the earth's curvature, so there must be an unobstructed view between transmitter and receiver. Thus, microwave stations need to be placed within 25–30 miles of each other, with no obstructions in between. A string of microwave relay stations will each receive incoming messages, boost the signal strength, and relay the signal to the next station.*

modem (p. 208, KQ 5.1) Short for "**mo**dulate/**dem**odulate"; device that converts digital signals into a representation of analog form (modulation) to send over phone lines; a receiving modem then converts the analog signal back to a digital signal (demodulation). *Why it's important: The modem provides a means for computers to communicate with one another using the standard copper-wire telephone network, an analog system that was built to transmit the human voice but not computer signals.*

MPEG (p. 221, KQ 5.3) Stands for Motion Picture Experts Group. MPEG is the leading compression standard for video images. *Why it's important: The file extension that identifies video (and sound) files compressed in this format is .mpeg or .mpg. People who work with videos are mainly interested in storing or transmitting an enormous amount of visual information in economical form; preserving details is a secondary consideration. The Motion Picture Experts Group sets standards for weeding out redundancies between neighboring images in a stream of video. Three MPEG standards have been developed for compressing visual information—MPEG-1, MPEG-2, and MPEG-4.*

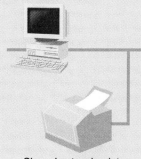

network (p. 222, KQ 5.4) Also called *communications network;* system of interconnected computers, telephones, or other communications devices that can communicate with one another and share applications and data. *Why it's important: The tying together of so many communications devices in so many ways is changing the world we live in.*

node (p. 223, KQ 5.4) Any device that is attached to a network. *Why it's important: A node may be a microcomputer, terminal, storage device, or peripheral device, any of which enhance the usefulness of the network.*

Shared network printer

Telecommunications

peer-to-peer LAN (p. 224, KQ 5.4) Type of local area network; all microcomputers on the network communicate directly with one another without relying on a server. Why it's important: *Peer-to-peer networks are less expensive than client-server networks and work effectively for up to 25 computers. Beyond that, they slow down under heavy use. They are appropriate for small networks.*

privacy (p. 228, KQ 5.5) The right of people not to reveal information about themselves. Why it's important: *Information technology presents constant threats to this right.*

radio frequency spectrum (p. 215, KQ 5.3) Refers to all the fields of electrical energy and magnetic energy that travel in waves. Why it's important: *These fields carry communications signals. The waves vary according to frequency—the number of times a wave repeats, or makes a cycle, in a second. The radio spectrum ranges from low-frequency waves, such as those used for aeronautical and marine navigation equipment, through the medium frequencies for CB radios, cordless phones, and baby monitors, to ultrahigh frequency bands for cell phones and also microwave bands for communications satellites.*

router (p. 225, KQ 5.4) Special computer that directs communicating messages when several networks are connected together. Why it's important: *High-speed routers can serve as part of the Internet backbone, or transmission path, handling the major data traffic.*

---

telecommuting (p. 213, KQ 5.2) Working at home while in telecommunication with the office. Why it's important: *Telecommuting has many benefits. The advantages to society are reduced traffic congestion, energy consumption, and air pollution. Productivity increases, because telecommuters experience fewer distractions at home and can work flexible hours. Teamwork improves, and the labor pool is expanded, because hard-to-get employees don't have to uproot themselves from where they want to live. A disadvantage is that people may feel isolated.*

twisted-pair wire (p. 218, KQ 5.3) Two strands of insulated copper wire, twisted around each other. Why it's important: *Twisted-pair wire has been the most common channel or medium used for telephone systems. It is relatively slow and does not protect well against electrical interference.*

videoconferencing (p. 213, KQ 5.2) Also called *teleconferencing;* use of television video and sound technology as well as computers to enable people in different locations to see, hear, and talk with one another. Why it's important: *Videoconferencing may eliminate the need for some travel for the purpose of meetings and allow people who cannot travel to visit "in person."*

virtual office (p. 214, KQ 5.2) Nonpermanent and mobile office run with computer and communications technology. Why it's important: *Employees work from their homes, cars, and other new work sites. They use pocket pagers, portable computers, fax machines, and various phone and network services to conduct business.*

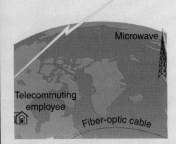

wide area network (WAN) (p. 223, KQ 5.4) Communications network that covers a wide geographical area, such as a country or the world. Why it's important: *Most long-distance and regional Bell telephone companies are WANs. A WAN may use a combination of satellites, fiber-optic cable, microwave, and copper wire connections and link a variety of computers, from mainframes to terminals.*

workgroup computing (p. 213, KQ 5.2) Also called *collaborative computing;* technology that enables teams of co-workers to use networks of microcomputers to share information and to cooperate on projects. Workgroup computing is made possible not only by networks and microcomputers but also by groupware. Why it's important: *Workgroup computing allows co-workers to collaborate with colleagues, suppliers, and customers and to tap into company information through computer networks.*

Chapter Review

Self-Test Questions

1. A(n) _modem_ converts digital signals into analog signals for transmission over phone lines.
2. A(n) _WAN_ _wide area network_ network covers a wide geographical area, such as a state or a country.
3. _fiber optic_ cable transmits data as pulses of light rather than as electricity.
4. _analog_ refers to waves continuously varying in strength and/or quantity; _digital_ refers to communications signals or information in a binary form.
5. _digital data compression_ is a method of removing repetitive elements from a file so that the file requires less storage space.
6. A(n) _file server_ is a computer that acts as a disk drive, storing programs and data files shared by users on a LAN.
7. The _module_ is the system software that manages the activities of a network.
8. _Modem_ is short for _modulate demodulate_
9. The leading compression standard for still images is _JPEG_.
10. _Filtering_ programs can screen out objectionable material on the Internet.

Multiple-Choice Questions

1. Which of the following best describes the telephone line that is used in most homes today?
   a. coaxial cable
   b. modem cable
   c. twisted-wire pair
   d. fiber-optic cable
   e. LAN
2. Which of the following do local area networks enable?
   a. sharing of peripheral devices
   b. sharing of programs and data
   c. better communications
   d. access to databases
   e. all of the above
3. Which of the following is not a data compression standard/method?
   a. lossless
   b. JPEG
   c. MPEG
   d. lossy
   e. NOS
4. Which of the following is not a type of server?
   a. file server
   b. print server
   c. mail server
   d. disk server
   e. database server

True/False Questions

1. T F In a LAN, a bridge is used to connect the same type of networks, whereas a gateway is used to enable dissimilar networks to communicate.
2. T F Frequency and amplitude are two characteristics of analog carrier waves.
3. T F A range of frequencies is called a _spectrum_.
4. T F Twisted-pair wire commonly connects residences to external telephone systems.
5. T F A cookie is an ID number used to visit specific Web sites.

Short-Answer Questions

1. What is the difference between an intranet and an extranet?
2. What is workgroup computing?
3. What is the difference between a LAN and a WAN?
4. Why is bandwidth a factor in data transmission?
5. What is a firewall?

Concept Mapping

On a separate sheet of paper, draw a concept map, or visual diagram, linking concepts. Show how the following terms are related.

analog	JPEG
backbone	LAN
bandwidth	MAN
coaxial cable	modem
communications channel	MPEG
compression	network
digital	node
fiber-optic cable	privacy
firewall	satellite
GPS	twisted-pair wire
host computer	WAN

Knowledge in Action

1. Are the computers at your school connected to a network? If so, what kind of network(s)? What types of computer are connected? What hardware and software allows the network to function? What department(s) did you contact to find the information you needed to answer these questions?

2. Using current articles, publications, and/or the Web, research cable modems. Where are they being used? What does a residential user need to hook up to a cable modem system? Do you think you will use a cable modem in the near future? Write a short report.

3. Research the FCC's role in regulating the communications industry. How are new frequencies opened up for new communications services? How are the frequencies determined? Who gets to use new frequencies?

4. On the Web, go to *www.trimble.com* and work through some of the tutorial on "How GPS Works." Then write a short report on the applications of a GPS system.

5. In your planned career or profession, can you imagine working as a telecommuter? Describe how telecommuting might work for you.

E-Commerce, Files, & Databases

Digital Engines for the New Economy

Key Questions

You should be able to answer the following questions.

6.1   **Databases & the New Economy** What are databases, and how are e-commerce, data mining, and business-to-business systems using them?

6.2   **All Databases Great & Small** What are four ways of organizing databases, and what are the benefits of databases?

6.3   **Managing Files: Basic Concepts** What are the data storage hierarchy, the key field, types of files, ways of keeping track of files, sequential versus direct access, and offline versus online storage?

6.4   **The Ways Databases Are Organized** What are four types of database organization?

6.5   **The Ethics of Using Storage & Databases: Concerns about Accuracy & Privacy** What are some ethical concerns about the uses of databases?

Graphical Interface

1 Databases & the New Economy: E-Commerce, Data Mining, & B2B Systems. Databases are organized collections of related files. Databases underpin the so-called New Economy of computer, telecommunications, and Internet companies in three ways: e-commerce, data mining, and business-to-business (B2B) systems. **E-commerce,** or electronic commerce, is the buying and selling of products and services through computer networks; an example is Amazon.com. **Data mining** is the computer-assisted process of sifting through and analyzing vast amounts of data in order to extract meaning and discover new knowledge. **Business-to-business (B2B) systems** allow businesses to sell to other businesses, using the Internet to cut transaction costs and increase efficiencies.

2 All Databases Great & Small. Databases can be classified as four types. (1) An **individual** database is a collection of integrated files used by one person. It could be a personal information manager, which helps people keep track of information they use daily. (2) A **shared database,** or company database, is shared by users in one organization in one location; a **database administrator** may coordinate its activities and needs. (3) A **distributed database** is one that is stored on different computers in different locations connected by a client/server network. (4) A **public databank** is a compilation of data available on a subject; many such databanks are Web sites.

The benefits of databases are file sharing, reduced data redundancy, improved data integrity, and increased security.

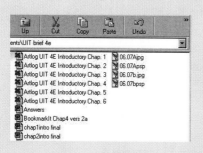

3 Managing Files: Basic Concepts. Data is organized in a **data storage hierarchy** of increasingly complex levels: bits, bytes (characters), fields, records, files, and databases. A **character** is a letter, number, or special character. A **field** consists of one or more characters (bytes). A **record** is a collection of related fields. A **file** is a collection of related records. A *database* is, as mentioned, an organized collection of integrated files. Important to data organization is the **key field,** a field used to uniquely identify a record so that it can be easily retrieved and processed.

Files are given names—**filenames.** Filenames also have *extension names,* three-letter additions such as *.doc* and *.com.* Among the types of files are the following. (1) **Program files** are files containing software instructions. The two most important are *source program files,* which contain instructions in the form written by the programmer, and *executable files,* which contain instructions that tell a computer how to perform a particular task. (2) **Data files** are files that contain data. (3) Other common files are *ASCII files,* which are text only; *image files* for digitized graphics; *audio files,* which contain digitized sound; *animation/video files,* used for conveying moving images; and *Web files,* which are files carried over the World Wide Web.

Microcomputer operating systems keep track of files using a filing system that uses drive letters, folders (directories), and filenames with extension. (1) Storage devices are assigned drive letters: A and B, floppy disk drives; C, hard drive; D, CD-ROM drive; E, Zip drive; F, tape backup drive. (2) A **directory,** or **folder,** is a storage place for files on your drive; directories or folders are organized in a hierarchy called a tree structure—root directory, directory, and subdirectory. (3) To identify a particular file, the operating system uses a **file specification** or **path,** which specifies the sequence of directories the computer must follow to locate a file. Example: root directory (root folder), directory (folder), subdirectory (subfolder), and filename with extension.

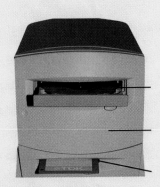

Two main ways in which a storage device accesses stored data are sequential access and direct access. **Sequential storage** means that data is stored and retrieved in sequence, as is the case with magnetic-tape storage. **Direct access storage** means that the computer can go directly to the information you want, resembling a CD player; hard disks and other types of disks are of this nature.

Whether on magnetic tape or disk, data may be stored offline or online. **Offline storage** means that data is not directly accessible for processing until the tape or disk has been loaded onto an input device. **Online storage** means that stored data is randomly (directly) accessible for processing.

4 **The Ways Databases Are Organized.** Databases can be organized in four ways. (1) In a **hierarchical** database, fields or records are arranged in related groups resembling a family tree, with child (lower-level) records subordinate to parent (higher-level) records. (2) A **network base** is similar to a hierarchical database but each child record can have more than one parent record. (3) A **relational database** relates, or connects, data in different files through the use of a key field. (4) An **object-oriented database** uses objects, software written in small, reusable chunks, as elements within database files. An **object** consists of data in any form and instructions on the action to be taken on the data.

5 **The Ethics of Using Storage & Databases: Concerns about Accuracy & Privacy.** In **morphing,** a film image is altered pixel by pixel, so that the image becomes something else. This manipulation of digitized images and sounds raises some ethical issues. Sound performances can be misrepresented, photos may be manipulated, and video and TV images may be altered in undetectable ways and all stored in a database.

Databases are also limited in accuracy and completeness, since not all facts can be found in a database, nor are all data items true. In addition, databases raise several concerns about privacy. Finally, those who own databases may be in a position to monopolize information.

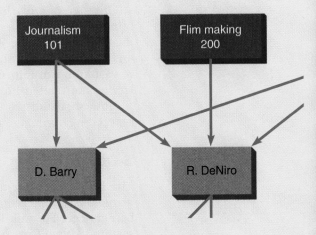

I

f information exists in one place, it exists in more than one place."

So says Carole A. Lane, a database expert.[1] This is what she calls "Lane's First Law of Information." Perhaps it could stand as a summary of one of the most important developments of the Digital Age.

How does information in one place get to be in more than one place? The answer has to do with databases. A database is not just the computerization of what used to go into manila folders and a filing cabinet. A database is an organized collection of related files, a technology for pulling together facts that allows the slicing and dicing and mixing and matching of data in all kinds of ways. As a result, the arrival of databases—especially when linked to the Internet—has stood many of our business and social institutions on their heads.

Your name and some facts about you can probably be found in scores if not hundreds of far-flung databases. How is that data being used? That is a very interesting question. In this chapter, we will examine the importance of databases and how they work.

6.1 Databases & the New Economy: E-Commerce, Data Mining, & B2B Systems

KEY QUESTIONS

What are databases, and how are e-commerce, data mining, and business-to-business systems using them?

It used to be there was a difference between the Old Economy and the New Economy. The first consisted of traditional companies—car makers, pharmaceuticals, retailers, publishers. The second consisted of computer, telecommunications, and Internet companies (AOL, Amazon, eBay, and a raft of "dot-com" firms). Until recently, New Economy companies were fast growing. From 1996 to 1997, for example, the U.S. Internet-based economy more than doubled, from $15.5 billion to $39 billion. By 2001, it was predicted to reach $350 billion.[2] Now, however, Old Economy companies have begun to absorb the new Internet-driven technologies, and the differences between the two sectors are dwindling.

One sign of growth is that Internet host computers have been almost doubling every year. But the mushrooming of computer networks and the booming popularity of the World Wide Web are only the most obvious signs of the digital economy. Behind them lies something equally important: the growth of vast stores of information in the form known as *databases*, organized collections of related files.

How are databases underpinning the New Economy? Let us consider three aspects: *e-commerce, data mining*, and *business-to-business (B2B) systems*.

E-Commerce

The Internet might have remained a text-based realm, the province of academicians and researchers, had it not been for the contributions of Tim Berners-Lee. He was the computer scientist who came up with the coding system (HyperText Markup Language), linkages, and addressing scheme (URLs) that debuted in 1991 as the graphical and multimedia World Wide Web. "It's hard to overstate the impact of the global system he created," writes *Time* technology writer Joshua Quittner. "He took a powerful communications system [the Internet] that only the elite could use, and turned it into a mass medium."[3]

The arrival of the Web quickly led to **_e-commerce_, or electronic commerce, the buying and selling of products and services through computer networks.**

By 2003, total U.S. e-commerce sales to consumers are expected to reach $108 billion, or 6% of consumer retail spending.[4] Indeed, online shopping is growing even faster than the increase in computer use, which has been fueled by the falling price of personal computers. (Half of American households now have a PC.)[5] Among the best-known e-firms are bookseller Amazon.com; auction network eBay; and Priceline, which lets you name the price you're willing to pay for airline tickets and hotel rooms.

Probably the foremost example of e-commerce is Amazon.com.[6] In 1994, seeing the potential for electronic retailing on the World Wide Web, Jeffrey Bezos left a successful career on Wall Street to launch an online bookstore called Amazon.com. Why the name *Amazon*?

"Earth's biggest river, Earth's biggest bookstore," said Bezos in a 1996 interview. "The Amazon River is ten times as large as the next largest river, which is the Mississippi, in terms of volume of water. Twenty percent of the world's fresh water is in the Amazon River Basin, and we have six times as many titles as the world's largest physical bookstore."[7] A more hard-headed reason is that, according to consumer tests, words starting with "A" show up on search-engine lists first.[8]

E-commerce
This Irish retailer of woolens and clothing goods sells not only in its store but also over the World Wide Web.

Still, Bezos realized that no bookstore with four walls could possibly stock the more than 2.5 million titles that are now active and in print. Moreover, he saw that an online bookstore wouldn't have to make the same investment in retail clerks, store real estate, or warehouse space (in the beginning, Amazon.com ordered books from the publisher *after* it took an order), so it could pass savings along to customers in the form of discounts. In addition, he appreciated that there would be opportunities to obtain demographic information about customers in order to offer personalized services. For example, Amazon could let customers know of books that might be of interest to them. Such personalized attention is difficult for traditional bookstores. Finally, Bezos saw that there could be a good deal of online interaction: Customers could post reviews of books they read and could reach authors by e-mail to provide feedback. All this was made possible not only by the Web but by the recording of information on giant databases.

Amazon.com sold its first book in July 1995 and by early 2000 had 1.1 million customers and a market capitalization of $18.3 billion, with Bezos owning about a third of that.[9] The firm also had expanded into the online retailing of music CDs, toys, electronics, drugs, cosmetics, pet supplies, and other goods and also into online auctions. Old Economy "brick and mortar" companies have since followed Amazon into the online sector.

Databases for marketing
Harrah's casinos use data mining to identify gambling patron preferences for marketing and special promotions.

Data Mining

A personal database, such as the address list of friends you have on your microcomputer, is generally small-scale. But some databases are almost unimaginably large-scale, involving records for millions of households and trillions of bytes of data. Some of these activities require the use of so-called massively parallel database computers that cost $1 million or more. "These machines gang together scores or even hundreds of the fastest microprocessors around," says one

description, "giving them the oomph to respond in minutes to complex database queries."[10]

These large-scale efforts go under the name *data mining*. **_Data mining (DM)_ is the computer-assisted process of sifting through and analyzing vast amounts of data in order to extract meaning and discover new knowledge.** The purpose of DM is to describe past trends and predict future trends. Thus, data-mining tools might sift through a company's immense collections of customer, marketing, production, and financial data and identify what's worth noting and what's not.

Data mining has come about because companies find that, in today's fiercely competitive business environment, they need to turn the gazillions of bytes of raw data at their disposal to new uses for further profitability. However, nonprofit institutions have also found DM methods useful, as in the pursuit of scientific and medical discoveries.

Some applications of data mining are as follows:[11]

- **Marketing:** Marketers use DM tools (such as one called Spotlight) to mine point-of-sale databases of retail stores, which contain facts (such as prices, quantities sold, dates of sale) for thousands of products in hundreds of geographic areas. By understanding customer preferences and buying patterns, marketers hope to target consumers' individual needs.

- **Health:** A coach in the U.S. Gymnastics Federation is using a DM system (called IDIS) to discover what long-term factors contribute to an athlete's performance, so as to know what problems to treat early on. A Los Angeles hospital is using the same tool to see what subtle factors affect success and failure in back surgery. Another system helps health-care organizations pinpoint groups whose costs are likely to increase in the near future, so that medical interventions can be made.

- **Science:** DM techniques are being employed to find new patterns in genetic data, molecular structures, global climate changes, and more. For instance, one DM tool (called SKICAT) is being used to catalog more than 50 million galaxies, which will be reduced to a 3-terabyte galaxy catalog.

Clearly, short-term payoffs can be dramatic. One telephone company, for instance, mined its existing billing data to identify 10,000 supposedly "residential" customers who spent more than $1000 a month on their phone bills. When it looked more closely, the company found these customers were really small businesses trying to avoid paying the more expensive business rates for their telephone service.[12]

However, the payoffs in the long term could be truly astonishing. Sifting medical-research data or subatomic-particle information may reveal new treatments for diseases or new insights into the nature of the universe.[13]

Business-to-Business (B2B) Systems

A **_business-to-business (B2B) system_ is the activity of a business selling to other businesses, using the Internet to cut transaction costs and increase efficiencies.** Business-to-business activity is expected to balloon to a $1 trillion industry by 2002 and a $2.8 trillion industry by 2004.[14]

One of the most famous examples of B2B is the auto industry online exchange developed by the big three U.S. automakers—General Motors, Ford, and DaimlerChrysler. The companies are putting their entire system of purchasing, involving more than $250 billion in parts and materials and 60,000 suppliers, on the Internet. Already so-called "reverse auctions," in which suppliers bid to provide the lowest price, have driven down the cost of parts such as tires and window sealers. This system replaces the old-fashioned bureaucratic procurement process built on phone calls and fax machines, and provides substantial cost savings.[15]

Online B2B exchanges have been developed to serve a variety of businesses, from manufacturers of steel and airplanes to convenience stores to olive oil producers.[16] To get a feel for how this works, you can participate yourself in a name-your-price exchange. Priceline.com, for example, has developed a system that allows a shopper to assemble a basket of groceries online and name a price; if it is accepted, the shopper can prepay on the Web site and pick up the groceries at a participating store.[17]

B2B exchanges are expected to revolutionize business by moving beyond pricing mechanisms and encompassing product quality, customer support, credit terms, and shipping reliability, which often count for more than price. The name given to this system is the *business web*, or *b-web*, in which suppliers, distributors, customers, and e-commerce service providers use the Internet for communications and transactions. In addition, b-webs are expected to provide extra revenue from ancillary services, such as financing and logistics.[18] *(See ● Panel 6.1.)*

None of these sectors of the New Economy is possible without databases.

CONCEPT CHECK

Describe what e-commerce is.

What is data mining, and how is it used?

What is a B2B system?

● PANEL 6.1

B2B exchanges
B2B exchanges, which draw on data from various databases, act as centralized online markets for buyers and sellers in specific fields, such as car parts or olive oil. Exchanges are expected to evolve into "b-webs," or business webs, encompassing other factors besides price.

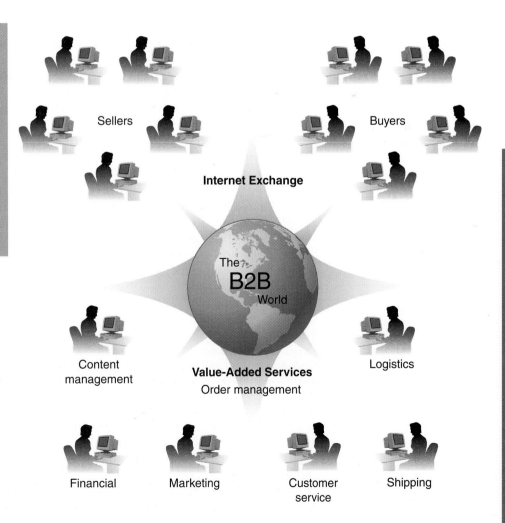

Sellers

Buyers

Internet Exchange

The
B2B
World

Content management

Value-Added Services
Order management

Logistics

Financial   Marketing   Customer service   Shipping

A *database,* we said, is an organized collection of integrated files. A database may be small, contained entirely within your own personal computer. Or, as we've seen, it may be massive, available through online connections. Such massive databases are of particular interest to us in this book because they offer phenomenal resources that until recently were unavailable to most ordinary computer users.

Databases are what make possible interactivity and multimedia. As we stated earlier, *interactivity* refers to the user's back-and-forth interaction with a computer. In other words, the user's actions and choices affect what the computer does. *Multimedia* refers to the use of a variety of media—text, sound, video—to deliver information. Interactivity and multimedia require systems that can store and make available enormous amounts of data—that is, databases.

Four Types of Databases

We may classify databases into four types: *individual, shared (company),* and *distributed databases* and what might be called *public databanks. (See* ● *Panel 6.2.)*

- Individual databases: **Individual databases are collections of integrated files used by one person.** As we discussed in Chapter 3, microcomputer users can set up their own individual databases using popular database management software; the information is stored on the hard drives of their personal computers. Today the principal database programs are Microsoft Access, Corel Paradox, and Lotus Approach. Such programs are used, for example, by graduate students to conduct research, by salespeople to keep track of clients, by purchasing agents to monitor orders, and by coaches to keep watch on other teams and players.

  In addition, types of individual databases known as *personal information managers (PIMs)* can help you keep track of and manage information you use on a daily basis, such as addresses, telephone numbers, appointments, to-do lists, and miscellaneous notes. Popular PIMs are Microsoft Outlook, Lotus Organizer, and Act.

- Shared (company) databases: A **shared database or company database is shared by users in one company or organization in one location.** The organization owns the database, which may be stored on a server such as a mainframe. Users are linked to the database via a local area or wide area network; the users access the network through terminals or microcomputers.

● PANEL 6.2

Four types of databases

Database	Description
Individual database	Collection of integrated files used by one person
Shared database	Database shared by users in one organization in one location
Distributed database	Database stored on different computers in different locations connected by a client/server network
Public databank	Compilation of data available to the public

Often a shared database is managed by a specialist called a database administrator. The *__database administrator (DBA)__* **coordinates all related activities and needs for an organization's database.** The DBA determines user access privileges; sets standards, guidelines, and control procedures; assists in establishing priorities for requests; prioritizes conflicting user needs; and develops user documentation and input procedures. He or she is also concerned with security—setting up and monitoring a system for preventing unauthorized access and making sure that the system is regularly backed up and that data can be recovered should a failure or disaster occur.

Shared databases, such as those you find when surfing the Web, are the foundation for a great deal of electronic commerce, particularly B2B commerce, as we have seen.

- Distributed databases: A *__distributed database__* **is one that is stored on different computers in different locations connected by a client/server network.** For example, Cisco Systems, which calls itself the fastest-growing company in the history of the computing industry (it supplies the vast network that connects computers to the Internet), uses its worldwide intranet to connect its distributed databases, even those located overseas. As a result, it is able to execute what is called a "virtual close"—defined as the ability to close the financial books with a one-hour notice. "By connecting an entire company via intranet, even one with operations in dozens of countries," explains Cisco CEO John Chambers, "what was once done quarterly can now be done anytime."[19]

- Public databanks: If you're looking for very specific information, you can use your browser to do a Web search, investigating hundreds or thousands of Web sites. Many of these Web sites represent *__public databanks__*, **compilations of data that are available to the public.**

Some public databanks are fee-based, such as Dialog Information Services, which offers scientific and technical information, and Dow Jones Interactive Publishing, which provides business information. Certain fee-based public databases are specialized, such as Lexis, which gives lawyers access to local, state, and federal laws, or Nexis, which gives journalists access to published articles in a range of newspapers.

Other public databanks are free. For instance, the U.S. government provides a great deal of free information, such as economic figures from the Bureau of Labor Statistics. Finally, many public databanks, such as Yahoo! or Amazon.com, are supported by advertising or online sales. The most popular revenue-producing Web sites—those supported by selling products or ads—are devoted to shopping, news and media, entertainment, online games, sweepstakes and lotteries, travel, finance, health and family, sports, and home and food.[20]

The Benefits of Databases

In the 1950s, when commercial use of computers was just beginning, a large organization would have different files for different purposes. For example, a university might have one file for course grades, another for student records, another for tuition billing, and so on. In a corporation, people in the accounting, order-entry, and customer-service departments all had their own separate files. Thus, if an address had to be changed, for example, each file would have to be updated separately. The database files were stored on magnetic tape and had to be accessed in sequence. Later magnetic disk technology

PRACTICAL ACTION BOX

Good Habits: Protecting Your Computer System, Your Data, & Your Health

Whether you set up a desktop computer and never move it or tote a portable PC from place to place, you need to be concerned about protection. You don't want to lose the use of your computer. You don't want to lose your data. And you certainly don't want to lose your health for computer-related reasons. Here are some tips for taking care of these vital areas.

Protecting Your Computer System

Computers are easily stolen, particularly portables. They also don't take kindly to fire, flood, or being dropped. Finally, a power surge through the power line can wreck the insides.

Guarding Against Hardware Theft & Loss Portable computers—laptops, notebooks, and subnotebooks—are easy targets for thieves. Obviously, anything conveniently small enough to be slipped into your briefcase or backpack can be slipped into someone else's. Never leave a portable computer unattended in a public place.

It's also possible to simply lose a portable. You might, for example, forget that it's in the overhead-luggage bin on an airplane. To help in its return, use a wide piece of clear tape to affix a card with your name and address to the outside of the machine. You should tape a similar card to the inside.

Desktop computers are also easily stolen. However, for under $25, you can buy a cable and lock, like those used for bicycles, and secure the computer, monitor, and printer to your work area. In addition, if your hardware does get stolen, its recovery may be helped if you have inscribed your driver's license number, Social Security number, or home address on each piece. Some campus and city police departments lend inscribing tools for such purposes.

Finally, insurance to cover computer theft or damage is surprisingly cheap. Look for advertisements in computer magazines. (If you have standard tenants' or homeowners' insurance, it may not cover your computer. Ask your insurance agent.)

Guarding Against Heat, Cold, Spills, & Drops "We fried them. We froze them. We hurled them. We even tried to drown them," proclaimed *PC Computing,* in a story about its sixth annual "torture test" of notebook computers. "And only about half survived."

The magazine put 16 notebook computers through durability trials. One test was equivalent to putting these machines in a car trunk in the desert heat (2 hours at 180 degrees), another with leaving them outdoors in a Chicago winter (2 hours at 0 degrees). A third test simulated sloshing coffee on a keyboard, and a fourth dropped them 29 inches to the floor. All

passed the bake and freeze tests. Some keyboards failed the coffee-spill test, although most revived after drying out and/or cleaning. Seven failed one of the two drop tests (flat drop and edge drop). Of the 16, nine ultimately survived (although some required reformatting after the drop tests).

This gives you an idea of how durable computers are. Designed for portability, notebooks may be hardier than desktop machines. Even so, you really don't want to tempt fate by dropping your computer, which could cause your hard-disk drive to fail. And you really shouldn't sip beverages around your notebook. Or, if you do, take your coffee black, since sugar and cream do the most damage.

Guarding Against Power Fluctuations Electricity is supposed to flow to an outlet at a steady voltage level. No doubt, however, you've noticed instances when the lights in your house suddenly brighten or, because a household appliance kicks in, dim momentarily. Such power fluctuations can cause havoc with your computer system, although most computers have some built-in protection. An increase in voltage may be a spike, lasting only a fraction of a second, or a surge, lasting longer. A surge can burn out the power supply circuitry in the system unit. A decrease may be a momentary voltage sag, a longer brownout, or a complete failure or blackout. Sags and brownouts can produce a slowdown of the hard-disk drive or a system shutdown.

Power problems can be handled by plugging your computer, monitor, and other devices into a surge protector or surge suppressor or an uninterruptible power supply (UPS) unit. If you're concerned that a lightning storm might send a surge to your system, simply unplug all your hardware until the storm passes.

Guarding Against Damage to Software System software and application software generally come on CD-ROM disks or floppy disks. The unbreakable rule is simply this: Copy the original disk, either onto your hard-disk drive or onto another floppy. Then store the original disk in a safe place. If your computer gets stolen or your software destroyed, you can retrieve the original and make another copy.

Protecting Your Data

Computer hardware and commercial software are nearly always replaceable, although perhaps with some expense and difficulty. Replacing data, however, may be a major problem, if it's possible at all. (A report of an eyewitness account, say, or a complex spreadsheet project might not come out the same way when you try to reconstruct it.) The following are some precautions to take.

Backup, Backup, Backup Sooner or later, almost every microcomputer user has the experience of accidentally wiping out or losing material and having no backup copy. After such an experience, you will probably become a true believer in backing up your data. If you're working on a research paper, for example, it's fairly easy to copy your work onto a diskette at the end of each work session. You can then store that disk in another location. If your computer is destroyed by fire, at least you'll still have the data (unless you stored your disk right next to the computer).

Treating Floppy Disks with Care Diskettes can be harmed by any number of enemies, including spills, dirt, heat, moisture, weights, and magnetic fields and magnetized objects. Some tips for taking care of your floppies are shown in the adjacent box. *(See ● Panel 6.3.)*

● PANEL 6.3
Taking care of your floppies

- Insert the floppy disk *carefully* into the disk drive.
- Do not manipulate the metal shutter on the diskette; it protects the surface of the magnetic material inside.
- Do not place heavy objects on the diskette.
- Do not expose the floppy disk to excessive heat or light.
- Do not use or place floppies near a magnetic field, such as a telephone or paper clips stored in magnetic holders. Magnetic fields can destroy data.
- Instead of leaving disks scattered on your desk, where they can be harmed by dust or beverage spills, store them in their boxes.

Guarding Against Viruses Computer viruses are programs—"deviant" programs—that can cause destruction to computers. They are spread from computer to computer in two ways: (1) They can be passed by way of an "infected" diskette—perhaps a floppy disk with a pirated videogame acquired from a friend. (2) They can be passed over a network—for example, as an e-mail attachment. They can then attach themselves to software on your hard disk, adding garbage to your files, erasing them, or wreaking havoc with the system software. They may display messages on your screen (such as "Jason Lives") or they may evade detection and spread their influence elsewhere in your system.

Each day, viruses are getting more sophisticated and harder to detect. The best protection is to install antivirus software. Some programs prevent viruses from infecting your system, others detect the viruses that have slipped through, and still others remove viruses or institute damage control. (Some of the major antivirus programs are Norton AntiVirus, Dr. Solomon's Anti-Virus Toolkit, McAfee Associates' Virus Scan, and Webscan.)

Protecting Your Health

More important than any computer system and (probably) any data is your health. What adverse effects might computers cause? The most serious are painful hand and wrist injuries, eyestrain and headache, and back and neck pains. Let's see what you can do to avoid these problems.

Protecting Your Hands & Wrists To avoid difficulties, consider the following factors:

- **Work-area setup:** With a computer, it's important to set up your work area so that you sit with both feet on the floor, thighs at right angles to your body. The chair should be adjustable and support your lower back. Your forearms should be parallel to the floor. You should look down slightly at the screen. This setup is particularly important if you are going to be sitting at a computer for hours.

- **Wrist position:** To avoid wrist and forearm injuries, you should keep your wrists straight and hands relaxed as you type. Instead of putting the keyboard on top of a desk, therefore, you should put it on a low table or in a keyboard drawer under the desk. Otherwise the nerves in your wrists will rub against the sheaths surrounding them, possibly leading to RSI pains. Some experts also suggest using a padded, adjustable wrist rest, which attaches to the keyboard.

  Various kinds of ergonomic keyboards are also available, such as those hinged in the middle.

Guarding Against Eyestrain, Headaches, & Back & Neck Pains Eyestrain and headaches usually arise because of improper lighting, screen glare, and long shifts staring at the screen. Make sure your windows and lights don't throw a glare on the screen and that your computer is not framed by an uncovered window.

Back and neck pains occur because furniture is not adjusted correctly or because of constant computer use. Adjustable furniture and frequent breaks should provide relief.

came along, allowing any file to be accessed randomly. This permitted the development of the relational database, which we will discuss in more detail shortly. In a relational database, an address change need be entered only once, and the updated information is then available in any relevant file.

The advantages of relational databases are as follows:

- **File sharing:** This is perhaps the biggest benefit. Now all authorized users can work with the same set of files.
- **Reduced data redundancy:** *Data redundancy* means that the same data fields (a person's address, say) appear in many different files and often in different formats. In the old file system, separate files would repeat the same data, wasting storage space. In a database, the information appears just once, freeing up more storage capacity. Moreover, the *same* information is available to *different* users.
- **Improved data integrity:** *Data integrity* means that data is accurate, consistent, and up to date. In the old system, when a change was made in one file, it might not get made in another. The result was that some reports were produced with erroneous information. In a relational database, reduced redundancy increases the chances of data integrity—the chances that the data is accurate, consistent, and up to date—because each updating change is made in only one place.
- **Increased security:** Although various departments may share data, access to specific information can be limited to selected users. Thus, through the use of passwords, a student's financial, medical, and grade information in a university database is made available only to those who have a legitimate need to know.

CONCEPT CHECK

Distinguish among four types of databases.

What are the advantages of databases?

6.3 Managing Files: Basic Concepts

KEY QUESTIONS
What are the data storage hierarchy, the key field, types of files, ways of keeping track of files, sequential versus direct access, and offline versus online storage?

When you create a letter or a spreadsheet or an address file on a computer, you often want to save it for future reference. Let's look at what the mechanics of this process are.

How Data Is Organized: The Data Storage Hierarchy

Data can be grouped into a hierarchy of categories, each increasingly more complex. The **data storage hierarchy** **consists of the levels of data stored in a computer: bits, bytes (characters), fields, records, files, and databases.** *(See ● Panel 6.4.)*

Computers, we have said, are based on the principle that electricity may be on or off. Thus, individual items of data are represented by the bits 0 for off and 1 for on. Bits and bytes are the building blocks for representing data, whether it is being processed, stored, or telecommunicated. The computer deals with the bits and bytes; you, however, will need to deal with characters, fields, records, files, and databases.

- **Characters:** A ***character*** **is a letter, number, or special character.** A, B, C, 1, 2, 3, #, $, % are all examples of single characters.
- **Field:** A ***field*** **is a unit of data consisting of one or more characters (bytes).** An example of a field is your name, your address, or your Social Security number.

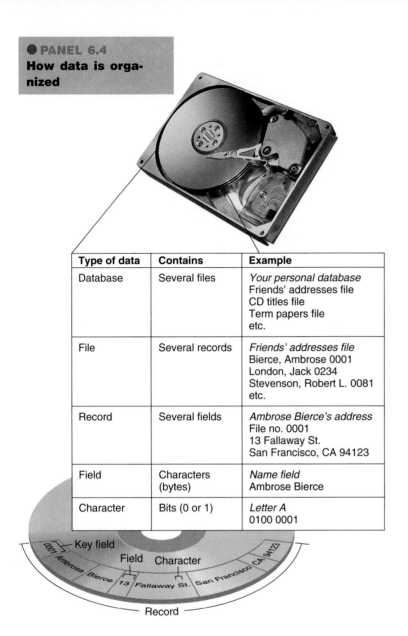

Type of data	Contains	Example
Database	Several files	*Your personal database* Friends' addresses file CD titles file Term papers file etc.
File	Several records	*Friends' addresses file* Bierce, Ambrose 0001 London, Jack 0234 Stevenson, Robert L. 0081 etc.
Record	Several fields	*Ambrose Bierce's address* File no. 0001 13 Fallaway St. San Francisco, CA 94123
Field	Characters (bytes)	*Name field* Ambrose Bierce
Character	Bits (0 or 1)	*Letter A* 0100 0001

- **Record:** A **_record_ is a collection of related fields.** An example of a record would be your name *and* address *and* Social Security number.
- **File:** A **_file_ is a collection of related records.** An example of a file is data collected on everyone employed in the same department of a company, including all names, addresses, and Social Security numbers. You use files a lot because the file is the collection of data or information that is treated as a unit by the computer.
- **Database:** A **_database_ is an organized collection of integrated files.** A company database might include files on all past and current employees in all departments. There would be various files for each employee: payroll, retirement benefits, sales quotas and achievements (if in sales), and so on.

The Key Field

An important concept in data organization is that of the *key field*. A **_key field_ is a field that is chosen to uniquely identify a record so that it can be easily retrieved and processed.** The key field is often an identification number, Social Security number, customer account number, or the like. The primary characteristic of the key field is that it is *unique*. Thus, numbers are clearly preferable to names as key fields because there are many people with common names like James Johnson, Susan Williams, Ann Wong, or Roberto Sanchez, whose records might be confused.

Types of Files: Program Files, Data Files, & Others

As we said, the *file* is the collection of data or information that is treated as a unit by the computer. *Files are given names—**filenames.*** If you're using a word processing program to write a psychology term paper, you might name it Psychreport.

Filenames also have *extension names*. These extensions of up to three letters are added after a period following the filename—for example, the *.doc* in Psychreport.doc is recognized by Microsoft Word as a "document." Extensions are usually inserted automatically by the application software.

When you look up the filenames listed on your hard drive (on the directory, as we will explain), you will notice a number of extensions, such as *.doc, .exe,* and *.com.* There are many kinds of files, but perhaps the two principal ones are *program files* and *data files.*

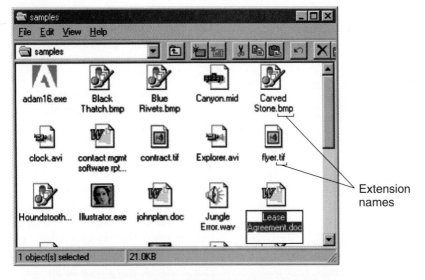

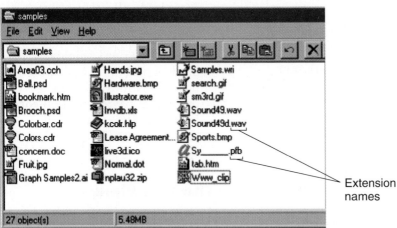

Extension names

Extension names

- **Program files:** ***Program files*** **are files containing software instructions.** Examples are word processing or spreadsheet programs, which are made up of several different program files. The two most important are source program files and executable files.

  Source program files contain high-level computer instructions in the original form written by the programmer. Some source program files have the extension of the language in which they are written, such as *.bas* for BASIC, *.pas* for Pascal, or *.jav* for Java.

  For the processor to use source program instructions, they must be translated into an *executable file*, which contains the instructions that tell the computer how to perform a particular task. You can identify an executable file by its extension, *.exe.* You use an executable file by running it—as when you select Microsoft Excel from your onscreen menu and run it. (There are some executable files that you cannot run—another computer program causes them to execute. These are identified by such extensions as *.dll, .drv, ocx, .sys,* and *.vbx.*)

- **Data files:** ***Data files*** **are files that contain data**—words, numbers, pictures, sounds, and so on. Unlike program files, data files don't instruct the computer to do anything. Rather, data files are there to be acted upon by program files. Examples of common extensions in data files are *.txt* (text) and *.xls* (spreadsheets). Certain proprietary software programs have their own extensions, such as *.ppt* for PowerPoint and *.mdb* for Access.

Other common types of files are *ASCII files, image files, audio files, animation/video files,* and *Web files.*

- **ASCII files:** ASCII is a common binary coding scheme used to represent data in a computer. *ASCII ("as-key") files* are text-only files that contain no graphics and no formatting, such as boldface or italics. This format is used to transfer documents between incompatible computers, such as PC and Macintosh. Such files may use the *.txt* extension.

- **Image files:** If ASCII files are for text, *image files* are for digitized graphics, such as art or photographs. They are indicated by such extensions as *.bmp, .gif, .jpg, .pcx, .tif,* and *.wmf.*

- **Audio files:** *Audio files* contain digitized sound and are used for conveying sound in CD-ROM multimedia and over the Internet. They have extensions such as *.wav* and *.mid.*

- **Animation/video files:** *Video files,* used for such purposes as conveying moving images over the Internet, contain digitized video images. Common extensions are *.avi, flc, .fli,* and *.mpg.*

- **Web files:** *Web files* are files carried over the World Wide Web. Their extensions include *.html*, *.htm*, and *.xml*.

Keeping Track of Files: Drive Letters, Directories/Folders, & Paths

How do you keep track of the files on your computer? On PC microcomputers, the operating system has a filing system that uses *drive letters, folders (directories)*, and *filenames* with *extensions*.

- **Drive letters:** Whether desktop or laptop, your computer probably has more than one storage device. Each device is assigned a letter, usually as follows:

  A for first floppy-disk drive.
  B for second floppy-disk drive (if any).
  C for hard drive.
  D for CD-ROM or DVD-ROM drive.
  E for Zip drive (if any).
  F for tape backup drive (if any).

  Some of these are shown in the illustration below. *(See ● Panel 6.5.)* Today there's a high probability you will have only *A* (one floppy drive), *C* (hard drive), and *D* (CD-ROM or DVD-ROM drive) on your PC. (Macintoshes don't use this letter system, although you can assign them to the drives if you wish.)

- **Directories or folders:** A *directory*, more commonly called a *folder*, is a storage place for files in one of your drives. These are analogous to manila folders in a file drawer, with each folder holding several files. All files on disk go into different folders. Directories or folders are organized in a hierarchy called a *tree structure:* root directory (root folder), directory (folder), and subdirectory (subfolder). The directories and subdirectories all start at the root.

  For each disk or CD-ROM, the Windows operating system provides a *root directory* or *root folder*, which is the top-level directory or main folder on that drive. When you open a disk drive in My

● PANEL 6.5
Letters assigned to microcomputer drives

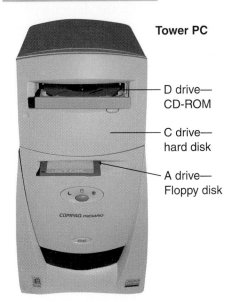

Tower PC

D drive—
CD-ROM

C drive—
hard disk

A drive—
Floppy disk

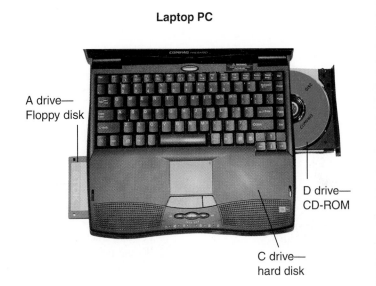

Laptop PC

A drive—
Floppy disk

D drive—
CD-ROM

C drive—
hard disk

Computer, the files and folders are all stored in the root directory (root folder). Beneath the root directory (root folder) are directories (folders), and beneath them subdirectories (subfolders). A Windows list-style directory screen will show the directory or subdirectory, filenames with extensions, their size in bytes, and the date and time they were created or last modified. Windows also provides an icon-style directory. *(See ● Panel 6.6.)*

- **File specification or path:** To identify a particular file, the operating system uses a *__file specification__*, or *path*, **which specifies the sequence of directories the computer must follow to locate a file.** When you specify the drive letter and various directories, it's important they be followed by a backslash symbol (\). An example is as follows:

C:\My Documents\Papers\Psychreport.doc

Root directory   Directory   Subdirectory   Filename   Extension
(root folder)   (folder)   (subfolder)

Two Types of Data Files: Master File & Transaction File

Among the several types of data files, two are commonly used to update data: a master file and a transaction file.

- **Master file:** The *__master file__* **is a data file containing relatively permanent records that are generally updated periodically.** An example of a master file would be the address-label file for all students currently enrolled at your college.

- **Transaction file:** The *__transaction file__* **is a temporary holding file that holds all changes to be made to the master file: additions, deletions, revisions.** For example, in the case of the address labels for your college, a transaction file would hold new names and addresses to be added (because over time new students enroll) and names and addresses to be deleted (because students leave). It would also hold revised names and addresses (because students change their names or move). Each month or so, the master file would be *updated* with the changes called for in the transaction file.

Data Access Methods: Sequential versus Direct Access

The way that a secondary-storage device allows access to the data stored on it affects its speed and its usefulness for certain applications. The two main types of data access are sequential and direct.

- **Sequential storage:** *__Sequential storage__* **means that data is stored and retrieved in sequence,** such as alphabetically. Tape storage falls in the category of sequential storage. Thus, if you are looking for employee number 8888 on a tape, the computer will have to start with 0001, then go past 0002, 0003, and so on, until it finally comes to 8888. This data access method is less expensive than other methods because it uses magnetic tape, which is cheaper than disks. The disadvantage of sequential file organization is that searching for data is slow.

- **Direct access storage:** *__Direct access storage__* **means that the computer can go directly to the information you want**—just as a CD player can go directly to a particular track on a music CD. The data is retrieved (accessed) according to a unique data identifier called a *key field*, as we will discuss. It also uses a *file allocation table (FAT)*, a hidden on-disk table that records exactly where the parts of a given file are stored.

List Style Directory

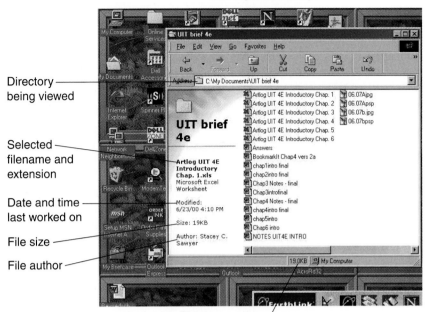

Directory being viewed

Selected filename and extension

Date and time last worked on

File size

File author

Storage space used
by all files in directory

Icon Style Directory

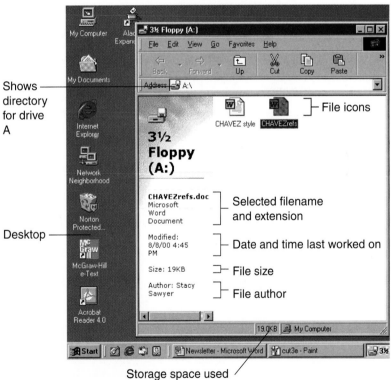

Shows directory for drive A

Desktop

File icons

Selected filename and extension

Date and time last worked on

File size

File author

Storage space used
by all files in directory

This method of file organization is used with hard disks and other types of disks. It is ideal for applications where there is no fixed pattern to the requests for data—for example, in airline reservation systems or computer-based directory-assistance operations.

If you need to find specific data, direct file access is much faster than sequential access. However, direct file access is also more expensive, for two reasons: (1) the complexity involved in maintaining a file allocation table and (2) the need to use hard-disk technology, rather than cheaper magnetic tape technology.

E-Commerce, Files, & Databases

251

Offline versus Online Storage

Whether it's on magnetic tape or on some form of disk, data may be stored either offline or online.

- **Offline:** ___Offline storage___ **means that data is not directly accessible for processing until the tape or disk it's on has been loaded onto an input device.** That is, the storage is not under the direct, immediate control of the central processing unit.

- **Online:** ___Online storage___ **means that stored data is randomly (directly) accessible for processing.** That is, storage is under the direct, immediate control of the central processing unit. You need not wait for a tape or disk to be loaded onto an input device.

For processing to be online, the storage must be online and *fast*. This nearly always means storage on disk (direct access storage) rather than magnetic tape (sequential storage).

6.4 The Ways Databases Are Organized

KEY QUESTION
What are four types of database organization?

Just as files can be organized in different ways (sequentially, for example), so databases can be organized in ways to best fit their use. The four most common arrangements are *hierarchical, network, relational,* and *object-oriented.*

Hierarchical Database

In a ___hierarchical database___, **fields or records are arranged in related groups resembling a family tree, with child (lower-level) records subordinate to parent (higher-level) records.** The parent record at the top of the database is called the *root parent.* (See ● *Panel 6.7.)*

Used principally on mainframes, hierarchical databases are the oldest of the four types. They are still used in some types of passenger reservation systems. In hierarchical databases, accessing or updating data is very fast, because the relationships have been predefined. However, because the structure must be defined in advance, it is quite rigid. There may be only one par-

● PANEL 6.7

Hierarchical database

Example of a cruise ship reservation system

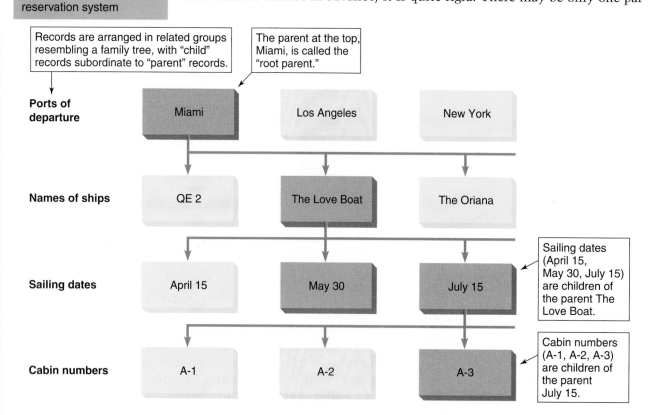

Records are arranged in related groups resembling a family tree, with "child" records subordinate to "parent" records.

The parent at the top, Miami, is called the "root parent."

Ports of departure	Miami	Los Angeles	New York
Names of ships	QE 2	The Love Boat	The Oriana
Sailing dates	April 15	May 30	July 15
Cabin numbers	A-1	A-2	A-3

Sailing dates (April 15, May 30, July 15) are children of the parent The Love Boat.

Cabin numbers (A-1, A-2, A-3) are children of the parent July 15.

ent per child, and no relationships among the child records are possible. Moreover, adding new fields to database records requires that the entire database be redefined.

Network Database

A **_network database_** **is similar to a hierarchical database, but each child record can have more than one parent record.** *(See ● Panel 6.8, next page.)* Thus, a child record, which in network database terminology is called a *member*, may be reached through more than one parent, called an *owner*.

Also used principally with mainframes, the network database is more flexible than the hierarchical arrangement, because different relationships may be established between different branches of data. However, it still requires that the structure be defined in advance. Moreover, there are limits to the number of possible links among records.

Relational Database

More flexible than hierarchical and network database models, **the _relational database_ relates, or connects, data in different files through the use of a key field, or common data element.** *(See ● Panel 6.9, next page.)* In this arrangement there are no access paths down through a hierarchy. Instead, data elements are stored in different tables made up of rows and columns. In database terminology, the tables are called *relations* (files), the rows are called *tuples* (records), and the columns are called *attributes* (fields). All related tables must have a key field that uniquely identifies each row; that is, the key field must be in *all* tables.

The advantage of relational databases is that the user does not have to be aware of any structure. Thus, they can be used with little training. Moreover, entries can easily be added, deleted, or modified. A disadvantage is that some searches can be time-consuming. Nevertheless, the relational model has become popular for microcomputer database management programs, such as Paradox and Access.

Object-Oriented Database

An **object-oriented database** *uses "objects," software written in small, reusable chunks, as elements within database files.* An *object* consists of (1) data in any form, including graphics, audio, and video, and (2) instructions on the action to be taken on the data.

A hierarchical or network database might contain only numeric and text data about a student—identification number, name, address, and so on. By contrast, an object-oriented database might also contain the student's photograph, a "sound bite" of his or her voice, and even a short piece of video. Moreover, the object would store operations, called *methods*, the programs that objects use to process themselves. For example, these programs might indicate how to calculate the student's grade-point average or how to display or print the student's record.

CONCEPT CHECK

Describe the data storage hierarchy and the concept of a key field.

Distinguish program files from data files.

List other types of files.

Explain how to keep track of files on your computer.

Distinguish sequential from direct access.

Discuss offline and online storage.

Describe the four types of database organization.

E-Commerce, Files, & Databases

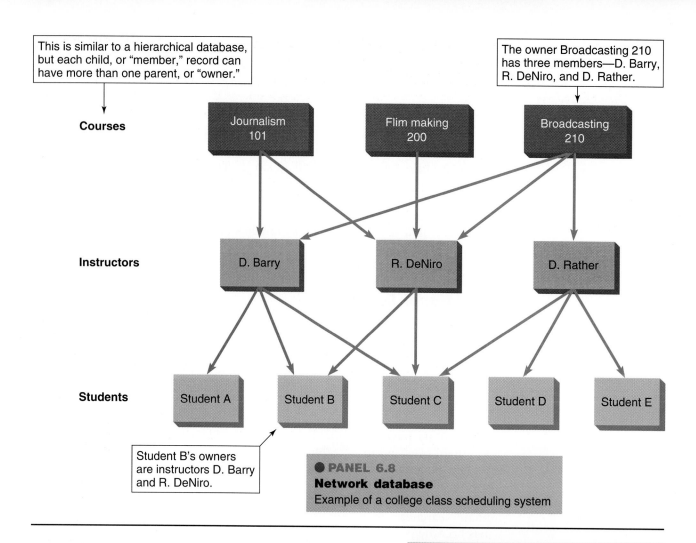

This is similar to a hierarchical database, but each child, or "member," record can have more than one parent, or "owner."

The owner Broadcasting 210 has three members—D. Barry, R. DeNiro, and D. Rather.

Courses

Journalism 101

Flim making 200

Broadcasting 210

Instructors

D. Barry

R. DeNiro

D. Rather

Students

Student A

Student B

Student C

Student D

Student E

Student B's owners are instructors D. Barry and R. DeNiro.

● PANEL 6.8
Network database
Example of a college class scheduling system

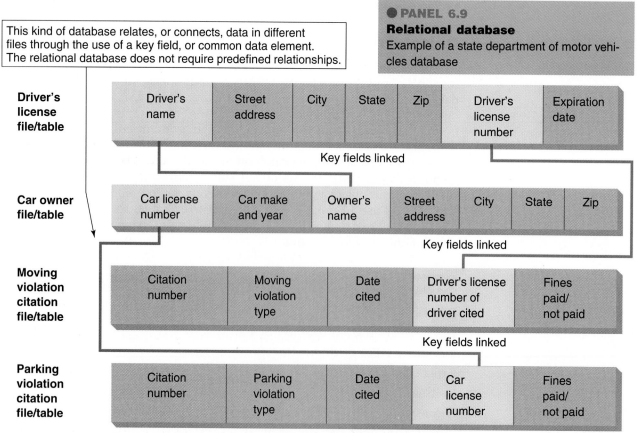

● PANEL 6.9
Relational database
Example of a state department of motor vehicles database

This kind of database relates, or connects, data in different files through the use of a key field, or common data element. The relational database does not require predefined relationships.

Driver's license file/table

| Driver's name | Street address | City | State | Zip | Driver's license number | Expiration date |

Key fields linked

Car owner file/table

| Car license number | Car make and year | Owner's name | Street address | City | State | Zip |

Key fields linked

Moving violation citation file/table

| Citation number | Moving violation type | Date cited | Driver's license number of driver cited | Fines paid/ not paid |

Key fields linked

Parking violation citation file/table

| Citation number | Parking violation type | Date cited | Car license number | Fines paid/ not paid |

6.5 The Ethics of Using Storage & Databases: Concerns about Accuracy & Privacy

KEY QUESTION

What are some ethical concerns about the uses of databases?

Ethics

The enormous capacities of today's storage devices have given photographers, graphics professionals, and others a new tool—the ability to manipulate images at the pixel level. For example, photographers can easily do *morphing*—transforming one image into another. In **_morphing_, a film or video image is displayed on a computer screen and altered pixel by pixel, or dot by dot.** As a result, the image metamorphoses into something else—a pair of lips into the front of a Toyota, for example.

The ability to manipulate digitized output—images and sounds—has brought a wonderful new tool to art. However, it has created some big new problems in the area of credibility, especially for journalism. How can we know that what we're seeing or hearing is the truth? Consider the following.

Manipulation of Sound

Frank Sinatra's 1994 album *Duets* paired him through technological tricks with singers like Barbra Streisand, Liza Minnelli, and Bono of U2. Sinatra recorded solos in a recording studio. His singing partners, while listening to his taped performance on earphones, dubbed in their own voices. These second voices were recorded not only at different times but often, through distortion-free phone lines, from different places. The illusion in the final recording is that the two singers are standing shoulder to shoulder.

Newspaper columnist William Safire called *Duets* "a series of artistic frauds." Said Safire, "The question raised is this: When a performer's voice and image can not only be edited, echoed, refined, spliced, corrected, and enhanced—but can be transported and combined with others not physically present—what is performance? . . . Enough of additives, plasticity, virtual venality; give me organic entertainment."[21] Some listeners feel that the technology changes the character of a performance for the better. Others, however, think the practice of assembling bits and pieces in a studio drains the music of its essential flow and unity.

Whatever the problems of misrepresentation in art, however, they pale beside those in journalism. What if, for example, a radio station were to edit a stream of digitized sound so as to misrepresent what actually happened?

Manipulation of Photos

When O. J. Simpson was arrested in 1994 on suspicion of murder, the two principal American newsmagazines both ran pictures of him on their covers.[22] *Newsweek* ran the mug shot unmodified, as taken by the Los Angeles Police Department. At *Time*, an artist working with a computer modified the shot with special effects as a "photo illustration." Simpson's image was darkened so that it still looked like a photo but, some critics said, with a more sinister cast to it.

Should a magazine that reports the news be taking such artistic license? Should *National Geographic* in 1982 have photographically moved two Egyptian pyramids closer together so that they would fit on a vertical cover? Was it even right for *TV Guide* in 1989 to run a cover showing Oprah Winfrey's head placed on Ann-Margret's body? In another case, to show what can be done, a photographer digitally manipulated the famous 1945 photo showing the meeting of the leaders of the wartime Allied powers at Yalta. Joining Stalin, Churchill, and Roosevelt are some startling newcomers: Sylvester Stallone and Groucho Marx. The additions are so seamless that it is impossible to tell the photo has been altered. *(See* ● *Panel 6.10, next page.)*

The potential for abuse is clear. "For 150 years, the photographic image has been viewed as more persuasive than written accounts as a form of 'evidence,'" says one writer. "Now this authenticity is breaking down under the

assault of technology."[23] Asks a former photo editor of the *New York Times Magazine*, "What would happen if the photograph appeared to be a straightforward recording of physical reality, but could no longer be relied upon to depict actual people and events?"[24]

Many editors try to distinguish between photos used for commercialism (advertising) versus for journalism, or for feature stories versus for news stories. However, this distinction implies that the integrity of photos is only important for some narrow category of news. In the end, it can be argued, altered photographs pollute the credibility of all of journalism.

Manipulation of Video & Television

The technique of morphing, used in still photos, takes a massive jump when used in movies, videos, and television commercials. Digital image manipulation has had a tremendous impact on filmmaking. Director and digital pioneer Robert Zemeckis *(Death Becomes Her)* compares the new technology to the advent of sound in Hollywood.[25] It can be used to erase jet contrails from the sky in a western and to make digital planes do impossible stunts. It can even be used to add and erase actors.

Virtual advertising
The oil company "76" doesn't really appear on this wall but looks as though it does.

Films and videotapes are widely thought to accurately represent real scenes (as evidenced by the reaction to the amateur videotape of the Rodney King beating by police in Los Angeles). Thus, the possibility of digital alterations raises some real problems. Videotapes supposed to represent actual events could easily be doctored.

Another concern is for film archives: Because digital videotapes suffer no loss in resolution when copied, there are no "generations." Thus, it will be impossible for historians and archivists to tell whether the videotape they're viewing is the real thing or not.[26]

Indeed, it is possible to create virtual images during live television events. These images—such as a Coca-Cola logo in the center of a soccer field—don't exist in reality but millions of viewers see them on their TV screens.[27]

Accuracy & Completeness

Databases—including public data banks such as Nexis/Lexis—can provide you with *more* facts and *faster* facts but not always *better* facts. Penny

Williams, professor of broadcast journalism at Buffalo State College in New York and formerly a television anchor and reporter, suggests five limitations to bear in mind when using databases for research:[28]

- **You can't get the whole story:** For some purposes, databases are only a foot in the door. There may be many facts or facets of the topic that are not in a database. Reporters, for instance, find a database is a starting point. It may take intensive investigation to get the rest of the story.
- **It's not the gospel:** Just because you see something on a computer screen doesn't mean it's accurate. Numbers, names, and facts must be verified in other ways.
- **Know the boundaries:** One database service doesn't have it all. For example, you can find full text articles from the *New York Times* on Lexis/Nexis, from *The Wall Street Journal* on Dow Jones News Retrieval, and from the *San Jose Mercury News* on America Online, but no service carries all three.
- **Find the right words:** You have to know which keywords (search words) to use when searching a database for a topic. As Lynn Davis, a professional researcher with ABC News, points out, if you're searching for stories on guns, the keyword "can be guns, it can be firearms, it can be handguns, it can be pistols, it can be assault weapons. If you don't cover your bases, you might miss something."[29]
- **History is limited:** Most public databases, Davis says, have information going back to 1980, and a few into the 1970s, but this poses problems if you're trying to research something that happened or was written about earlier.

Matters of Privacy

Privacy **is the right of people to not reveal information about themselves.** Who you vote for in a voting booth and what you say in a letter sent through the U.S. mail arπe private matters. However, the ease of pulling together and disseminating information via databases and communications lines has put privacy under extreme pressure.

As you've no doubt discovered, it's no trick at all to get your name on all kinds of mailing lists. Theo Theoklitas, for instance, has received applications for credit cards, invitations to join video clubs, and notification of his finalist status in Ed McMahon's $10 million sweepstakes. Theo is a black cat who's been getting mail ever since his owner sent in an application for a rebate on cat food. A whole industry has grown up of professional information gatherers and sellers, who collect personal data and sell it to fundraisers, direct marketers, and others.

In the 1970s, the Department of Health, Education, and Welfare developed a set of five Fair Information Practices regarding the use and misuse of information. These rules have since been adopted by a number of public and private organizations and have led to the enactment of a number of laws to protect individuals from invasion of privacy. *(See ● Panel 6.11, next page.)*

CONCEPT CHECK

Discuss how computers can manipulate sounds and images.

What are the limitations of using databases for research?

Discuss concerns about privacy.

Fair Information Practices

1. There must be no personal data record-keeping systems whose existence is a secret from the general public.

2. People have the right to access, inspect, review, and amend data about them that is kept in an information system.

3. There must be no use of personal information for purposes other than those for which it was gathered without prior consent.

4. Managers of systems are responsible and should be held accountable and liable for the reliability and security of the systems under their control, as well as for any damage done by those systems.

5. Governments have the right to intervene in the information relationships among private parties to protect the privacy of individuals.

Important Federal Privacy Laws

Freedom of Information Act (1970): Gives you the right to look at data concerning you that is stored by the federal government. A drawback is that sometimes a lawsuit is necessary to pry it loose.

Fair Credit Reporting Act (1970): Bars credit agencies from sharing credit information with anyone but authorized customers. Gives you the right to review and correct your records and to be notified of credit investigations for insurance or employment. A drawback is that credit agencies may share information with anyone they reasonably believe has a "legitimate business need." Legitimate is not defined.

Privacy Act (1974): Prohibits federal information collected about you for one purpose from being used for a different purpose. Allows you the right to inspect and correct records. A drawback is that exceptions written into the law allow federal agencies to share information anyway.

Family Educational Rights and Privacy Act (1974): Gives students and their parents the right to review, and to challenge and correct, students' school and college records; limits sharing of information in these records.

Right to Financial Privacy Act (1978): Sets strict procedures that federal agencies must follow when seeking to examine customer records in banks; regulates financial industry's use of personal financial records. A drawback is that the law does not cover state and local governments.

Privacy Protection Act (1980): Prohibits agents of federal government from making unannounced searches of press offices if no one there is suspected of a crime.

Cable Communications Policy Act (1984): Restricts cable companies in the collection and sharing of information about their customers.

Computer Fraud and Abuse Act (1986): Allows prosecution for unauthorized access to computers and databases. A drawback is that people with legitimate access can still get into computer systems and create mischief without penalty.

Electronic Communications Privacy Act (1986): Makes eavesdropping on private conversations illegal without a court order.

Computer Security Act (1987): Makes actions that affect the security of computer files and telecommunications illegal.

Computer Matching and Privacy Protection Act (1988): Regulates computer matching of federal data; allows individuals a chance to respond before government takes adverse actions against them. A drawback is that many possible computer matches are not affected, such as those done for law-enforcement or tax reasons.

Video Privacy Protection Act (1988): Prevents retailers from disclosing video-rental records without the customer's consent or a court order.

Visual Summary

business-to-business (B2B) system (p. 240, KQ 6.1) Activity of a business selling to other businesses, using the Internet to cut transaction costs and increase efficiencies. *Why it's important: Business-to-business activity is expected to balloon to a $1 trillion industry by 2002 and a $2.8 trillion industry by 2004.*

character (p. 246, KQ 6.3) A single letter, number, or special character. *Why it's important: A, B, C, 1, 2, 3, #, $, % are all examples of single characters. Characters are part of the data storage hierarchy.*

database (p. 247, KQ 6.3) Organized collection of related (integrated) files. *Why it's important: Businesses and organizations build databases to help them keep track of and manage their affairs. In addition, online database services put enormous research resources at the user's disposal.*

database administrator (DBA) (p. 243, KQ 6.2) Person who coordinates all related activities and needs for an organization's database. *Why it's important: The DBA determines user access privileges; sets standards, guidelines, and control procedures; assists in establishing priorities for requests; prioritizes conflicting user needs; and develops user documentation and input procedures. He or she is also concerned with security—setting up and monitoring a system for preventing unauthorized access and making sure that the system is regularly backed up and that data can be recovered should a failure or disaster occur.*

data files (p. 248, KQ 6.3) Files that contain data—words, numbers, pictures, sounds, and so on. *Why it's important: Unlike program files, data files don't instruct the computer to do anything. Rather, data files are there to be acted on by program files. Examples of common extensions in data files are .txt (text) and .xls (spreadsheets). Certain proprietary software programs have their own extensions, such as .ppt for PowerPoint and .mdb for Access.*

data mining (DM) (p. 240, KQ 6.1) Computer-assisted process of sifting through and analyzing vast amounts of data in order to extract meaning and discover new knowledge. *Why it's important: The purpose of DM is to describe past trends and predict future trends. Thus, data-mining tools might sift through a company's immense collections of customer, marketing, production, and financial data and identify what's worth noting and what's not.*

data storage hierarchy (p. 246, KQ 6.3) Consists of the levels of data stored in a computer: bits, bytes (characters), fields, records, files, and databases. *Why it's important: Understanding the data storage hierarchy is necessary to understand how to use a database.*

direct access storage (p. 250, KQ 6.3) Storage system that allows the computer to go directly to the desired information. The data is retrieved (accessed) according to a unique data identifier called a key field. It also uses a file allocation table (FAT), a hidden on-disk table that records exactly where the parts of a given file are stored. *Why it's important: This method of file organization is used with hard disks and other types of disks. It is ideal for applications where there is no fixed pattern to the requests for data—for example, in airline reservation systems or computer-based directory-assistance operations. Direct access storage is much faster than sequential access storage.*

directory (p. 249, KQ 6.3) More commonly called a *folder;* storage place for files in one of your drives. *Why it's important: Directories are analogous to manila folders in a file drawer, with each folder holding several files. All files on disk go into different folders.*

distributed database (p. 243, KQ 6.2) Database that is stored on different computers in different locations connected by a client/server network. *Why it's important: Data need not be centralized in one location.*

Customer   Shipping
service

e-commerce (p. 238, KQ 6.1) Electronic commerce; the buying and selling of products and services through computer networks. Why it's important: *By 2003, total U.S. e-commerce sales to consumers are expected to reach $108 billion, or 6% of consumer retail spending; online shopping is growing even faster than the increase in computer use, which has been fueled by the falling price of personal computers.*

field (p. 246, KQ 6.3) Unit of data consisting of one or more characters (bytes). An example of a field is your name, your address, or your Social Security number. Why it's important: *A collection of fields makes up a record. Also see key field.*

file (p. 247, KQ 6.3) Collection of related records. An example of a file is data collected on everyone employed in the same department of a company, including all names, addresses, and Social Security numbers. Why it's important: *A file is the collection of data or information that is treated as a unit by the computer; a collection of related files makes up a database.*

file specification (p. 250, KQ 6.3) Also called *path;* specifies the sequence of directories the computer must follow to locate a file. Why it's important: *The operating system uses file specifications to identify a particular file. When you specify the drive letter and various directories, they must be followed by a backslash symbol (\).*

filename (p. 247, KQ 6.3) A file is the collection of data or information that is treated as a unit by the computer. Why it's important: *Files are given names so that they can be differentiated. Filenames also have extension names. These extensions of up to three letters are added after a period following the filename—for example, the .doc in Psychreport.doc is recognized by Microsoft Word as a "document." Extensions are usually inserted automatically by the application software.*

hierarchical database (p. 252, KQ 6.4) Database in which fields or records are arranged in related groups resembling a family tree, with child (lower-level) records subordinate to parent (higher-level) records. The parent record at the top of the database is called the *root record.* Why it's important: *The hierarchical database is one of the common database structures.*

individual database (p. 242, KQ 6.2) Collection of integrated files used by one person. Why it's important: *Microcomputer users can set up their own individual databases using popular database management software; the information is stored on the hard drives of their personal computers. Today the principal database programs are Microsoft Access, Corel Paradox, and Lotus Approach. In addition, types of individual databases known as personal information managers (PIMs) can help users keep track of and manage information used on a daily basis, such as addresses, telephone numbers, appointments, to-do lists, and miscellaneous notes. Popular PIMs are Microsoft Outlook, Lotus Organizer, and Act.*

key field (p. 247, KQ 6.3) Field that is chosen to uniquely identify a record so that it can be easily retrieved and processed. The key field is often an identification number, Social Security number, customer account number, or the like. Why it's important: *The primary characteristic of the key field is that it is unique and thus can be used to identify one specific record.*

master file (p. 250, KQ 6.3) Data file containing relatively permanent records that are generally updated periodically. Why it's important: *Master files contain relatively permanent information used for reference purposes. They are updated through the use of transaction files.*

morphing (p. 255, KQ 6.5) Altering a film or video image displayed on a computer screen pixel by pixel, or dot by dot. Why it's important: *Morphing and other techniques of digital manipulation can produce images that misrepresent reality.*

network database (p. 253, KQ 6.4) Database similar in structure to a hierarchical database; however, each child record can have more than one parent record. Thus, a child record, which in network database terminology is called a *member,* may be reached through more than one parent, which is called an *owner.* Why it's important: *The network database is one of the common database structures.*

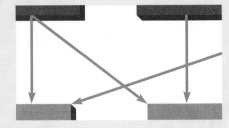

object-oriented database (p. 253, KQ 6.4) Database that uses "objects," software written in small, reusable chunks, as elements within database files. An object consists of (1) data in any form, including graphics, audio, and video, and (2) instructions on the action to be taken on the data. *Why it's important: A hierarchical or network database might contain only numeric and text data. By contrast, an object-oriented database might also contain photographs, sound bites, and video clips. Moreover, the object would store operations, called methods, the programs that objects use to process themselves.*

offline storage (p. 252, KQ 6.3) System in which stored data is not directly accessible for processing until the tape or disk it's on has been loaded onto an input device. *Why it's important: The storage medium and data are not under the direct, immediate control of the central processing unit.*

online storage (p. 252, KQ 6.3) System in which stored data is randomly (directly) accessible for processing. *Why it's important: The storage medium and data is under the direct, immediate control of the central processing unit. One need not wait for a tape or disk to be loaded onto an input device.*

program files (p. 248, KQ 6.3) Files containing software instructions. *Why it's important: Contrast data files.*

Ethics

privacy (p. 257, KQ 6.5) Right of people to not reveal information about themselves. *Why it's important: The ease of pulling together and disseminating information via databases and communications lines has put privacy under extreme pressure.*

public databanks (p. 243, KQ 6.2) Compilations of data that are available to the public. *Why it's important: Public databanks are one of the basic types of database.*

record (p. 247, KQ 6.3) Collection of related fields. An example of a record would be your name and address and Social Security number. *Why it's important: Related records make up a file.*

relational database (p. 253, KQ 6.4) Common database structure that relates, or connects, data in different files through the use of a key field, or common data element. In this arrangement there are no access paths down through a hierarchy. Instead, data elements are stored in different tables made up of rows and columns. In database terminology, the tables are called *relations* (files), the rows are called *tuples* (records), and the columns are called *attributes* (fields). All related tables must have a key field that uniquely identifies each row; that is, the key field must be in all tables. *Why it's important: The*

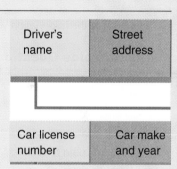

relational database is one of the common database structures; it is more flexible than hierarchical and network database models.

sequential storage (p. 250, KQ 6.3) Storage system whereby data is stored and retrieved in sequence, such as alphabetically. *Why it's important: Sequential storage is the only type of storage provided by tape, which is used mostly for archiving and backup. The disadvantage of sequential file organization is that searching for data is slow. Compare direct access storage.*

shared database (p. 242, KQ 6.2) Company database that is shared by users in one company or organization in one location. The organization owns the database, which may be stored on a server such as a mainframe. Users are linked to the database via a local area or wide area network; the users access the network through terminals or microcomputers. *Why it's important: Shared databases, such as those you find when surfing the Web, are the foundation for a great deal of electronic commerce, particularly B2B commerce.*

transaction file (p. 250, KQ 6.3) Temporary holding file that holds all changes to be made to the master file: additions, deletions, revisions. *Why it's important: The transaction file is used to periodically update the master file.*

Chapter Review

Self-Test Questions

1. According to the data storage hierarchy, databases are composed of _____, _____, _____, _____, and _____.

2. An individual piece of data within a record is called a _____.

3. A(n) _____ coordinates all activities related to an organization's database.

4. _____ is the right of people not to reveal information about themselves.

5. The four types of databases are _____, _____, _____, and _____.

6. The buying and selling of products and services through computer networks is called _____.

7. _____ files contain software instructions; _____ files contain data.

8. _____ storage means that the computer can go directly to the information you want.

Multiple-Choice Questions

1. Which of the following is not an advantage of a relational database?
   a. file sharing
   b. reduced data redundancy
   c. increased data redundancy
   d. improved data integrity
   e. increased security

2. Which of the following database models stores data in any form, including graphics, audio, and video?
   a. hierarchical
   b. network
   c. object-oriented
   d. relational
   e. offline

3. Which of the following is not a type of database?
   a. distributed
   b. combination
   c. public
   d. shared
   e. individual

True/False Questions

T F 1. The use of key fields makes it easier to locate a record in a database.

T F 2. A transaction file contains permanent records that are periodically updated.

T F 3. A database is an organized collection of integrated files.

T F 4. A shared database is stored on different computers in different locations connected by a client/server network.

T F 5. A directory (folder) is a storage place for files in one of your drives.

Short-Answer Questions

1. Name three responsibilities of a database administrator.

2. What is the difference between master files and transaction files?

3. What is data mining?

4. What is the difference between offline and online storage?

5. Briefly explain what *data mining* is.

6. What is an ASCII file?

7. What drive/device does each of the following letters indicate? A, B, C, D, E, F

8. Give an example of a file specification path.

Concept Mapping

On a separate sheet of paper, draw a concept map, or visual diagram, linking concepts. Show how the following terms are related.

character	individual database
database	key field
DBA	master file
data files	network database
data storage hierarchy	object-oriented
directory	database
distributed database	program files
field	record
file	relational database
filename	shared database
hierarchical database	transaction file

Knowledge in Action

Interview someone who works with or manages an organization's database. What types of records make up the database? Which departments use it? What database structure is used? What are the types and sizes of storage devices? Are servers used? Was the database software custom written?

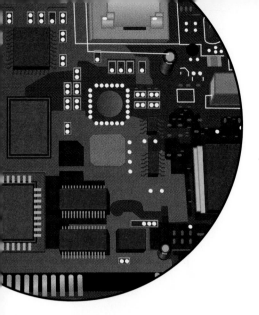

The Challenges of the Digital Age

Society & Information Technology Today

Key Questions

You should be able to answer the following questions.

7.1 The Digital Environment: Is There a Grand Design? What are the NII, the new Internet, the Telecommunications Act, the 1997 White House plan, and ICANN?

7.2 Security Issues: Threats to Computers & Communications Systems What are some characteristics of the key security issues for information technology?

7.3 Security: Safeguarding Computers & Communications What are the characteristics of the four components of security?

A Visual Overview of This Chapter

1 **The Digital Environment: Is There a Grand Design?** Some factors affecting the design of the digital environment are as follows. (1) The *National Information Infrastructure* is a grand design relying on private companies and the Internet. (2) The old Internet is being replaced by new Internet networks—VBNS, Internet2, and the NGI. The **VBNS** is the main government component to upgrade the backbone, or primary hubs of data transmission. **Internet2** is a cooperative university-business program to enable high-end users to quickly move data, using VBNS. The **Next Generation Internet (NGI)** is designed to help tie campus high-performance backbones to the federal infrastructure. (3) The *1996 Telecommunications Act* was designed to increase competition between telecommunications businesses, so different companies are no longer restricted from offering different services. (4) The 1997 White House plan stresses that government should stay out of the way of Internet commerce. (5) **ICANN** is the Internet Corporation for Assigned Names and Numbers, which is a nonprofit corporation established to regulate Internet domain names.

2 **Security Issues: Threats to Computers & Communications Systems.** Among the threats to the security of computers are the following. (1) Errors and accidents, such as human errors, procedural errors, software errors, electromechanical problems, and "dirty data" problems. (2) Natural hazards, such as earthquakes, hurricanes, and fires, and civil strife and terrorism. (3) **Computer crimes,** which can be either an illegal act perpetrated against computers or the use of computers to accomplish an illegal act. Crimes against computers include theft of hardware, software, time and services, or information, or crimes of malice and destruction. (4) Crimes using computers include credit card theft, investment fraud, and **plagiarism,** or representation of another writer's material as one's own. (5) **Worms,** programs that copy themselves repeatedly into a computer's memory or disk drive, and **viruses,** deviant programs stored on hard drives that can destroy data. Worms and viruses are passed by exchange of infected floppy disks or infected data sent over a network; **antivirus software** can detect viruses. (6) Computer criminals may be an organization's employees, outside users, hackers and crackers, and professional criminals. **Hackers** gain unauthorized access to computers often just for the challenge, **crackers** for malicious purposes.

3 **Security: Safeguarding Computers & Communications. Security,** the system of safeguards for protecting computers against disasters, failure, and unauthorized access, has four components. (1) Computer systems try to determine authorized users by three criteria: by what they have (keys, badges, signatures); by what they know (as with **PINs,** or personal identification numbers, and **passwords,** or special words or codes); and by who they are (as by physical traits, as determined perhaps through biometrics, the science of measuring individual body characteristics). (2) **Encryption,** altering data so it is not usable unless the changes are undone, tries to make computer messages more secure. (3) Software and data are protected by controlling access to files, by audit controls that track the programs used, and by people controls that screen job applicants and other users. (4) **Disaster-recovery plans** are methods for restoring computer operations after destruction or accident.

I

f . . . the Internet is on its way to becoming the dominant mode of information exchange, then it is no longer a luxury but, like the telephone, a necessity."

And it follows, in this analyst's opinion, that "anyone without it is in danger of being shut out."[1]

This begins to suggest the "digital divide," which is growing steadily wider as the Information Age continues to expand productivity and wealth. Addressing that divide is one of the most important challenges of our time. Internationally, the digital divide is between rich countries and poor countries. In the United States, 10% of the economy is devoted to the purchase of hardware and software, as compared with, say, Bangladesh, where it is one-tenth of 1%.[2] Of the world's approximately 1 billion Web pages, more than 80% are in English.[3] With only 5% of the world's population, the U.S. has 50% of the world's Internet-linked home computers.[4] Even in the United States, however, great numbers of people are shut out of the new system of information exchange, on account of their low income and/or limited knowledge of English.[5]

The digital divide is only one of the many challenges confronting us as infotech sweeps the world. Elsewhere in the book we have considered privacy and ergonomics. In this chapter, we consider some other major issues:

- Security issues—accidents, hazards, crime, viruses—and security safeguards
- Quality-of-life issues—environment, mental health, the workplace
- Economic issues—employment and the haves/have-nots

7.1 The Digital Environment: Is There a Grand Design?

KEY QUESTIONS

What are the NII, the new Internet, the Telecommunications Act, the 1997 White House plan, and ICANN?

In Chapter 1, we mentioned the *information superhighway*, a term that since has lost its luster in favor of other coinages such as the *digital environment*. The presumed goal of this worldwide system of computers and telecommunications is to give us lightning-fast (high-bandwidth) voice and data exchange, multimedia, interactivity, and near-universal, low-cost access—and to do so reliably and securely. Whether you're a Russian astronaut aloft in a spacecraft, a Bedouin tribesman in the desert with a PDA/cell phone, or a Canadian work-at-home mother with her office in a spare bedroom, you'll be able, it is hoped, to connect with nearly anything or anybody anywhere. You'll be able to participate in telephony, teleconferencing, telecommuting, teleshopping, telemedicine, tele-education, tele-voting, and even tele-psychotherapy (already available now), to name a few possibilities. What shape will this digital environment take? Some government officials hope it will follow a somewhat orderly model, such as that envisioned in the National Information Infrastructure (of which new versions of the Internet are a part). Others hope it will evolve out of competition intended by the passage of the 1996 Telecommunications Act. Still others hope that a White House document, *A Framework for Global Electronic Commerce*, offers a realistic policy. Finally, a nonprofit organization, ICANN, is concerned with Internet addressing issues. What can these attempts to create an all-encompassing design do?

The National Information Infrastructure

As portrayed by government officials, the *National Information Infrastructure (NII)* is a kind of grand vision for today's existing networks and technologies, as well as technologies yet to be deployed. Services would be delivered by telecommunications companies, cable-television companies, and the Internet. Applications would be varied—education, health care, information access, electronic commerce, and entertainment.

Who would put the pieces of the NII together? The current national policy is to let private industry do it, with the government trying to ensure fair competition among the carriers—phone, cable, and satellite companies—and compatibility among various technological systems. In addition, NII envisions open access to people of all income levels.

The New Internet: VBNS, Internet2, & NGI

Call your office
A traveling salesman contacts his office via cell phone and notebook computer.

Lately less is being said about NII and more about new Internet networks: *VBNS, Internet2,* and the *Next Generation Internet.*

What are these, and does this mean three new networks will be built? Actually, all three names refer to the same network. This high-speed Internet is designed to relieve the congested electronic highway of the older Internet. Here's what the three efforts represent:

- VBNS: **Linking supercomputers and other banks of computers across the nation, _VBNS (Very-High-Speed Backbone Network Service)_ is the main U.S. government component to upgrade the "backbone," or primary hubs of data transmission.** Speeds are 1000 times conventional Internet speeds.

  Financed by the National Science Foundation and managed by the telecommunications giant MCI, VBNS is somewhat exclusive. It is being used by just 101 top universities and other research institutions for network-intensive applications, such as transmitting high-quality video for distance education.[6] (Internet2, by contrast, will eventually touch a great many more members.) VBNS has been underway since 1996, and most of the present members are also members of Internet2.

- Internet2: **_Internet2_ is a cooperative university-business program to enable high-end users to quickly and reliably move huge amounts of data, using VBNS as the official backbone.** Whereas VBNS provides data transfer at 1000 times commercial Web speeds, Internet2 operates at only 100 times those speeds.

  In effect, Internet2 adds "toll lanes" to the older Internet to speed things up. The purpose is to advance videoconferencing, research, and academic collaboration—to enable a kind of "virtual university." Presently more than 150 universities in partnership with companies such as Cisco Systems and IBM are participants.

- Next Generation Internet: **The _Next Generation Internet (NGI)_ is the U.S. government's program to parallel the university-business–sponsored effort of Internet2. It is designed to provide money to six government agencies to help tie the campus high-performance backbones into the broader federal infrastructure.** The technical goals for NGI include connecting at least 100 sites, including universities, federal national laboratories, and other research organizations. Speeds are 100 times those of the older Internet, and 10 sites are connected at speeds that are 1000 times as fast.

In general, all of these are modeled after the original Internet, except that they will use high-speed fiber-optic circuits and more sophisticated software. NGI and Internet 2 should be available to the public by 2003.

The 1996 Telecommunications Act

After years of legislative attempts to overhaul the 1934 Communications Act, in February 1996 President Bill Clinton signed into law the *Telecommunications Act of 1996*, undoing 60 years of federal and state communications regulations. The act is designed to let phone, cable, and TV businesses compete and combine more freely. The purpose of the law was to cultivate greater competition between local and long-distance telephone companies, as well as between the telephone and cable industries. Under this legislation, different carriers can offer the same services—for example, cable companies can offer telephone services, phone companies can offer cable.

Is the law successful? "The only point on which all parties agree," says Laurence Tribe, Harvard professor of constitutional law, "is that the law isn't working as intended, and that American consumers are still waiting for free and healthy competition in communications services."[7]

No batteries needed
Boeing's new 777 airplanes feature a digital phone service, allowing laptop users to send and receive faxes and log onto the Internet. Power outlets on the back of the seats eliminate users' need to change batteries when their laptops run low on power during the flight.

The 1997 White House Plan for Internet Commerce

In 1997, the Clinton administration unveiled a document, authored by a White House group, that is significant because it endorses a governmental hands-off approach to the Internet—or, as the report more grandly calls it, "the Global Information Infrastructure."[8] Behind the title *A Framework for Global Electronic Commerce*, was a plan whose gist is this: Government should stay out of the way of Internet commerce.

"Where government is needed," it states, "its aim should be to support and enforce a predictable, minimalist, consistent, and legal environment for commerce."[9] Otherwise, the plan states that private companies, not government, should take the lead in promoting the Internet as an electronic marketplace, in adopting self-regulation, and in devising ratings systems to help parents guide their children away from objectionable online content.

The policy "literally follows the first rule of the Hippocratic oath: Do no harm," says one observer. "It lets businesses and consumers determine the Internet's growth."[10]

Eli Noam, professor of finance and economics at the Columbia Business School, remains a skeptic, believing that cyberspace will be regulated no matter what the White House says:

> For all the rhetoric of an Internet "free trade zone," will the United States readily accept an Internet that includes Thai child pornographers, Albanian tele-doctors, Cayman Island tax dodges, Monaco gambling, Nigerian blue sky stock schemes, Cuban mail-order catalogues? Or, for that matter, American violators of privacy, purveyors of junk e-mail or "self-regulating" price-fixers?

Unlikely. And other countries will feel the same on matters they care about.[11]

We consider some of these concerns in the following sections.

ICANN: The Internet Corporation for Assigned Names & Numbers

Acting on the belief that the Internet moves too quickly to be regulated by government, the White House then decided to let Internet users govern themselves. In June 1998, it proposed the creation of a series of nonprofit corporations to manage such complex issues as fraud prevention, privacy, and intellectual property protection. The first such group, **_ICANN (Internet Corporation for Assigned Names & Numbers)_ was established to regulate Internet domain names,** those addresses ending with .com, .org, .net, and so on, that identify a Web site. ICANN got off to a rocky start, but if it succeeds, according to one report, "it could evolve into the preeminent regulatory body on the Internet, with power extending far beyond the arena of domain names."[12]

CONCEPT CHECK

What is the National Information Infrastructure?

Distinguish among VBNS, Internet2, and the Next Generation Internet.

What is the purpose of the 1996 Telecommunications Act?

Describe the 1997 White House plan for Internet commerce and its offshoot, ICANN.

7.2 Security Issues: Threats to Computers & Communications Systems

KEY QUESTION
What are some characteristics of the key security issues for information technology?

Security issues go right to the heart of the workability of computer and communications systems. Here several threats to computers and communications systems are discussed.

Errors & Accidents

In general, errors and accidents in computer systems may be classified as human errors, procedural errors, software errors, electromechanical problems, and "dirty data" problems.

- **Human errors:** Quite often, when experts speak of the "unintended effects of technology," what they are referring to are the unexpected things people do with it. Among the ways in which people can complicate the workings of a system are the following:[13]

    (1) Humans often are not good at assessing their own information needs. For example, many users will acquire a computer and communications system that either is not sophisticated enough or is far more complex than they need.

    (2) Human emotions affect performance. For example, one frustrating experience with a computer is enough to make some people abandon the whole system. But throwing your computer out the window isn't going to get you any closer to learning how to use it better.

    (3) Humans act on their perceptions, which in modern information environments are often too slow to keep up with the equipment. Decisions influenced by information overload, for example, may be just as faulty as those based on too little information.

- **Procedural errors:** Some spectacular computer failures have occurred because someone didn't follow procedures. In 1999, the $125 million Mars Climate Orbiter was fed data expressed in pounds, the English unit of force, instead of newtons, the metric unit (about 22% of a pound). As a result, the spacecraft flew too close to the surface of Mars and broke apart.[14] A few years earlier, Nasdaq, the nation's second largest stock market, was shut down for 2½ hours by an effort, ironically, to make the computer system more user friendly. Technicians were phasing in new software, adding technical improvements a day at a time. A few days into this process, technicians trying to add more features to the software flooded the data-storage capability of the computer system. The result was to delay the opening of the stock market and shorten the trading day.[15]

- **Software errors:** We are forever hearing about "software glitches" or "software bugs." A software bug is an error in a program that causes it not to work properly. In a column lamenting the absence of a bug-free Web browser, *Business Week* technology journalist Stephen Wildstrom wrote: "I have used every browser produced by Netscape, . . . and every one of them has been buggy." But its major rival is no better. "Where Netscape is crash-prone," Wildstrom said, Microsoft Internet Explorer 5.0 "has gotten downright scary," requiring numerous software "patches" to close the security holes in the software that allow strangers to access your computer.[16]

- **Electromechanical problems:** Mechanical systems, such as printers, and electrical systems, such as circuit boards, don't always work. They may be faultily constructed, get dirty or overheated, wear out, or become damaged in some other way. Power failures (brownouts and blackouts) can shut a system down. Power surges can also burn out equipment.

- **"Dirty data" problems:** When keyboarding a research paper, you undoubtedly make a few typing errors (which, hopefully, you clean up). So do all the data-entry people around the world who feed a continual stream of raw data into computer systems. A lot of problems are caused by this kind of "dirty data." *Dirty data* is incomplete, outdated, or otherwise inaccurate data.

Natural & Other Hazards

Some disasters do not merely lead to temporary system downtime; they can wreck the entire system. Examples are natural hazards, and civil strife and terrorism.

Natural hazards

Tornadoes can inflict major damage to both people and computer systems.

- **Natural hazards:** Whatever is harmful to property (and people) is harmful to computers and communications systems. This certainly includes natural disasters: fires, floods, earthquakes, tornadoes, hurricanes, blizzards, and the like. If they inflict damage over a wide area, as have ice storms in eastern Canada or hurricanes in Florida, natural hazards can disable all the electronic systems we take for granted. Without power and communications connections, automated teller machines, credit-card verifiers, and bank computers are useless.

- **Civil strife and terrorism:** We may take comfort in the fact that wars and insurrections seem to take place in other parts of the world. Yet we are not immune to civil unrest, such as the riot in Los Angeles in 2000 after the hometown Lakers won the National Basketball Association championship. Nor are we immune, apparently, to acts of terrorism, such as the 1993 bombing of New York's World Trade Center. In that case, companies found themselves frantically moving equipment to new offices and reestablishing their computer networks. The Pentagon (which has 650,000 terminals and workstations, 100 WANs, and 10,000 LANs) has been taking steps to reduce its own systems' vulnerability to intruders, although hackers have managed to penetrate top-secret operations.

Crimes against Computers & Communications

Ethics

A _computer crime_ can be of two types: (1) It can be an illegal act perpetrated against computers or telecommunications, or (2) it can be the use of computers or telecommunications to accomplish an illegal act.

Crimes against information technology include theft—of hardware, of software, of computer time, of cable or telephone services, or of information. Other illegal acts are crimes of malice and destruction. Some examples are as follows:

Terrorism

The kind of terrorist bombing that occurred in 1995 with the federal building in Oklahoma City obviously can be a major threat to computer systems.

- **Theft of hardware:** Hardware theft can range from shoplifting an accessory in a computer store to removing a laptop or cellular phone from someone's car. Professional criminals may steal shipments of microprocessor chips off a loading dock or even pry cash machines out of shopping-center walls.

    Eric Avila, 26, a history student at the University of California at Berkeley, had his doctoral dissertation—involving six years of painstaking research—stored on the hard drive of his Macintosh laptop when a thief stole it out of his apartment. Although he had copied an earlier version of his dissertation (70 pages entitled "Paradise Lost: Politics and Culture in Post-War Los Angeles") onto a diskette, the thief stole that, too. "I'm devastated," Avila said. "Now it's gone, and there is no way I can recover it other than what I have in my head." To make matters worse, he had no choice but to pay off the $2000 loan for a computer he did not have anymore.[17] The moral, as we've tried to stress in this book: _Always make backup copies of your important data, and store them in a safe place—away from your computer._

Laptop left on a car seat

Computers left in cars are temptations for smash-and-grab thieves.

- **Theft of software:** Generally, software theft involves illegally copying programs, rather than physically taking someone's floppy disks. Software makers secretly prowl electronic bulletin boards in search of purloined products, then try to get a court order to shut down the bulletin boards. They also look for organizations that "softlift"—companies, colleges, or other institutions that buy one copy of a program and make copies for many computers.

    Many such so-called pirates are reported by coworkers or fellow students to the "software police," the Software Publishers Association. The SPA has a toll-free number (800–388–7478) for reporting illegal copying of software. In the 1990s, two New England college students were indicted for allegedly using the Internet to encourage the exchange of copyrighted software.[18]

Another type of software theft is copying or counterfeiting of well-known software programs. These pirates often operate in China, Taiwan, Mexico, Russia, and various parts of Asia and Latin America. In some countries, most of the U.S. microcomputer software in use is thought to be illegally copied.

- **Theft of time and services:** The theft of computer time is more common than you might think. Probably the biggest instance of it is people using their employer's computer time to play games, do online shopping or stock trading, or dip into Web pornography. Some people even operate sideline businesses.

  For years "phone phreaks" have bedeviled the telephone companies. For example, they have found ways to get into company voice-mail systems, then use an extension to make long-distance calls at the company's expense. They have also found ways to tap into cellular phone networks and dial for free.

- **Theft of information:** "Information thieves" have infiltrated the files of the Social Security Administration, stolen confidential personal records, and sold the information. On college campuses, thieves have snooped on or stolen private information such as grades. Thieves have also broken into computers of the major credit bureaus and stolen credit information. They have then used the information to charge purchases or have resold it to other people. One thief broke into the systems of Internet retailer CD Universe and stole 300,000 customers' credit-card numbers. When the company's executives refused to pay a ransom demand, he sold them off piecemeal on the Internet (for others to illegally charge on) until he was stopped.

- **Crimes of malice and destruction:** Sometimes criminals are more interested in abusing or vandalizing computers and telecommunications systems than in profiting from them. For example, a student at a Wisconsin campus deliberately and repeatedly shut down a university computer system, destroying final projects for dozens of students. A judge sentenced him to a year's probation, and he left the campus.

Ethics

Crimes Using Computers & Communications

Just as a car can be used to perpetrate or assist in a crime, so can information technology. For example, four college students on New York's Long Island who met via the Internet used a specialized computer program to steal credit-card numbers, then, according to police, went on a one-year, $100,000 shopping spree. When arrested, they were charged with grand larceny, forgery, and scheming to defraud.[19]

In addition, investment fraud has come to cyberspace. Many people now use online services to manage their stock portfolios through brokerages hooked into the services. Scam artists have followed, offering nonexistent investment deals and phony solicitations, and manipulating stock prices.

Finally, there is the kind of fraud known as **_plagiarism_, the expropriation of another writer's text, findings, or interpretations, and presentation of this material as one's own.** Plagiarism has always been a great problem in educational institutions, but the rise of the Internet has aggravated the problem. The box on the next page considers this matter further.

Worms & Viruses

Worms and viruses are forms of high-tech maliciousness. **A _worm_ is a program that copies itself repeatedly into a computer's memory or onto a disk drive.** Sometimes it will copy itself so often it will cause a computer to crash. An example was Melissa, a 1999 worm program that infected perhaps a million PCs. Melissa launched a file of porno Web sites via e-mail and sent them on to people on the recipient's e-mail address list.

PRACTICAL ACTION BOX

Web Research, Term Papers, & Plagiarism

Ethics

No matter how much some students may be able to rationalize cheating in college—for example, trying to pass off someone else's term paper as their own (plagiarism)—ignorance of the consequences is not an excuse. Most instructors announce the penalties for cheating at the beginning of their course—usually a failing grade in the course and possible suspension or expulsion from school.

Even so, every student becomes aware before long that the World Wide Web contains sites that offer term papers, either for free or for a price. Some dishonest students may download papers and just change the author's name to their own, although others are more likely to use the papers for ideas.

How the Web Can Lead to Plagiarism

Two types of term-paper Web sites are:

- **Sites offering papers for free:** Such a site requires users to fill out a membership form, then provides at least one free student term paper.

- **Sites offering papers for sale:** Commercial sites may charge $6–$10 a page, which users may charge to their credit card.

"These paper-writing scams present little threat to serious students or savvy educators," says one writer. "No self-respecting student will hand in purchased papers; no savvy educator will fail to spot them and flunk the student outright."[a] English professor Bruce Leland, director of writing at Western Illinois University, points out that many Web papers are so bad that no one is likely to benefit from them in any case.[b]

How do instructors detect and defend against student plagiarism? Leland says professors are unlikely to be fooled if they tailor term-paper assignments to work done in class, monitor students' progress—from outline to completion—and are alert to papers that seem radically different from a student's past work.

Eugene Dwyer, a professor of art history at Kenyon College, requires that papers in his classes be submitted electronically, along with a list of World Wide Web site references. "This way I can click along as I read the paper. This format is more efficient than running around the college library, checking each footnote."[c]

The World Wide Web, points out Mitchell Zimmerman, also provides the means for detecting cheaters. "Faster computer search programs will make it possible for teachers to locate texts containing identified strings of words from amid the millions of pages found on the Web."[d] Indeed, there exists a Web site that scans student papers against millions of pages on the Internet. The system can lock on to a stolen phrase as short as eight words. It can even detect material that has been changed slightly to appear original. In 1999, a University of California at Berkeley professor using the site caught 45 of his 320 neurobiology students using term papers plagiarized from online sources.[e]

How the Web Can Lead to Low-Quality Papers

Besides tempting some students to plagiarism, the Web also creates another problem in term-paper writing, points out William Rukeyser, coordinator for Learning in the Real World, a nonprofit information clearinghouse. It enables students "to cut and paste together reports or presentations that appear to have taken hours or days to write but have really been assembled in minutes with no actual mastery or understanding by the student."[f] Students use hypertext to skip to isolated phrases or paragraphs in source material, which they then pull together, complains Rukeyser, without making the effort to gain any real knowledge of the material.

How does an instructor spot a term paper based primarily on Web research? Philosophy professor David Rothenberg, of New Jersey Institute of Technology, offers four clues:[g]

- **No books cited:** The student's bibliography cites no books, just articles or references to Web sites. Sadly, says Rothenberg, "one finds few references to careful, in-depth commentaries on the subject of the paper, the kind of analysis that requires a book, rather than an article, for its full development."

- **Outdated material:** Much of the material in the bibliography is strangely out of date. Says Rothenberg. "A lot of stuff on the Web that is advertised as timely is actually at least a few years old."

- **Unrelated pictures and graphs:** Students may intersperse the text with a lot of impressive-looking pictures and graphs that bear little relation to the precise subject of the paper. "Cut and pasted from the vast realm of what's out there for the taking, they masquerade as original work."

- **Superficial references:** "Too much of what passes for information [online] these days is simply *advertising* for information," points out Rothenberg. "Screen after screen shows you where you can find out more, how you can connect to this place or that." And a lot of other kinds of information are detailed but often superficial: "pages and pages of federal documents, corporate propaganda, snippets of commentary by people whose credibility is difficult to assess."

A _virus_ is a "deviant" program, stored on a computer hard drive, that can cause unexpected and often undesirable effects, such as destroying or corrupting data. _(See_ ● _Panel 7.1.)_ The famous e-mail Love Bug (its subject line was I LOVE YOU), which originated in the Philippines in May 2000 and did perhaps as much as $10 billion in damage worldwide, was both a worm and a virus. A variation on the Melissa worm, it spread faster and caused more damage than any other bug before it.[20] The Love Bug was followed almost immediately by a variant virus. This new Love Bug didn't reveal itself with an ILOVEYOU line but changed to a random word or words each time a new computer was infected.[21]

Worms and viruses are passed in two ways:

- **By diskette:** The first way is via an infected diskette, perhaps obtained from a friend or a repair person.
- **By network:** The second way is via a network, as from e-mail or an electronic bulletin board. This is why, when taking advantage of all the freebie games and other software available online, you should use virus-scanning software to check downloaded files.

The virus usually attaches itself to your hard disk. It might then display annoying messages ("Your PC is stoned—legalize marijuana") or cause Ping-Pong balls to bounce around your screen and knock away text. More seriously, it might add garbage to your files, then erase or destroy your system

---

● PANEL 7.1
Types of worms and viruses

- **Boot-sector virus:** The boot sector is that part of the system software containing most of the instructions for booting, or powering up, the system. The boot sector virus replaces these boot instructions with some of its own. Once the system is turned on, the virus is loaded into main memory before the operating system. From there it is in a position to infect other files. Any diskette that is used in the drive of the computer then becomes infected. When that diskette is moved to another computer, the contagion continues. Examples of boot-sector viruses: AntCMOS, AntiEXE, Form.A, NYB (New York Boot), Ripper, Stoned.Empire.Monkey.

- **File virus:** File viruses attach themselves to executable files—those that actually begin a program. (In DOS these files have the extensions .com and .exe.) When the program is run, the virus starts working, trying to get into main memory and infecting other files.

- **Multipartite virus:** A hybrid of the file and boot-sector types, the multipartite virus infects both files and boot sectors, which makes it better at spreading and more difficult to detect. Examples of multipartite viruses are Junkie and Parity Boot.

  A type of multipartite virus is the _polymorphic virus,_ which can mutate and change form just as human viruses can. Such viruses are especially troublesome because they can change their profile, making existing antiviral technology ineffective.

  A particularly sneaky multipartite virus is the _stealth virus,_ which can temporarily remove itself from memory to elude capture. An example of a multipartite, polymorphic stealth virus is One Half.

- **Macro virus:** Macro viruses take advantage of a procedure in which miniature programs, known as macros, are embedded inside common data files, such as those created by e-mail or spreadsheets, which are sent over computer networks. Until recently, such documents have typically been ignored by antivirus software. Examples of macro viruses are Concept, which attaches to Word documents and e-mail attachments, and Laroux, which attaches to Excel spreadsheet files. Fortunately, the latest versions of Word and Excel come with built-in macro virus protection.

- **Logic bomb:** Logic bombs, or simply bombs, differ from other viruses in that they are set to go off at a certain date and time. A disgruntled programmer for a defense contractor created a bomb in a program that was supposed to go off two months after he left. Designed to erase an inventory tracking system, the bomb was discovered only by chance.

- **Trojan horse:** The Trojan horse covertly places illegal, destructive instructions in the middle of a legitimate program, such as a computer game. Once you run the program, the Trojan horse goes to work, doing its damage while you are blissfully unaware. An example of a Trojan horse is FormatC.

software. It may evade your detection and spread its havoc elsewhere, since an infected hard disk will infect every floppy disk used by the system.

If you look in the utility section of any software store, you'll see a variety of virus-fighting programs. **_Antivirus software_ scans a computer's hard disk, floppy disks, and main memory to detect viruses and, sometimes, to destroy them.** Such virus watchdogs operate in two ways. First, they scan disk drives for "signatures," characteristic strings of 1s and 0s in the virus that uniquely identify it. Second, they look for suspicious virus-like behavior, such as attempts to erase or change areas on your disks. Examples of antivirus programs are Norton Anti-Virus 2000, McAfee Virus Scan for Windows, and Virex for Macs.

Computer Criminals

What kind of people are perpetrating most of the information-technology crime? Over 80% may be employees, the rest are outside users, hackers and crackers, and professional criminals.

- **Employees:** Says Michigan State University criminal justice professor David Carter, who surveyed companies about computer crime, "Seventy-five to 80% of everything happens from inside."[22] Most common frauds, Carter found, involved credit cards, telecommunications, employees' personal use of computers, unauthorized access to confidential files, and unlawful copying of copyrighted or licensed software.

  Workers may use information technology for personal profit, or steal hardware or information to sell. They may also use it to seek revenge for real or imagined wrongs, such as being passed over for promotion. Sometimes they may use the technology simply to demonstrate to themselves they have power over people.

- **Outside users:** Suppliers and clients may also gain access to a company's information technology and use it to commit crimes. In both cases, this becomes more of a possibility as electronic connections such as intranets and extranets become more commonplace.

- **Hackers and crackers:** *Hacker* is so overused it has come to be applied to anyone who breaks into a computer system. Some people think it means almost any computer lover. In reality, there is a difference between hackers and crackers, although the term cracker has never caught on with the general public.

  Hackers are people who gain unauthorized access to computer or telecommunications systems, often just for the challenge of it. Some hackers even believe they are performing a service by exposing security flaws. Whatever the motivation, network system administrators view any kind of unauthorized access as a threat, and they usually try to pursue offenders vigorously. The most flagrant cases of hacking are met with federal prosecution.

  Crackers are people who illegally break in to computers for malicious purposes—to obtain information for financial gain, shut down hardware, pirate software, or alter or destroy data. Sometimes you hear of "white hat" hackers (who aren't malicious) versus "black-hat" hackers (who are). The Federal Bureau of Investigation estimates that cybercrime costs Americans more than $20 billion a year—and that more than 60% of such computer crime goes unreported. Says one article:

  > On the low end, Web vandals, mostly teenagers derided by older [crackers] as "script kiddies," have attacked and shut down hundreds of busy Web sites, including those of the NASDAQ stock exchange, ABC, the White House, the Senate, and the FBI itself. On the more sophisticated end, a single program triggered by a New Jersey hacker—[the] Melissa virus—caused an estimated $80 million in damage to computer users.[23]

- **Professional criminals:** Members of organized crime rings don't just steal information technology. They also use it the way that legal businesses do—as a business tool, but for illegal purposes. For instance, databases can be used to keep track of illegal gambling debts and stolen goods. Not surprisingly, the old-fashioned illegal bookmaking operation has gone high-tech, with bookies using computers and fax machines in place of betting slips and paper tally sheets.

As information-technology crime has become more sophisticated, so have the people charged with preventing it and disciplining its outlaws. Campus administrators are no longer being quite as easy on offenders and are turning them over to police. Industry organizations such as the Software Publishers Association are going after software pirates large and small. (Commercial software piracy is now a felony, punishable by up to five years in prison and fines of up to $250,000 for anyone convicted of stealing at least 10 copies of a program, or more than $2500 worth of software.) Police departments as far apart as Medford, Massachusetts, and San Jose, California, now have police patrolling a "cyber beat." They regularly cruise online bulletin boards and chat rooms looking for pirated software, stolen trade secrets, child molesters, and child pornography.

In 1988, after the last widespread Internet break-in, the U.S. Defense Department created the Computer Emergency Response Team (CERT). Although it has no power to arrest or prosecute, CERT provides round-the-clock international information and security-related support services to users of the Internet. Whenever it gets a report of an electronic snooper, whether on the Internet or on a corporate e-mail system, CERT stands ready to lend assistance. It counsels the party under attack, helps thwart the intruder, and evaluates the system afterward to protect against future break-ins.

CONCEPT CHECK

Explain some of the errors, accidents, and hazards that can affect computers.

What are the principal crimes against computers?

What are some crimes using computers?

Describe some types of computer criminals.

7.3 Security: Safeguarding Computers & Communications

KEY QUESTION
What are the characteristics of the four components of security?

The ongoing dilemma of the Digital Age is balancing convenience against security. **Security** **is a system of safeguards for protecting information technology against disasters, systems failure, and unauthorized access that can result in damage or loss.** We consider four components of security—identification and access, encryption, protection of software and data, and disaster-recovery plans.

Identification & Access

Are you who you say you are? The computer wants to know.

There are three ways a computer system can verify that you have legitimate right of access. Some security systems use a mix of these techniques. The systems try to authenticate your identity by determining (1) what you have, (2) what you know, or (3) who you are.

- **What you have—cards, keys, signatures, badges:** Credit cards, debit cards, and cash-machine cards all have magnetic strips or built-in computer chips that identify you to the machine. Many require you to display your signature, which may be compared with any future signature you write. Computer rooms are always kept locked, requiring a key. Many people also keep a lock on their personal computers. A computer room may also be guarded by security officers, who may need to see an authorized signature or a badge with your photograph before letting you in.

Of course, credit cards, keys, and badges can be lost or stolen. Signatures can be forged. Badges can be counterfeited.

- **What you know—PINs, passwords, and digital signatures:** To gain access to your bank account through an automated teller machine (ATM), you key in your PIN. **A _PIN (personal identification number)_ is the security number known only to you that is required to access the system.** Telephone credit cards also use a PIN. If you carry either an ATM or a phone card, never carry the PIN written down elsewhere in your wallet (even disguised).

   A _password_ is a special word, code, or symbol required to access a computer system. Passwords are one of the weakest security links, says AT&T security expert Steven Bellovin. Passwords (and PINs, too) can be guessed, forgotten, or stolen. To reduce a stranger's guessing, Bellovin recommends never choosing a real word or variations of your name, birthdate, or those of your friends or family. Instead you should mix letters, numbers, and punctuation marks in an oddball sequence of no fewer than eight characters.[24] Some good passwords: 2b/orNOT2b%. Alfred!E!Newman7. You can also choose an obvious and memorable password but shift the position of your hands on the keyboard, creating a meaningless string of characters—the best kind of password. (Thus, ELVIS becomes R;BOD when you move your fingers one position right on the keyboard.)

- **Who you are—physical traits:** Some forms of identification can't be easily faked—such as your physical traits. Biometrics tries to use these in security devices. **_Biometrics_ is the science of measuring individual body characteristics.**

   For example, before University of Georgia students can use the all-you-can-eat plan at the campus cafeteria, they must have their hands read. As one writer describes the system, "a camera automatically compares the shape of a student's hand with an image of the same hand pulled from the magnetic strip of an ID card. If the patterns match, the cafeteria turnstile automatically clicks open. If not, the would-be moocher eats elsewhere."[25]

   Besides handprints, other biological characteristics read by biometric devices are fingerprints (computerized "finger imaging"), voices, the blood vessels in the back of the eyeball (retinal scan), the lips, and even the entire face.

Some computer security systems have a "call-back" provision. In a _call-back system_, the user calls the computer system, punches in the password and hangs up. The computer then calls back a certain preauthorized number. This measure will block anyone who has somehow got hold of a password but is calling from an unauthorized telephone.

Ethics

Encryption

PGP is a computer program written for encrypting computer messages—putting them into secret code. **_Encryption_ is the altering of data so it is not usable unless the changes are undone.** _PGP_ (for _Pretty Good Privacy_) is so good that it is practically unbreakable; even government experts can't crack it. Another encryption system is _DES_ (for _Data Encryption Standard_), adopted as a federal standard in 1976.

Encryption is clearly useful for some organizations, especially those concerned with trade secrets, military matters, and other sensitive data. Some maintain that encryption will determine the future of e-commerce, because transactions cannot flourish over the Internet unless they are secure.[26] However, from the standpoint of our society, encryption is a two-edged sword. For instance, police in Sacramento, California, found that PGP blocked them from reading the computer diary of a convicted child molester and finding links to a suspected child pornography ring. Should the government be allowed to read the encoded e-mail of its citizens? What about government surveillance of overseas terrorists, drug dealers, and other enemies?

Protection of Software & Data

Organizations go to tremendous lengths to protect their programs and data. As might be expected, this includes educating employees about making backup disks, protecting against viruses, and so on. Other security procedures include the following:

- **Control of access:** Access to online files is restricted to those who have a legitimate right to access—because they need them to do their jobs. Many organizations have a system of transaction logs for recording all accesses or attempted accesses to data.

- **Audit controls:** Many networks have *audit controls* for tracking which programs and servers were used, which files opened, and so on. This creates an *audit trail*, a record of how a transaction was handled from input through processing and output.

- **People controls:** Because people are the greatest threat to a computer system, security precautions begin with the screening of job applicants. Résumés are checked to see if people did what they said they did. Another control is to separate employee functions, so people are not allowed to wander freely into areas not essential to their jobs. Manual and automated controls—input controls, processing controls, and output controls—are used to check if data is handled accurately and completely during the processing cycle. Printouts, printer ribbons, and other waste that may contain passwords and trade secrets to outsiders are disposed of through shredders or locked trash barrels.

Disaster-Recovery Plans

A _disaster-recovery plan_ **is a method of restoring information processing operations after being halted by destruction or accident.** "Among the countless lessons that computer users have absorbed in the hours, days, and weeks after the [New York] World Trade Center bombing," wrote one reporter, "the most enduring may be the need to have a disaster-recovery plan. The second most enduring lesson may be this: Even a well-practiced plan will quickly reveal its flaws."[27]

Mainframe computer systems are operated in separate departments by professionals, who tend to have disaster plans. Mainframes are usually backed up. However, many personal computers, and even entire local area networks, are not backed up. The consequences of this lapse can be great. It has been reported that, on average, a company loses as much as 3% of its gross sales within eight days of a sustained computer outage. In addition, the average company struck by a computer outage lasting more than 10 days never fully recovers.[28]

A disaster-recovery plan is more than a big fire drill. It includes a list of all business functions and the hardware, software, data, and people to support those functions, as well as arrangements for alternate locations. The disaster-recovery plan includes ways for backing up and storing programs and data in another location, ways of alerting necessary personnel, and training for those personnel.

Visual Summary

antivirus software (p. 274, KQ 7.2) Program that scans a computer's hard disk, floppy disks, and main memory to detect viruses and, sometimes, to destroy them. Why it's important: *Computer users must find out what kind of antivirus software to install in their systems—and how to keep it up to date—for protection against damage or shutdown.*

biometrics (p. 276, KQ 7.3) Science of measuring individual body characteristics. Why it's important: *Biometrics is used in some computer security systems, to restrict user access.*

computer crime (p. 270, KQ 7.2) Crime of two types: (1) an illegal act perpetrated against computers or telecommunications; (2) the use of computers or telecommunications to accomplish an illegal act. Why it's important: *Crimes against information technology include theft—of hardware, of software, of computer time, of cable or telephone services, or of information. Other illegal acts are crimes of malice and destruction.*

crackers (p. 274, KQ 7.2) People who illegally break into computers for malicious purposes. Why it's important: *Crackers attempt to break into computers to obtain information for financial gain, shut down hardware, pirate software, or alter or destroy data.*

disaster-recovery plan (p. 277, KQ 7.3) Method of restoring information processing operations that have been halted by destruction or accident. Why it's important: *Such a plan is important if an organization desires to resume computer operations quickly.*

encryption (p. 276, KQ 7.3) The altering of data so it is not usable unless the changes are undone. Why it's important: *Encryption is clearly useful for some organizations, especially those concerned with trade secrets, military matters, and other sensitive data. Some maintain that encryption will determine the future of e-commerce, because transactions cannot flourish over the Internet unless they are secure.*

hackers (p. 274, KQ 7.2) People who gain unauthorized access to computer or telecommunications systems, often just for the challenge of it. Why it's important: *Hackers create problems not only for the institutions that are victims of break-ins but also for ordinary users of the systems.*

ICANN (Internet Corporation for Assigned Names & Numbers) (p. 268, KQ 7.1) Organization established to regulate Internet domain names, those addresses ending with .com, .org, .net, and so on, that identify a Web site. Why it's important: *ICANN could evolve into the preeminent regulatory body on the Internet, with power extending far beyond the arena of domain names.*

Internet2 (p. 266, KQ 7.1) Cooperative university-business program established to enable high-end users to quickly and reliably move huge amounts of data, using VBNS as the official backbone. Why it's important: *Whereas VBNS provides data transfer at 1000 times commercial Web speeds, Internet2 operates at only 100 times those speeds. In effect, Internet2 adds "toll lanes" to the older Internet to speed things up. The purpose is to advance videoconferencing, research, and academic collaboration—to enable a kind of "virtual university."*

Next Generation Internet (NGI) (p. 266, KQ 7.1) U.S. government's program to parallel the university-business–sponsored effort of Internet2. Why it's important: *NGI is designed to provide money to six government agencies to help tie the campus high-performance backbones into the broader federal infrastructure. The technical goals for NGI include connecting at least 100 sites, including universities, federal national laboratories, and other research organizations. Speeds are 100 times those of the older Internet, and 10 sites are connected at speeds that are 1000 times as fast.*

password (p. 276, KQ 7.3) Special word, code, or symbol required to access a computer system. Why it's important: *Passwords are one of the weakest security links; they can be guessed, forgotten, or stolen.*

PIN (personal identification number) (p. 276, KQ 7.3) Security number known only to the user; it is required to access a system. Why it's important: *PINS are required to access many computer systems, as well as automated teller machines.*

plagiarism (p. 271, KQ 7.2) The expropriation of another writer's text, findings, or interpretations, and presentation of this material as one's own. Why it's important: *Information technology offers plagiarists new opportunities for unauthorized copying.*

security (p. 275, KQ 7.3) System of safeguards for protecting information technology against disasters, systems failure, and unauthorized access that can result in damage or loss. Four components of security are identification and access, encryption, protection of software and data, and disaster-recovery plans. Why it's important: *With proper security, organizations and individuals can minimize information technology losses from disasters, system failures, and unauthorized access.*

VBNS (Very-High-Speed Backbone Network Service) (p. 266, KQ 7.1) Main U.S. government component to upgrade the "backbone," or primary hubs of data transmission, of the Internet. Why it's important: *Linking supercomputers and other banks of computers across the United States, speeds are1000 times conventional Internet speeds. Financed by the National Science Foundation and managed by the telecommunications giant MCI, VBNS is somewhat exclusive. It is being used by just 101 top universities and other research institutions for network-intensive applications, such as transmitting high-quality video for distance education.*

virus (p. 273, KQ 7.2) Deviant program that can cause unexpected and often undesirable effects, such as destroying or corrupting data. Why it's important: *Viruses can cause users to lose data and/or files or can shut down entire computer systems.*

worm (p. 271, KQ 7.2) Program that copies itself repeatedly into a computer's memory or onto a disk drive until no space is left. Why it's important: *Worms can shut down computers.*

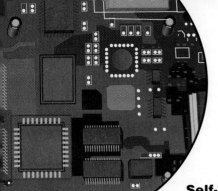

Chapter Review

Self-Test Questions

1. The purpose of _____ is to scan a computer's disk devices and memory to detect viruses and, sometimes, to destroy them.

2. So that information processing operations can be restored after destruction or accident, organizations should adopt a _____.

3. _____ is the altering of data so that it is not usable unless the changes are undone.

4. _____ is incomplete, outdated, or otherwise inaccurate data.

5. _____ is the expropriation of another writer's text, findings, or interpretations and presentation of this material as one's own.

Multiple-Choice Questions

1. People who illegally break into computers for malicious purposes are called _____.

   a. hackers

   b. encrypters

   c. crackers

   d. analysts

   e. viruses

2. A firewall is generally intended to protect against _____.

   a. software bugs

   b. employee use of e-mail

   c. unauthorized access via the Internet

   d. natural disasters

   e. encryption

3. Which of the following is not a new form of the Internet?

   a. NGI

   b. Internet2

   c. ICANN

   d. VBNS

True/False Questions

T F   1. Viruses cannot be passed from computer to computer through a network.

T F   2. One of the weakest links in a security system is biometrics.

T F   3. Humans are always very good at assessing their own information needs.

T F   4. It's necessary to back up files only when one is working with financial data.

T F   5. ICANN regulates Internet domain names.

Short-Answer Questions

1. What is the difference between a hacker and a cracker?

2. What does a worm do?

3. Briefly describe and differentiate VBNS, Internet2, and Next Generation Internet.

4. What was the 1996 U.S Telecommunications Act intended to do?

5. What does ICANN stand for, and what does it do?

6. Name five threats to computers and communications systems.

7. What are the two types of computer crime?

Concept Mapping

On a separate sheet of paper, draw a concept map, or visual diagram, linking concepts. Show how the following terms are related.

antivirus software	password
biometrics	PIN
crackers	plagiarism
disaster-recovery plan	security
encryption	VBNS
ICANN	virus
Internet2	worm
NGI	

Knowledge in Action

1. What, in your opinion, are the most significant disadvantages of using computers? What do you think can be done about these problems?

2. What's your opinion about the issue of free speech on an electronic network? Research some recent legal decisions in various countries, as well some articles on the topic. Should the contents of messages be censored? If so, under what conditions?

3. Research the problems of stress and isolation experienced by computer users in the United States, Japan, and one other country. Write a brief report on your findings.

chapter 8

The Promises of the Digital Age

Society & Information Technology Tomorrow

Key Questions

You should be able to answer the following questions.

8.1 **Some Emerging Technology: Global Telecommunications & Artificial Intelligence** What are the two models of telecommunications, and what are the main areas of artificial intelligence?

8.2 **Information & Education** How is information technology being used in education?

8.3 **Health & Medicine** How is information technology being used in health and medicine?

8.4 **Commerce & Money** How is information technology being used in sales and marketing, retailing, banking, and stock trading?

8.5 **Entertainment & the Arts** How is information technology being used in music and movies?

8.6 **Jobs & Careers** How can you use the Internet to find a job, and what are some hot jobs in information technology?

A Visual Overview of This Chapter

1 **Some Emerging Technology: Global Communications & Artificial Intelligence.** Telecommunications can be arranged according to two models. In the *tree-and-branch model,* a centralized information provider sends out messages through many channels to many consumers; this model is followed by many mass media. In the *switched-network model,* people on the system are not only consumers of information but also possible providers of it; this model is that of the Internet, and it is much more participatory than the other model.

Artificial intelligence, technologies used for developing machines to emulate human-like qualities, has several areas, including the following: (1) **Virtual reality,** devices that project a person into a sensation of three-dimensional space, is used in arcade-type games and also in **simulators,** devices that represent the behavior of physical or abstract systems and are used in training, as of airplane pilots. (2) **Robotics,** the development and study of machines that can perform work normally done by people, has produced **robots,** automatic devices that perform functions usually performed by people. (3) **Expert systems** are interactive computer programs used to solve problems normally requiring the assistance of human experts. (4) **Natural language processing** is the study of ways for computers to recognize and understand human language.

Artificial life is the study of "creatures"—computer instructions, or pure information—that, like live organisms, are created, replicate, evolve, and die.

2 **Information & Education.** Helping the vast stores of information make sense is a challenge that is being addressed with **intelligent agents,** programs that roam networks and compile data and perform work tasks on your behalf. Students at all levels are finding computers helpful in education, including their use in distance learning over the Internet.

3 **Health & Medicine.** *Telemedicine*—medical care delivered via telecommunications—is one way computers and communications are changing health and medicine. Another way is through the way patients are tapping health-care databases. In addition, medical information is being digitized, with everything from psychotherapy to implants being affected.

4 **Commerce & Money.** One big thrust of information technology is that it erases boundaries in business between company departments, suppliers, and customers, so the idea of what constitutes an organization is changing. Changes are being found in sales and marketing and retailing, as with online sales, and in banking and cybercash, as with electronic payment systems.

5 **Entertainment & the Arts.** Two fields in which information technology is being used are music and movies. In music, computers are being used to give digitized instruments a superior range of sounds and to change the system of marketing songs. In movies, computers are used for all kinds of animation and other special effects, to do better film editing, and to enable amateurs to make movies cheaper.

6 **Jobs & Careers.** Job seekers can now use employer databases to get leads on jobs, and they can post résumés with electronic job registries so employers can find them. The five information-technology jobs projected to have the largest percentage increase in the near future are computer engineers, computer support specialists, systems analysts, database administrators, and desktop publishing specialists.

Is the computer just a convenient new gadget that makes things easier? Or is it a truly revolutionary tool whose long-range result will be to change all the rules?

In 1869, steam engines delivered 1.2 million horsepower to U.S. manufacturers. Seventy years later, the electric motor, which replaced the steam engine, delivered 45 million horsepower—a 40-fold increase in mechanical power.

The electric motor, suggests economist Stephen Cohen, was a *tool* as opposed to a *gadget*. The difference? Gadgets are inventions that make life easier but do not change a country's economic dynamics. An automatic transmission may make driving easier but, Cohen argues, "a car with an automatic transmission still does what a car does."[1] Startling innovations such as television in the 1950s and air transport in the 1960s did not change society's basic economic forces. The electric motor, by contrast, is a tool. It took 40 years for industry to learn how to take full advantage of it, but among the revolutionary results was mass production. "The long-run consequences—industrial, organizational, social—were enormous," says Cohen.

In the late 1950s, the world's 2000 computers had an average processing power of 10,000 instructions per second, points out Cohen's University of California, Berkeley colleague Brad DeLong. Today there are 200 million computers with an average processing power of 100 million instructions per second. "This is a million-fold increase in 40 years," writes DeLong.

So are computers and communications technology gadgets or tools? The tools of the industrial revolution amplified muscle power, so that we don't have to rely on a human or a horse anymore. Today's information technology amplifies brain power. "We no longer need to rely on human memory or on human eyes for scanning, or human brains for organizing and computing information and data," DeLong and Cohen suggest. Because the forces of this technological tool thread through everything, they are "important in every single economic activity." Thus, the two economists think, a century from now, people will look back and conclude that today's innovations in information technology constituted life-changing tools.

What are the promises of this future Digital Age? We consider the following areas:

- Some emerging technology: global telecommunications and artificial intelligence
- Information and education
- Health and medicine
- Commerce and money
- Entertainment and the arts
- Jobs and careers

8.1 Some Emerging Technology: Global Telecommunications & Artificial Intelligence

KEY QUESTIONS

What are the two models of telecommunications, and what are the main areas of artificial intelligence?

Stephen L. Talbott is editor of NetFuture, an online newsletter that examines how people's lives have changed as tasks become automated. A former programmer and technical writer, Talbott lives in a wooded area of rural upstate New York. But he found that writing his newsletter on a word processor caused him to withdraw from his pleasant surroundings. Accordingly, he now divides his time between writing on the computer and writing longhand on a yellow pad in front of his picture window so that he can look at birds and trees.

"The computer is our hope if we can accept it as our enemy," he says. "As our friend, it will destroy us."[2]

The computer is our enemy? What Talbott means by this is that we must learn to distance ourselves from it. The act of pulling away can bring a feeling of freedom, he says, and a new appreciation of the personal, the creative, and the natural. "The computer gives us the opportunity to assert ourselves against it," Talbott says. "This is a gift."

Such is the riddle we need to keep in mind—the computer is both enemy and friend—as we contemplate its benefits. Here let us consider two important emerging technologies: global telecommunications and artificial intelligence.

Global Telecommunications: The Internet as a Participatory System

The hallmark of great civilizations has been their systems of communications.

In the beginning, communications was based on transportation: the Roman Empire had its network of roads, the European powers had their far-flung navies, and the United States and Canada built their transcontinental railroads. Then transportation yielded to the electronic exchange of information. Beginning in 1844, the telegraph ended the short existence of the Pony Express. In 1876, the telephone came along as a competitor to the telegraph.

In 1889, electromagnetic radio waves were transmitted the width of the English Channel. The amplifying vacuum tube, invented in 1906, led to commercial radio, beginning with Pittsburgh's KDKA in 1920. Television came into being in England in 1925; one of the first scheduled TV shows, broadcast from New York on July 21, 1931, introduced Kate Smith singing "When the Moon Comes Over the Mountain."

During the 1950s and 1960s, as TV sets appeared seemingly everywhere, Canadian communications philosopher Marshall McLuhan posed the notion of a "global village." The global village refers to the "shrinking of the world society because of the ability to communicate." This shrinking has resulted from the global reach of mass media such as television. People throughout the world can now view similar shows and so have common points of reference.

Recently, technology has become portable, giving individuals even more power over communications and making the global village even more closely knit. Two decades ago, cellphones, pagers, and portable computers with communications links barely existed; now they are commonplace. Moreover, as computer and communications technology go digital, the computer, communications, consumer electronics, entertainment, and mass media industries are undergoing a technological convergence.

Telecommunications can be organized through two kinds of arrangements: the "tree-and-branch" model or the "switched-network" model.

- Tree-and-branch model: In the *tree-and-branch model*, a centralized information provider sends out messages through many channels to thousands of consumers. This is the model of most mass media, such as radio and TV broadcasting. It is also the model envisioned by some cable and entertainment companies looking to provide movies and other services from a centralized library or a core of "content providers."

- Switched-network model: In the *switched-network model*, people on the system are not only consumers of information ("content") but also possible providers of it. This is the model of the telephone system, of course, and also of most computer networks, including, most certainly, the Internet.

Videoconferencing
Army Lieutenant Frank Holmes, stationed in Sarajevo, Bosnia, talks to his wife, Amanda, and daughter, Morgan, 5000 miles away at Fort Bragg, N.C., using ProShare Videoconferencing technology.

Whereas the tree-and-branch allows only few-to-many communications, the switched-network allows many-to-many communications. Thus, the more the switched-network becomes prominent, the more participatory our systems of communications become.

For example, the file-sharing software Napster has tapped into the participatory potential of the switched-network model. Napster, like Gnutella and similar programs enables users to go on the Web and download songs that reside in MP3 format on the hard drives of other music lovers. (MP3 is a simple way of condensing a sound file.) These music files, which are free, can be shared regardless of their location on the Internet. There is no way to police this file swapping, and many such files were originally taken— pirated—from copyrighted music released by music companies on CDs. (Warning: When you share music using Napster, this means that nearly anyone can access your hard disk through the phone line, making you vulnerable to worms and viruses.)[3] The few-to-many system has given way to a many-to-many system.

More than 70% of college students, according to one survey, use the controversial Napster music Web service at least once a month.[4] Because most Napster users get their downloaded music free from other fans' hard disks, not from record company servers, they have changed the economics of music distribution. Students may be unaware that most of the 7000 new CD titles made available each year by record companies are by not-yet-famous bands. These bands are subsidized by mega-sellers such as Metallica and Dr. Dre. Record companies may charge $16 for a CD that costs only 60 cents to manufacture. But their net profit is only about 59 cents per CD after they subtract out the costs of supporting all the artists they carry.[5]

Artificial Intelligence

You're having trouble with your new software program. You call the customer "help desk" at the software maker. Do you get a busy signal or get put on hold to listen to music (or, worse, advertising) for several minutes? Technical support lines are often swamped, and waiting is commonplace. Or, to deal with your software difficulty, do you find yourself dealing with . . . other software?

The odds are good that you will. Programs that can walk you through a problem and help solve it are called *expert systems*. As the name suggests, these are systems that are imbued with knowledge by a human expert. Expert systems are one of the most useful applications of artificial intelligence.

<u>**Artificial intelligence (AI)**</u> **is a group of related technologies used for developing machines to emulate human-like qualities, such as learning, reasoning,**

communicating, seeing, and hearing. Today the main areas of AI are *virtual reality, robotics, expert systems, and natural language processing.*

- Virtual-reality and simulation devices: ***Virtual reality (VR),*** **a computer-generated artificial reality, projects a person into a sensation of three-dimensional space.** *(See* ● *Panel 8.1.)* To put yourself into virtual reality, you need software and special headgear; then you can add gloves, and later perhaps a special suit. The headgear—which is called a *head-mounted display*—has two small video display screens, one for each eye, for creating the sense of three-dimensionality. Headphones pipe in stereophonic sound or even 3-D sound. Three-dimensional sound makes you think you are hearing sounds not only near each ear but also in various places all around you. The glove has sensors for collecting data about your hand movements. Once you are wearing this equipment, software gives you interactive sensory feelings similar to real-world experiences.

  You may have seen virtual reality used in arcade-type games, such as Atlantis, a computer simulation of The Lost Continent. However, there are far more important uses—for example, in simulators for training. ***Simulators*** **are devices that represent the behavior of physical or abstract systems.** Virtual-reality simulation technologies are applied a great deal in training. For instance, to train bus drivers, they create lifelike bus control panels and various scenarios such as icy road conditions. They are used to train pilots on various aircraft and to prepare air-traffic controllers for equipment failures. Surgeons-in-training can develop their skills through simulation on "digital patients." Virtual-reality therapy has been used for autistic children and in the treatment of phobias, such as extreme fear of public speaking or of being in public places or high places.

- Robotics: More than 40 years ago, in the film *Forbidden Planet,* Robby the Robot could sew, distill bourbon, and speak 187 languages. We haven't caught up with science-fiction movies, but maybe we'll get there yet.

  Robotics **is the development and study of machines that can perform work normally done by people.** The machines themselves are called robots. *(See* ● *Panel 8.2.)* Basically, a ***robot*** **is an automatic device that performs functions ordinarily executed by human beings or that operates with what appears to be almost human intelligence.** ScrubMate—a robot equipped with computerized controls, ultrasonic "eyes," sensors, batteries, three different cleaning and scrubbing tools, and a self-squeezing mop—can clean bathrooms. Rosie the HelpMate

WABOT-2

This robot, created in 1984, can read sheet music well enough to play simple tunes on an electric organ.

Robots
Left: The Human Extender Robot allows people to pick up very heavy packages. *Right:* NASA remote-controlled research robot.

delivers special-order meals from the kitchen to nursing stations in hospitals. Robodoc is used in surgery to bore the thighbone so that a hip implant can be attached. A driverless harvester, guided by satellite signals and artificial vision system, is used to harvest alfalfa and other crops.

Robots are also used for more exotic purposes such as fighting oil-well fires, doing nuclear inspections and cleanups, and checking for mines and booby traps. An eight-legged, satellite-linked robot called Dante II was used to explore the inside of Mount Spurr, an active Alaskan volcano, sometimes without human guidance. A six-wheeled robot vehicle named Sojourner was used in NASA's 1997 Pathfinder exploration of Mars to sample the planet's atmosphere and soil and to radio data and photos back to Earth.

- **Expert systems:** An *expert system* **is an interactive computer program used in solving problems that would otherwise require the assistance of a human expert.** Fundamental to an expert system is a knowledge base constructed by experts that can "learn" by adding new knowledge. The expert system MYCIN helps diagnose infectious diseases. PROSPECTOR assesses geological data to locate mineral deposits. DENDRAL identifies chemical compounds. Home-Safe-Home evaluates the residential environment of an elderly person. Business Insight helps businesses find the best strategies for marketing a product. REBES (Residential Burglary Expert System) helps detectives investigate crime scenes.

- **Natural language processing:** Natural languages are ordinary human languages, such as English. ***Natural language processing*** **is the study of ways for computers to recognize and understand human language,** whether in spoken or written form.

Think how challenging it is to make a computer translate English into another language. In one instance, the English sentence "The spirit is willing, but the flesh is weak" came out in Russian as "The wine is agreeable, but the meat is spoiled." The problem with human language is that it is often ambiguous; different listeners may arrive at different interpretations.

Today you can buy a handheld computer that will translate a number of English sentences—principally travelers' phrases ("Please take me to the airport")—into another language. This trick is similar to teaching an English-speaking child to sing "Frere Jacques." More complex is the work being done by AI scientists trying to endow the computer with an "understanding" of how human language works. These scientists are wrestling with ideas about the instinctual instructions or genetic code that allows us to understand language. Because the context of a phrase is critical in understanding human language, this work requires a very large associated database.

Still, some natural-language systems are already in use. Intellect, a software product, employs a limited English vocabulary to help users orally query databases. LUNAR, developed to help analyze moon rocks, answers questions about geology on the basis of an extensive database. Verbex, used by the U.S. Postal Service, lets mail sorters read aloud an incomplete address and will reply with the correct ZIP code.

Ethics

• Artificial life, the Turing test, and AI ethics: **_Artificial life_, or A-life, is a field of study concerned with "creatures"—computer instructions, or pure information—that are created, replicate, evolve, and die as if they were living organisms.** Thus, A-life software (such as LIFE) tries to simulate the responses of a human being.

How can we know when we have reached the point where computers have achieved human intelligence? How will you know, say, whether you're talking to a human being on the phone or to a computer? Clearly, with the strides made in the fields of artificial intelligence and artificial life, this question is no longer just academic.

Interestingly, this matter was addressed back in 1950 by Alan Turing, an English mathematician and computer pioneer. Turing predicted that by the end of the century computers would be able to mimic human thinking and converse so naturally that their communications would be indistinguishable from a person's. Out of these observations came the Turing test. The Turing test is intended to determine whether a computer possesses "intelligence" or "self-awareness."

In the Turing test, a human judge converses by means of a computer terminal with two entities hidden in another location. One entity is a person typing on a keyboard. The other is a software program. As the judge types in and receives messages on the terminal, he or she must decide whether the entity is human. In this test, intelligence and the ability to think is demonstrated by the computer's success in fooling the judge.

Judith Anne Gunther participated as one of eight judges in the third annual Loebner Prize Competition, which is based on Turing's ideas.[6] (There have been other competitions since.) The "conversations" are restricted to predetermined topics, such as baseball. This is because today's best programs have neither the databases nor the syntactical ability to handle an unlimited number of subjects. Conversations with each entity are limited to 15 minutes. At the end of the contest, the program that fools the judges most is the winner.

Gunther found that she wasn't fooled by any of the computer programs. The winning program, for example, relied as much on deflection and wit as it did on responding logically and conversationally. (For example, to a judge trying to discuss a federally funded program, the computer said: "You want logic? I'll give you logic: shut up, shut up, shut up, shut up, shut up, now go away! How's that for logic?") However, Gunther was fooled by one of the five humans, a real person discussing abortion. "He was so uncommunicative," wrote Gunther, "that I pegged him for a computer."

Behind everything to do with artificial intelligence and artificial life—just as it underlies everything we do—is the whole matter of ethics. William A. Wallace, professor of decision sciences at

Rensselaer Polytechnic Institute and author of *Ethics in Modeling*, points out that many users are not aware that computer software, such as expert systems, is often subtly shaped by the ethical judgments and assumptions of the people who create them.[7] In one instance, he points out, a bank had to modify its loan-evaluation software after it discovered that it tended to reject some applications because it unduly emphasized old age as a negative factor. Clearly, there is no such thing as completely "value-free" technology. Human beings build it, use it, and have to live with the results.

CONCEPT CHECK

Distinguish between the tree-and-branch model and the switched-network model.

Distinguish among the main areas of AI.

What is artificial life?

8.2 Information & Education

KEY QUESTION

How is information technology being used in education?

Getting the right kind of information is one of our greatest challenges. Can everything in the Library of Congress be made available online to citizens and companies? What about government records, patents, contracts, and other legal documents? Or geographical maps photographed from satellites?

An even greater challenge is getting information to make sense. What the online world really needs, therefore, is a terrific librarian. "What bothers me most," says Christine Borgman, chair of the UCLA Department of Library and Information Science, "is that computer people seem to think that if you have access to the Web, you don't need libraries."[8] But what's in a library is standardized and well organized, whereas what's on the Web is often overwhelming, unstandardized, and chaotic.

As one solution, computer scientists have been developing so-called intelligent agents to find information on computer networks and filter it. An ***intelligent agent program*** **performs work tasks on your behalf, including roaming networks and compiling data.** A software agent is a kind of electronic assistant that will filter messages, scan news services, and perform similar secretarial chores. An agent will also travel over communications lines to computer databases, collecting files to add to a personalized database. As you might suspect, one popular use for such agents is as "shopping bots" that consumers can use to find online bargains.

With their potential for finding and processing information, computers are pervasive on campus. More than a third of college courses require the use of e-mail, and many have their own dedicated Web pages.[9] College students spend an average of 5.6 hours a week on the Internet.[10] And as computer prices drop, more and more students have their own PCs. (For the rest, colleges often make public microcomputers available.)

Many students have been exposed to computers since the lower grades. In 1997, for example, 75% of elementary schools (and 89% of secondary schools) were wired to the Internet, according to the U.S. Department of Education.[11] The results have not always been productive, especially in elementary schools. However, when used selectively by trained teachers in middle schools, according to research, computers can significantly enhance performance in math learning.[12] In art classes, computers can be used to encourage creativity without mess, as children learn to make art digitally.[13]

Thousands of schoolchildren have tapped into a Web site by a Wake Forest University biology researcher that tracks albatrosses across the Pacific Ocean. The Web site offers the students a chance to test their own hypotheses

PRACTICAL ACTION BOX

Distance Learning

Distance learning, one writer points out, is "a contemporary version in some ways of the correspondence [education-by-mail] schools of the 1950s."[a] Students take courses given at a distance, using television or a computer network. The teacher may be close by in another room or on the other side of the world. (For testing, however, students may have to go to a classroom for an examination in the presence of the instructor or proctors.)

Studying online has its advantages: You can log on from anywhere, at any time of day or night, and participate in discussions to whatever extent you choose. Of course, there are drawbacks. "Some courses just don't translate well at a distance," says Pam Dixon, author of *Virtual College*. "Sometimes it's better to get your hands dirty and experience the course in a physical setting. Would you have wanted your dentist to have learned his or her craft entirely by videoconferencing?"[b] But, Dixon points out, courses in business, writing, computers, mathematics, and library science, to name just a few, are well suited to teaching from afar. Indeed, students may get more attention from teachers online than they would from teachers in a large classroom with hundreds of students. In addition, says tele-education entrepreneur Glenn Jones, distance learning can give students computer access to virtual libraries of print, photos, recordings, and movies.[c]

To be sure, distance learning may take a bit of getting used to, as Joseph Walter, professor of communication stud-ies at Northwestern University, found in studying the subject. One thing Walter learned is "that the computer changes the dynamics of communication. It takes four or five times as long as speaking when you have to type out what you want to say."[d] Moreover, as Jones points out, it is still the enthusiasm and wit and skill of caring teachers that make the difference in the learning experience.

But, in return for giving up the live classroom setting, students gain the convenience of being able to take courses not offered locally, often at times that suit their own schedules, not the logistical needs of an educational institution. This is a particular boon to the growing numbers of part-time and non-traditional (older) students. Part-timers presently make up about 45% of all college enrollments. This, says one writer, is "a group for whom 'anytime, anywhere' education holds special appeal."[e]

With convenience, however, come responsibility and accountability. "Online school is a venue for self-starters," says one writer, "which may be one reason that, though 10 people enrolled in the class I took, only half were ever heard from after the first few weeks."[f]

If you can't get the educational offerings you want locally, where do you start your search for the distance-learning equivalent? You might begin by looking at Distance Learning on the Net *(http://homepage.interaccess.com/~ghoyle)* and the Comprehensive Distance Education List of Resources *(http://talon.extramural.uiuc.edu/ramage/disted.html)*.

about the tracking data.[14] Students exposed to the Internet in high school say they think the Web has helped them improve the quality of their academic research and of their written work.[15] There is even software such as Intelligent Essay Assessor and E-Rater that grades essays, although its effectiveness is still debated.[16]

One of the greatest revolutions in education—before, during, and after the college years—is the advent of distance learning, or "cyberclasses," along with the explosion of Internet resources. The home-schooling movement, for example, has come of age, thanks to Internet resources.[17] Many colleges—both individually and in associations such as that of the Western Governors University (backed by 17 states and Guam) and the Community College Distance Learning Network—are offering a variety of Internet and/or video-based online courses.[18] Corporations are also offering training classes via the Internet or corporate intranet.[19]

Viktor Yazykov, competing in the perilous Around Alone solo sailing competition, found himself in the stormy South Atlantic with a seriously infected arm that needed emergency surgery. So, with the help of step-by-step instructions sent by e-mail from Boston-based Dr. Daniel Carlin to his solar-powered laptop computer, Yazykov operated on his own arm.[20]

Here we have a dramatic example of *telemedicine*—medical care delivered via telecommunications. For some time, physicians in rural areas lacking local access to radiologists have used "teleradiology" to exchange digital images such as X-rays via telephone-linked networks with expert physicians in metropolitan areas. Now telemedicine is moving to an exciting new level, as the use of digital cameras and sound, in effect, moves patients to doctors. Already telemedicine is being embraced by administrators in the American prison system, where by law inmates are guaranteed medical treatment—and where the 8% increase in prisoners every year has led to the need to control health costs.[21]

But, besides physicians, patients and health consumers are also going online. Although online health information can be misleading and even dangerous, growing numbers of consumers are seeking medical advice on the Internet—by tapping into health-care databases, e-mailing health professionals, or communicating with people who have similar conditions in online chat rooms. For instance, hours after 10-year-old Robert Lord of San Diego fractured his spine in a fall from a tree, his father found an experimental drug on the Internet, saving the boy from lifetime paralysis.[22] Often patients are already steeped in information about their conditions when they arrive in the offices of health-care professionals. "It's a fundamental shift of knowledge, and therefore power, from physicians to patients," says one consultant.[23] In addition, health-care consumers are able to share experiences and information with each other. Inquisitive mothers-to-be, for example, can find a gathering spot at the Pregnancy Today Web site.[24]

Computer technology is also radically changing the tools of medicine. "All medical information—X-ray or lab test or blood pressure or pulse monitor—can now be digitized," says surgeon Rick Satava. "It can all be brought to the physician or surgeon in digital format."[25] Although only a small percentage of such information is digitized at present, some breakthroughs are producing results that would have been science fiction only 10 years ago. Software is now able to compute a woman's breast cancer risk.[26] Mental-health researchers are using computers to screen troubled teenagers in need of psychotherapy.[27] And psychotherapy that is done online has been found to provide short-term relief.[28] Epidemiologists are using the Web to improve public health in developing nations.[29] MIT has developed a "micropill," a pharmacy on a chip that can be implanted in the body to deliver tiny amounts of medicine on a controlled-release basis.[30] Hydraulics and computers are being used to help artificial limbs get "smarter."[31] And a patient paralyzed by a stroke has received an implant that allows communication between his brain and a computer, enabling him to move a cursor across a screen by brainpower and convey simple messages—as in "Star Trek."[32]

Eye-gaze technology

In 1986, after receiving a diagnosis of Lou Gehrig's disease, Intel physicist and engineer Mike Ward began drawing blueprints for a computer system so he could continue working as the disease progressed. "Typing" an e-mail message using eye-gaze technology, Ward says, "I cannot move or speak, but I can still be productive."

Businesses clearly see the Internet as a way to enhance productivity and competitiveness. However, the changes are going well beyond this, as we have already seen.

The thrust of the original Industrial Revolution was separation—to break work up into its component parts so as to permit mass production. The effect of computer networks in the Digital Revolution, however, is unification—to erase boundaries between company departments, suppliers, and customers.[33] Indeed, the parts of a company can now as easily be global as down the hall from one another. Thus, designs for a new product can be tested and exchanged with factories in remote locations. With information flowing faster, goods can be sent to market faster and inventories reduced. Says an officer of the Internet Society, "Increasingly you have people in a wide variety of professions collaborating in diverse ways in other places. The whole notion of 'the organization' becomes a blurry boundary around a set of people and information systems and enterprises."[34]

Some areas of business in which we are seeing rapid changes are sales and marketing, retailing, banking, and stock trading.

Virtual shopping

In Shanghai, China, the New World Department Store saves on dressing-room space by letting shoppers try on virtual clothes right on the sales floor. After their video image is captured on screen, customers can see themselves in any outfit in any color at the click of a mouse.

- **Sales and marketing:** In the old days, "you could earn a good living" as a salesperson if you could talk and read a price list, says Ted Urpens, the head of sales training for Caradon Everest, a British manufacturer of replacement windows. Now sales representatives not only need to be better educated and more knowledgeable about their customers' businesses but also must be comfortable with computer technology.

  Caradon Everest, for example, equips its sales staff with laptops containing software that "configures" customized windows and calculates the prices on the spot, a process that was once handled by the company's technical people and took a week. "The company can also load on product images for multimedia presentations and training programs," says one account. "Using a digital camera, pictures of the customer's house can be loaded into the computer, which can superimpose the company's windows and print out a color preview."[35]

  Other companies are also taking advantage of information technology in marketing. An Atlanta company called The Mattress Firm, for instance, uses geographical information systems, or GIS, software called MapLinx that plots drop-off points for a driver's most efficient delivery of mattresses to customers. However, MapLinx also depicts sales by neighborhood, showing owner Darin Lewin where his mattresses are selling well and thus where he should spend money to market them. "It's like shooting a rifle instead of a shotgun," says Lewin. "We can pinpoint our customer."[36]

  The Internet and World Wide Web have also, of course, become popular marketing tools. Indeed, online sales, and the cost of tools necessary to generate and support them, will constitute a $2 trillion annual market in 2003, predicts Forester Research.[37] For example, Lou Ann Hammond's company, Car-List Inc., of San Francisco, uses telecommunications links to provide customers with access to a database describing features, prices, and availability of new and used cars at selected dealerships and among individual sellers. A similar service, Auto-By-Tel, in Irvine, California, which is free to consumers, slashes the marketing cost to $25 per car—as compared to the average of $425 and 3 to 5 salespeople that is required to market a car the old-fashioned way. "At traditional dealerships you talk to lots of people," says president Peter Ellis. "Through Auto-By-Tel, you're already predisposed to buying so all the dealer does is write up the contract."[38]

- **Retailing becomes e-tailing:** Are there any limitations to what can be sold on the Internet? Computers, books, CDs, and airline tickets seem obvious. But what about groceries?

"You can't exactly sniff cantaloupes over your modem," points out one writer.[39] Even so, it's predicted that, by 2007, online grocery shopping will be a $60 billion to $85 billion business, reaching some 15% of U.S. households.[40] Two pioneer online supermarkets, Peapod and NetGrocer, have been joined by HomeGrocer.com (partly owned by Amazon.com), Webvan, and WholeFoods.com.

The challenges for all these Internet food ventures are daunting: Even designing a Web site shopping menu can be difficult because the profusion of groceries makes them hard to categorize. Figuring out how to deliver items, particularly perishable ones (drop off at customer's workplace? put in refrigerated box in the garage?), is one of the biggest headaches. And many people don't want to pay the surcharges to support grocery delivery. Even filling orders can be an expensive task, since most online grocers employ clerks to wheel carts through a warehouse and fill individual customer orders. Webvan has tried to solve this problem by building a mechanized warehouse in which machines zip groceries around so that a single worker can fill several orders.[41]

- **Banking and e-money:** "This is the beginning of the end of cash," says San Francisco secretary Valerie Baptiste, as she hands over her smart card—plastic embedded with a microchip containing stored-up funds like an electronic purse—to pay for her morning coffee and bagel. The cashier inserts the smart card into a machine that, in less than five seconds deducts $2.15 from the stored value. Later, when Baptiste finds her card running low, she can stick it into a kind of modem and dial up her bank account to "refuel" the card.[42]

The world of cybercash has come to banking—not only smart cards, but Internet banking, electronic deposits, electronic bill paying, online stock and bond trading, and online insurance shopping

Some banks are backing an electronic-payment system that will allow Internet users to buy tiny online goods and services with "micropayments" of as little as 25 cents, from participating merchants. For instance, publishers could charge buyers a quarter to buy an article or listen to a song online, a small transaction that until now has not been practical. "It allows you to buy things by the sip rather than the gulp," says futurist Paul Saffo.[43]

More than 1000 banks now have World Wide Web sites, offering services that include account access, funds transfer, bill payment, loan and credit card applications, and investments. In one interactive application, you can apply for a Visa card called NextCard online and get approved (or turned down) within about two minutes. The system then shows you the account balances on your other credit cards so that, if you wish, you can transfer your balances to NextCard and benefit from a lower interest rate.[44]

Only about 46% of U.S. workers have their paychecks electronically deposited into their bank accounts (as opposed to 95% or more in Japan, Norway, and Germany, for example), but this is sure to change as Americans discover that direct deposit is actually safer and faster.[45] Getting them used to electronic bill paying will probably take longer, but that day too is arriving. For more than two decades, it has been possible to pay your bills online with special software and direct modem hookups to your bank, but only about 3% of U.S. households have tried it. The main delay has been that scarcely any bills are delivered online, but now banks have convinced corporations to electronically deliver bills to them, which will then be sent to customers via the Web.[46]

- **Stock trading:** Computer technology is unquestionably changing the nature of stock trading. Only a few years ago, hardly anyone had the occupation of "day trader," most of whom are individual investors who rely on quick market fluctuations to turn a profit. Now, points out technology observer Denise Caruso, "anyone with a computer, a connection to the global network, and the requisite ironclad stomach for risk has the information, tools, and access to transaction systems required to play the stock market, a game that was once the purview of an elite few."[47]

  Online trading has also changed the nature of trading institutions themselves. The nation's two largest stock exchanges—the New York Stock Exchange (NYSE) and the Nasdaq—have felt the pressure from ECNs, or electronic trading networks, with names such as Island and Instinet.[48] ECNs "offer investors a way to trade for a fraction of what it costs to trade on the NYSE or Nasdaq," explains one report. "Their secret is that computers post, match, and execute all trades. No specialists or market makers serve as middlemen, as they do on the floor of the NYSE or in Nasdaq trading rooms."[49] Both exchanges have taken steps such as extending their trading day into the evening hours to accommodate investors on the West Coast.[50]

  Full-service brokerage firms such as the 85-year-old Merrill Lynch & Co. have begun to offer online trading, bowing to competition from such electronic discount brokerage firms as E*trade, Ameritrade, Schwab Online, and DLJ direct. As a *USA Today* editorial observed, "With the Internet, investors can quickly amass boatloads of company information, pick stocks, and execute trades on their own and on the cheap. Who needs brokers, particularly when their advice can easily cost 100 times as much?"[51] But Allan Sloan, *Newsweek*'s Wall Street editor, begs to differ: "The more information that's available online, the more valuable competent advisers and middlemen become. . . . Someone has to filter the information—much of which may not be true—and do something useful with it."[52] Indeed, major scandals involving stock fraud show that anyone considering buying securities online should make sure the seller is trustworthy.

8.5 Entertainment & the Arts

KEY QUESTION
How is information technology being used in music and movies?

Information technology is being used for all kinds of entertainment, ranging from video games to tele-gambling. It is also being used in the arts, from painting to photography. Let's consider just two examples, music and film.

- **Music:** Yamaha's Disklavier GranTouch looks like an acoustic grand piano, and the keyboard action is the same as in concert grands. However, in this new instrument, there are no strings and hence no heavy iron frame to support them.[53] The sounds themselves are drawn from a huge 30-megabyte database of digitized sounds. Because the sounds are synthesized, the instrument never needs tuning. In addition, it can summon the voices of more than 700 instruments, from harp to soprano sax.

This is just one example of how information technology is changing music. More importantly, as we mentioned earlier, it is changing the financial underpinnings of the music industry. For example, Internet retailers compete with brick-and-mortar record stores by providing easy online shopping for music CDs. Spinner Networks, owned by America Online, "broadcasts" about 2 million popular songs each day through the Net.[54]

But the World Wide Web is standing the system of music recording and distribution on its head. Not only do Web sites offer guitar chords and sheet music, but promotional sites feature signed and unsigned artists, both garage bands and established professional musicians.[55] There are legal MP3 sites on the Internet but also many illegal ones, which operate in violation of copyright laws protecting ownership of music. Many of these will probably cease to exist as the recording industry develops new Web technologies that frustrate the pirating of music and instead offer fee-based services.[56]

Ethics

Some musicians want a revolution; the Internet, they believe, is a force that can democratize the market, bypassing record companies and radio stations and offering music directly to listeners.[57] Many people who avoid buying CDs by using Napster or other file-sharing software to download music as MP3 files think they are striking a blow against greedy record companies.[58] But if these companies are destroyed by changing economics, will buyers have lost the crucial function of *winnowing*? "With the demise of business as usual," writes one critic, "every garage band in existence will have equal standing in the undifferentiated mass of millions of titles thrown up on the Web. Yes, downloads will be inexpensive, but how will one find the good stuff?"[59]

- **Movies:** "In computers, we know there are going to continue to be bazillions of jobs," says Jan Millsapps, chair of the cinema department at San Francisco State University, where cinema arts programs lead to degrees in multimedia and animation. "But artists are expanding the possibilities. . . . More and more the art of these students will be the movie stars of the future."[60]

Now that blockbuster movies routinely meld live action and animation, computer artists are in big demand. *Star Wars: Episode I*, for instance, had fully 1965 digital shots out of about 2200 shots. Even when film was used, it was scanned into computers to be tweaked with animated effects, lighting, and the like. Entire beings were created on computers by artists working on designs developed by producer George Lucas and his chief artist.[61]

What is driving the demand for computer artists? One factor is that animation, though not cheap, looks more and more like a bargain, because hiring movie actors costs so much—some make $20 million a film. Moreover, special effects are readily understood by audiences in other countries, and major studios increasingly count on revenues from foreign markets to make a film profitable. With the soaring demand for computer-generated imagery—not only for movies but also for TV ads and video games—colleges and trade schools are expanding their digital-animation training programs.

But animation is not the only area in which computers are revolutionizing movies. Digital editing has radically transformed the way films are assembled. Says one report, "Traditional film editing was always a funky, hands-on proposition: reeling and unreeling spools of film, cutting and gluing pieces of celluloid together, working amid a sea of film that sometimes got trampled underfoot." Today, by contrast, "150 miles of film can be stored on hard drives and an editor with the press of a key or the click of a mouse can instantly access any visual or audio moment in the film."[62] Thus, hundreds of variations of a scene can be stored and called up for review and comparison. A crowd of extras, for example, can be multiplied into an army of thousands. Computer techniques have even been used to develop digitally created actors—called "synthespians." The late John Wayne was recruited for a Coors beer ad.[63]

Even nonprofessionals can get into movie making as new computer-related products come to market. Digital video capture and editing systems are now available for under $1000, enabling amateurs to turn home videotapes into digital data and edit them. Digital camcorders, which offer outstanding picture and sound quality, are dropping in price.

CONCEPT CHECK

Describe the benefits of an intelligent agent.

What are some uses of computers in education, health, and medicine?

Discuss some uses of information technology in commerce and money.

How are computers being used in music and movies?

8.6 Jobs & Careers

KEY QUESTIONS

How can you use the Internet to find jobs, and what are some hot jobs in information technology?

It's 3 A.M. Even at this hour you can, if you wish, find job-search advice, tips on interviewing and résumé writing, and postings of employment opportunities around the world. One means for doing so is to direct your Web browser to a portal such as Yahoo! *(www.yahoo.com)* to obtain a list of popular Web sites. In the menu, you can click on Business, then Employment, then Jobs. This will bring up a list of sites that offer career advice, résumé postings, job listings, research about specific companies, and other services. (Caution: As might be expected, there is also a fair amount of junk out there: get-rich-quick offers, résumé-preparation firms, and other attempts to separate you from your money.)

Advice about careers, occupational trends, employment laws, and job hunting is also available through on-line chat groups, and other portals—America Online, Alta Vista, Microsoft Network, and Prodigy.

Ways for You to Find Employers

As you might expect, the first to use cyberspace as a job bazaar were companies seeking people with technical backgrounds and technical people seeking employment. However, as the public's interest in commercial services and the Internet has exploded, the technical orientation of online job exchanges has changed. Now, says one writer, "interspersed among all the ads for programmers on the Internet are openings for English teachers in China, forest rangers in New York, physical therapists in Atlanta, and models in Florida."[64] Most Web sites are free to job seekers, although some may require you to fill out an online registration form. *(See ● Panel 8.3.)*

Ways for Employers to Find You

Posting your résumé on line for prospective employers to view is attractive because of its low (or zero) cost and wide reach. But does it have any disadvantages? Certainly it might if the employer happens to be the one you're already working for. In addition, you have to be aware that you lose control over anything broadcast into cyberspace. You're putting your credentials out there for the whole world to see, and you need to be somewhat concerned about who might gain access to them.

Posting your résumé with an electronic jobs registry is certainly worth doing if you have a technical background, since technology companies in particular find this an efficient way of screening and hiring. However, it may also benefit people with less technical backgrounds. Online recruitment "is popular with companies because it pre-screens applicants for at least basic computer skills," says one writer. "Anyone who can master the Internet is likely to know something about word processing, spreadsheets, or database searches, knowledge required in most good jobs these days."[65]

The latest wrinkle in job seeking is to prepare a résumé with hypertext links and/or clever graphics and multimedia effects and then put it on a Web site to entice employers to chase after you. If you don't know how to do this, there are many companies that—for a fee—can convert your résumé to HTML and publish it on their own Web sites. Some of these services can't dress it up with fancy graphics or multimedia, but since complex pages take longer for employers to download anyway, the extra pizzazz is probably not worth the effort. A number of Web sites allow you to post your résumé for free. *(See* ● *Panel 8.4, next page.)*

Companies are also beginning to replace their recruiters' campus visits with online interviewing. For example, the firm VIEWnet Inc. of Madison, Wisconsin, offers first-round screenings or interviews for summer internships through its teleconferencing "InterVIEW" technology, which allows video signals to be transmitted via telephone lines.

What Do You Want to Do?

Even low-tech people can find high-tech jobs—positions that don't require a technological background in fields such as human resources, product marketing, and contract management. But high tech leads the employment parade, offering the most opportunities—"jobs that are fun, well-paying, and with benefits galore," as career writer Belle Wise puts it.[66] The five jobs projected to have the largest percentage increase from 1998 to 2008, according to the Bureau of Labor Statistics, are as follows:

- Computer engineers—108% increase
- Computer support specialists—102% increase
- Systems analysts—94% increase
- Database administrators—77% increase
- Desktop publishing specialists—73% increase

General

America's Job Bank: *www.ajb.dni.us*

Career Magazine: *www.careermag.com*

Career Mosaic: *www.careermosaic.com*

CareerPath.com: *www.careerpath.com*

College Grad Job Hunter: *www.collegegrad.com*

The Employment Guide's CareerWeb: *www.cweb.com*

4Work: *www.4work.com*

Job-Hunt.Org: *www.job-hunt.org*

JobOptions: *www.joboptions.com*

JobTrak.com: *www.jobtrak.com/jobguide*

Monster.com: *www.monster.com*

The Occupational Outlook Handbook: *www.jobweb.org/occhandb.htm*

Specialized

eAttorney (jobs in field of law): *www.attorneysatwork.com*

EngineeringJobs.com: *www.engineeringjobs.com*

FINANCIALjobs.com (accounting and finance jobs): *www.financialjobs.com*

Health Search USA (health and medicine): *www.healthsearchusa.com*

Industrial Light & Magic (George Lucas): *www.ilm-jobs.com*

These hot jobs, those with the fastest employment growth, require a good education. Still, for most careers that you might want to pursue, the Knowledge Age demands nearly perpetual education.

"Knowledge is the hot ticket now," says Karyne Conley, a vice president of SBC Communications. "Although it's hard to stay on top of all the new information, it seems I at least have to try. Really, there aren't many excuses left not to."[67]

CONCEPT CHECK

What are ways for you to find employers?

What are ways employers can find you?

What are the five fastest-growing jobs in information technology?

Visual Summary

artificial intelligence (AI) (p. 285, KQ 8.1) Group of related technologies used for developing machines to emulate human-like qualities, such as learning, reasoning, communicating, seeing, and hearing. Why it's important: *Today the main areas of AI are virtual reality, robotics, expert systems, and natural language processing.*

artificial life (p. 288, KQ 8.1) Also called *A-life;* field of study concerned with "creatures"—computer instructions, or pure information—that are created, replicate, evolve, and die as if they were living organisms. Why it's important: *A-life software (such as LIFE) tries to simulate the responses of a human being.*

expert system (p. 287, KQ 8.1) Interactive computer program that can apply rules to input in such a way as to generate conclusions. Fundamental to an expert system is a knowledge base constructed by experts that can "learn" by adding new knowledge. Why it's important: *Expert systems help users solve problems that would otherwise require the assistance of a human expert.*

intelligent agent (p. 289, KQ 8.2) Program that performs work tasks on the user's behalf, including roaming networks and compiling data. Why it's important: *A software agent is a kind of electronic assistant that will filter messages, scan news services, and perform similar secretarial chores. An agent will also travel over communications lines to computer databases, collecting files to add to a personalized database.*

natural language processing (p. 287, KQ 8.1) Study of ways for computers to recognize and understand human language, whether in spoken or written form. Why it's important: *Natural languages make it easier for users to work with computers.*

---

robot (p. 286, KQ 8.1) Automatic device that performs functions ordinarily executed by human beings or that operates with what appears to be almost human intelligence. Why it's important: Robots are performing more and more functions in business and the professions.

robotics (p. 286, KQ 8.1) Development and study of machines that can perform work normally done by people. Why it's important: *See robot.*

simulator (p. 286, KQ 8.1) Device that represents the behavior of physical or abstract systems. Why it's important: *Virtual-reality simulation technologies are applied a great deal in training.*

virtual reality (VR) (p. 286, KQ 8.1) Computer-generated artificial reality, projects a person into a sensation of three-dimensional space. To experience virtual reality, the user needs software and special headgear, gloves, and perhaps a special suit. The headgear has two small video display screens, one for each eye, for creating the sense of three-dimensionality. Headphones pipe in stereophonic sound or even 3-D sound. The glove has sensors for collecting data about hand movements. The software provides interactive sensory feelings similar to real-world experiences. Why it's important: *With virtual reality, users can experience almost anything they want without ever leaving their chairs. Virtual reality is employed in simulators for training programs of many types.*

Chapter Review

Self-Test Questions

1. A(n) _____ is an automatic device that performs functions ordinarily performed by human beings.

2. The goal of _____ is to enable the computer to communicate with the user in the user's native language.

3. In the _____ model, used by most mass media, a centralized information provider sends out messages through many channels to thousands of consumers; the _____ model, in contrast, is used by the telephone system and computer networks.

4. _____ is a group of related technologies used for developing machines to emulate human-like qualities such as learning, reasoning, communicating, seeing, and hearing.

5. _____ is a computer-generated artificial reality.

6. A(n) _____ performs tasks on the user's behalf.

7. Taking college classes online is referred to as _____.

True/False Questions

T F   1. Expert systems are not interactive.

T F   2. Simulators create a type of virtual reality.

T F   3. "A-life" refers to robots.

T F   4. The thrust of the Digital Revolution is to break work up into its component parts.

Short-Answer Questions

1. What are intelligent agents used for?

2. What do you need to experience virtual reality?

3. What is the Turing test?

4. What are the four main areas of artificial intelligence?

Concept Mapping

On a separate sheet of paper, draw a concept map, or visual diagram, linking concepts. Show how the following terms are related.

AI	intelligent agent
artificial life	Internet
Digital Revolution	natural language
e-money	robot
e-tailing	simulator
expert system	VR

Knowledge in Action

1. If you could design a robot, what kind would you create? What would it do?

2. If you could build an expert system, what would it do? What kinds of questions would you ask experts in order to elicit the appropriate information?

3. Go to three of the Web sites listed on the top of page 297 and report on ten job listings in your planned job or profession.

Systems & Programming

Development, Programming, & Languages

Key Questions

You should be able to answer the following questions.

A.1 **Systems Development: The Six Phases of Systems Analysis & Design** What are the six phases of the systems development life cycle?

A.2 **Programming: A Five-Step Procedure** What is programming, and what are the five steps in accomplishing it?

A.3 **Programming Languages** What is a programming language, and how is a high-level language converted to a low-level language?

A.4 **Object-Oriented Programming & Visual Programming** How do OOP and visual programming work?

A.5 **Internet Programming: HTML, XML, VRML, Java, & ActiveX** What are the features of HTML, XML, VRML, Java, and ActiveX?

Organizations can make mistakes, of course, and big organizations can make *really big* mistakes.

California's state Department of Motor Vehicles' databases needed to be modernized, and in 1988 Tandem Computers said it could do the job. "The fact that the DMV's database system, designed around an old IBM-based platform, and Tandem's new system were as different as night and day seemed insignificant at the time to the experts involved," said one writer investigating the project later.[1] The massive driver's license database, containing the driving records of more than 30 million people, first had to be "scrubbed" of all information that couldn't be translated into the language used by Tandem computers. One such scrub yielded 600,000 errors. Then the DMV had to translate all its IBM programs into the Tandem language. "Worse, DMV really didn't know how its current IBM applications worked anymore," said the writer, "because they'd been custom-made decades before by long-departed programmers and rewritten many times since." Eventually the project became a staggering $44 million loss to California's taxpayers.

This example shows how important planning is, especially when an organization is trying to launch a new kind of system. How do you avoid such mistakes? By employing systems analysis and design.

A.1 Systems Development: The Six Phases of Systems Analysis & Design

KEY QUESTION

What are the six phases of the systems development life cycle?

You may not have to wrestle with problems on the scale of motor-vehicle departments. That's a job for computer professionals. You're mainly interested in using computers and communications to increase your own productivity. Why, then, do you need to know anything about systems analysis and design?

In many careers, you may find your department or your job the focus of a study by a systems analyst. Knowing how the procedure works will help you better explain how your job works or what goals your department is supposed to achieve. In progressive companies, management is always interested in suggestions for improving productivity. This is the method for expressing your ideas.

The Purpose of a System

A **system** is defined as a collection of related components that interact to perform a task in order to accomplish a goal. A system may not work very well, but it is nevertheless a system. The point of systems analysis and design is to ascertain how a system works and then take steps to make it better.

An organization's computer-based information system consists of hardware, software, people, procedures, and data, as well as communications setups. These work together to provide people with information.

Getting the Project Going: How It Starts, Who's Involved

To get a project rolling, all it takes is a single individual who believes that something badly needs changing. An employee may influence a supervisor. A customer or supplier may get the attention of someone in higher management. Top management on its own may decide to take a look at a system that looks inefficient. A steering committee may be formed to decide which of many possible projects should be worked on.

Participants in the project are of three types:

- **Users:** The system under discussion should *always* be developed in consultation with users, whether floor sweepers, research scientists, or customers. Indeed, inadequate user involvement in analysis and design can be a major cause of a system's failing for lack of acceptance.
- **Management:** Managers within the organization should also be consulted about the system.
- **Technical staff:** Members of the company's information systems (IS) department, consisting of systems analysts and programmers, need to be involved. For one thing, they may well have to carry out and execute the project. Even if they don't, they may have to work with outside IS people contracted to do the job.

Complex projects will require one or several systems analysts. A **_systems analyst_ is an information specialist who performs systems analysis, design, and implementation.** His or her job is to study the information and communications needs of an organization and determine what changes are required to deliver better information to people who need it. "Better" information means information that is summarized in the acronym "CART"—complete, accurate, relevant, and timely. The systems analyst achieves this goal through the problem-solving method of systems analysis and design.

The Six Phases of Systems Analysis & Design

Systems analysis and design is a six-phase problem-solving procedure for examining an information system and improving it. The six phases (actually the phases overlap, so the number varies according to practitioner) make up what is called the systems development life cycle. The **_systems development life cycle (SDLC)_ is the step-by-step process that many organizations follow during systems analysis and design.**

Whether applied to a Fortune 500 company or a three-person engineering business, systems analysis and design consists of six phases. *(See ● Panel A.1.)*

● **PANEL A.1**
Systems development life cycle
An SDLC typically includes six phases.

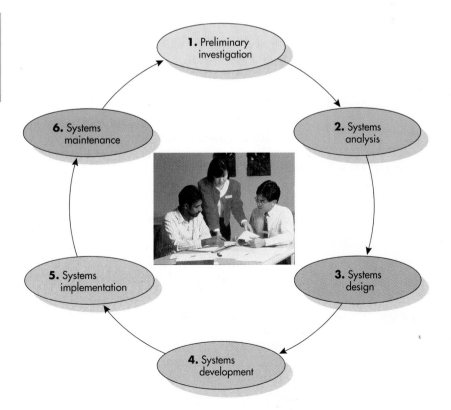

1. Preliminary investigation
2. Systems analysis
3. Systems design
4. Systems development
5. Systems implementation
6. Systems maintenance

Phases often overlap, and a new one may start before the old one is finished. After the first four phases, management must decide whether to proceed to the next phase. *User input and review is a critical part of each phase.*

The First Phase: Conduct a Preliminary Investigation

The objective of Phase 1, _preliminary investigation_, is to conduct a preliminary analysis, propose alternative solutions, describe costs and benefits, and submit a preliminary plan with recommendations.

- **Conduct the preliminary analysis:** In this step, you need to find out what the organization's objectives are and the nature and scope of the problem under study. Even if a problem pertains only to a small segment of the organization, you cannot study it in isolation. You need to find out what the objectives of the organization itself are. Then you need to see how the problem being studied fits in with them.

- **Propose alternative solutions:** In delving into the organization's objectives and the specific problem, you may have already discovered some solutions. Other possible solutions can come from interviewing people inside the organization, clients or customers affected by it, suppliers, and consultants. You can also study what competitors are doing. With this data, you then have three choices. You can leave the system as is, improve it, or develop a new system.

- **Describe the costs and benefits:** Whichever of the three alternatives is chosen, it will have costs and benefits. In this step, you need to indicate what these are. Costs may depend on benefits, which may offer savings. There are many possible kinds of benefits. A process may be speeded up, streamlined through elimination of unnecessary steps, or combined with other processes. Input errors or redundant output may be reduced. Systems and subsystems may be better integrated. Users may be happier with the system. Customers or suppliers may interact better with the system. Security may be improved. Costs may be cut.

- **Submit a preliminary plan:** Now you need to wrap up all your findings in a written report. The readers of this report will be the executives who decide in which direction to proceed—make no changes, change a little, or change a lot—and how much money to allow the project. You should describe the potential solutions, costs, and benefits and indicate your recommendations.

The Second Phase: Do an Analysis of the System

The objective of Phase 2, _systems analysis_, is to gather data, analyze the data, and write a report. In this second phase, you will follow the course selected by management on the basis of your Phase 1 feasibility report. We are assuming that they have ordered you to perform Phase 2—to do a careful analysis or study of the existing system in order to understand how the new system you proposed would differ. This analysis will also consider how people's positions and tasks will change if the new system is put into effect.

- **Gather data:** In gathering data, you will review written documents, interview employees and managers, develop questionnaires, and observe people and processes at work.

- **Analyze the data:** The next step is to come to grips with the data you have gathered and analyze it. Many analytical tools, or modeling tools, are available. *Modeling tools* enable a systems analyst to present graphic, or pictorial, representations of a system. Some of these tools involve creating flowcharts and diagrams on paper. Examples of modeling tools are *grid charts, decision tables, data flow diagrams, system flowcharts,* and *connectivity diagrams.*

- **Write a report:** Once you have completed the analysis, you need to document this phase, by writing a report to management. This report should have three parts. First, it should explain how the existing system works. Second, it should explain the problems with the existing system. Finally, it should describe the requirements for the new system and make recommendations on what to do next.

At this point, not a lot of money will have been spent on the systems analysis and design project. If the costs of going forward seem prohibitive, this is a good time for the managers reading the report to call a halt. Otherwise, you will be asked to move to Phase 3.

The Third Phase: Design the System

The objective of Phase 3, _systems design_, is to do a preliminary design and then a detailed design, and write a report. In this third phase of the SDLC, you will essentially create a "rough draft" and then a "detail draft" of the proposed information system.

- **Do a preliminary design:** A _preliminary design_ describes the general functional capabilities of a proposed information system. It reviews the system requirements and then considers major components of the system. Usually several alternative systems (called _candidates_) are considered, and the costs and the benefits of each are evaluated.

  Some tools that may be used in the design are _CASE tools_ and _project management software._

  CASE (computer-aided software engineering) tools are programs that automate various activities of the SDLC in several phases. This technology is intended to speed up the process of developing systems and to improve the quality of the resulting systems. CASE tools, also known as _automated design tools,_ may be used at other stages of the SDLC as well. Examples of such programs are Excelerator, Iconix, System Architect, and Powerbuilder.

  CASE tools may also be used to do prototyping. In _prototyping_, workstations, CASE tools, and other software applications are used to build working models of system components, so that they can be quickly tested and evaluated. Thus, a _prototype_ is a limited working system developed to test out design concepts. A prototype, which may be constructed in just a few days, allows users to find out immediately how a change in the system might benefit them.

  Project management software consists of programs used to plan, schedule, and control the people, costs, and resources required to complete a project on time.

- **Do a detail design:** A _detail design_ describes how a proposed information system will deliver the general capabilities described in the preliminary design. The detail design usually considers the following parts of the system in this order: output requirements, input requirements, storage requirements, processing requirements, and system controls and backup.

- **Write a report:** All the work of the preliminary and detail designs will end up in a large, detailed report.

The Fourth Phase: Develop the System

In Phase 4, _systems development_, the systems analyst or others in the organization develop or acquire the software, acquire the hardware, and then test the system. Depending on the size of the project, this is the phase that will probably involve the organization in spending substantial sums of money. It could also involve spending a lot of time. However, at the end you should have a workable system.

- **Develop or acquire the software:** During the design stage, the systems analyst may have had to address what is called the "make-or-buy" decision, but that decision certainly cannot be avoided now. In the *make-or-buy decision,* you decide whether you have to create a program—have it custom-written—or buy it, meaning simply purchase an existing software package. Sometimes programmers decide they can buy an existing program and modify it.

  If you decide to create a new program, then the question is whether to use the organization's own staff programmers or hire outside contract programmers (outsource it). Whichever way you go, the task could take many months.

  Programming is an entire subject unto itself, and we address it in the next major section.

- **Acquire hardware:** Once the software has been chosen, the hardware to run it must be acquired or upgraded. It's possible your new system will not require obtaining any new hardware. It's also possible that the new hardware will cost millions of dollars and involve many items: microcomputers, mainframes, monitors, modems, and many other devices. The organization may find it's better to lease rather than to buy some equipment, especially since, as we mentioned (Moore's law), chip capability has traditionally doubled every 18 months.

- **Test the system:** With the software and hardware acquired, you can now start testing the system. Testing is usually done in two stages: unit testing, then system testing.

  In *unit testing,* individual parts of the program are tested, using test (made-up, or sample) data. If the program is written as a collaborative effort by multiple programmers, each part of the program is tested separately.

  In *system testing,* the parts are linked together, and test data is used to see if the parts work together. At this point, actual organization data may be used to test the system. Tests with erroneous data and massive amounts of data check whether the system can be made to fail ("crash").

  At the end of this long process, the organization will have a workable information system, one ready for the implementation phase.

The Fifth Phase: Implement the System

Whether the new information system involves a few handheld computers, an elaborate telecommunications network, or expensive mainframes, the fifth phase will involve some close coordination in order to make the system not just workable but successful. **Phase 5, _systems implementation,_ consists of converting the hardware, software, and files to the new system and training the users.**

- Convert to the new system: *Conversion,* the process of converting from an old information system to a new one, involves converting hardware, software, and files. *Hardware conversion* may be as simple as taking away an old PC and plunking a new one down in its place. Or it may involve acquiring new buildings and putting in elaborate wiring, climate-control, and security systems.

  Software conversion means making sure the applications that worked on the old equipment can be made to work on the new.

  File conversion, or *data conversion,* means converting the old files to new ones without loss of accuracy. For example, can the paper contents from the manila folders in the personnel department be input to the system with a scanner? Or do they have to be keyed in manually, with the consequent risk of introducing errors?

There are four strategies for handling conversion: *direct, parallel, phased,* and *pilot.*

Direct implementation means the user simply stops using the old system and starts using the new one. The risk of this method should be evident: What if the new system doesn't work? If the old system has truly been discontinued, there is nothing to fall back on.

Parallel implementation means that the old and new systems are operated side by side until the new system has shown it is reliable, at which time the old system is discontinued. Obviously there are benefits to this cautious approach. If the new system fails, the organization can switch back to the old one. The difficulty with this method is the expense of paying for the equipment and people to keep two systems going at the same time.

Phased implementation means that parts of the new system are phased in separately—either at different times (parallel) or all at once in groups (direct).

Pilot implementation means that the entire system is tried out but only by some users. Once the reliability has been proved, the system is implemented with the rest of the intended users. The pilot approach still has its risks, since all of the users of a particular group are taken off the old system. However, the risks are confined to a small part of the organization.

In general, the phased and pilot approaches are the most favored methods. Phased is best for large organizations in which people are performing different jobs. Pilot is best for organizations in which all people are performing the same task (such as order takers at a direct-mail house).

- **Train the users:** Various tools are available to familiarize users with the new system. They run from documentation (instruction manuals) to videotapes to live classes to one-on-one, side-by-side teacher-student training. Sometimes the organization's own staffers do the training; sometimes it is contracted out.

The Sixth Phase: Maintain the System

Phase 6, <u>systems maintenance</u>, involves adjustment and improvement of the system by conducting system audits and periodic evaluations and by making changes based on new conditions. After conversion and the user training, the system won't just run itself. There is a sixth—and never-ending—phase in which the information system must be monitored to ensure that it is successful. Maintenance includes not only keeping the machinery running but also in updating and upgrading the system to keep pace with new products, services, customers, government regulations, and other requirements.

A.2 Programming: A Five-Step Procedure

KEY QUESTIONS
What is programming, and what are the five steps in accomplishing it?

To see how programming works, we must understand what a program is. A **<u>program</u> is a list of instructions that the computer must follow in order to process data into information.** The instructions consist of *statements* used in a programming language, such as BASIC. Examples are programs that do word processing, desktop publishing, or payroll processing.

The decision whether to buy or create a program forms part of Phase 4 in the systems development life cycle. *(See ● Panel A.2, next page.)* Once the decision is made to develop a new system, the programmer goes to work.

A program, we said, is a list of instructions for the computer. **<u>Programming</u>, also called software engineering, is a multistep process for creating that list of instructions.**

The five steps are as follows.

SDLC

● PANEL A.2
Where programming fits in the systems development life cycle
The fourth phase of the six-phase systems development life cycle includes a five-step procedure of its own. These five steps constitute the problem-solving process called *programming*.

1. **Clarify the problem**—include needed output, input, processing requirements.
2. **Design a solution**—use modeling tools to chart the program.
3. **Code the program**—use a programming language's syntax, or rules, to write the program.
4. **Test the program**—get rid of any logic errors, or "bugs," in the program ("debug" it).
5. **Document and maintain the program**—include written instructions for users, explanation of the program, and operating instructions.

Coding—sitting at the keyboard and typing words into a computer—is what many people imagine programming to be. As we see, however, it is only one of the five steps. Coding consists of translating the logic requirements into a programming language—the letters, numbers, and symbols that make up the program.

A.3 Programming Languages

KEY QUESTIONS

What is a programming language, and how is a high-level language converted to a low-level language?

A **_programming language_** is a set of rules and symbols that tells the computer what operations to do. Examples of well-known programming languages are BASIC, COBOL, and C. *(See* ● *Panel A.3.)* Not all languages are appropriate for all uses. Some, for example, have strengths in mathematical and statistical processing. Others are more appropriate for database management. Thus, in choosing a language, the programmer must consider what purpose the program is to serve and what languages are already being used in the organization or field.

The most important high-level languages that you may come across are the following, ranging from oldest to newest: *FORTRAN, COBOL, BASIC* (and *Visual Basic*), *Pascal, C* (and *C++*), and *Ada.*

- *FORTRAN—the language of mathematics and the first high-level language:* Developed in 1954 by IBM, FORTRAN (for *FORmula TRANslator*) was the first procedural language and is still the most widely used language for mathematical, scientific,and engineering problems. It is also useful for complex business applications, such as forecasting and modeling. However, because it cannot handle a large volume of input/output operations or file processing, it is not used for more typical business problems. The newest version of FORTRAN is FORTRAN 95.

- *COBOL—the language of business:* Formally adopted in 1960, COBOL (for *COmmon Business Oriented Language*) is the most frequently used business programming language for large computers. Its most significant attribute is that it is extremely readable. For example, a COBOL line might read: MULTIPLY HOURLY-RATE BY HOURS-WORKED GIVING GROSS PAY

    Writing a COBOL program resembles writing an outline for a research paper. The program is divided into four divisions, which in turn are divided into sections, which are divided into paragraphs, which are divided into sections, which are divided into statements.

- *BASIC—the easy language:* BASIC was developed by John Kemeny and Thomas Kurtch in 1965 for use in training their students at Dartmouth College. By the late 1960s, it was widely used in academic settings on all kinds of computers, from mainframes to PCs. Now its use has extended to business.

    BASIC (for *Beginner's All-purpose Symbolic Instruction Code*) has been the most popular microcomputer language and is considered the easiest programming language to learn. The interpreter form is popular with first-time and casual users because it is interactive, meaning that user and computer can communicate with each other during the writing and running of the program.

    Today there is no one version of BASIC, but one of the current evolutions is Visual Basic, which is the most popular visual programming language.

- *PASCAL—the simple language:* Created in 1970 and named after the 17th-century French mathematician Blaise Pascal, *Pascal* is an alternative to BASIC as a language for teaching purposes and is relatively easy to learn. A difference from BASIC is that Pascal uses structured programming. It also has extensive capabilities for graphics programming.

- *C and C++—for portability and applications software development:* "C" is this language's entire name, and it does not "stand" for anything. Developed at Bell Laboratories in the early 1970s, *C* is a general-purpose language that works well for microcomputers and is portable among many computers. It is widely used for writing operating systems, utilities, spreadsheet programs, database programs, and some scientific uses. C is also the programming language used most commonly in commercial software development, including games, robotics, and graphics.

    Developed in the early 1980s, *C++* ("C plus plus")—the plus signs stand for "more than C"—combines the traditional C with object-oriented programming (OOP) capability, a technique we discuss later in the chapter. C++ is used for developing applications software.

- *Ada—for weapons and commercial uses:* *Ada* is an extremely powerful structured programming language designed by the U.S. Department of Defense to ensure portability of programs from one application to another. Ada was named for Countess Ada Lovelace, considered the world's "first programmer." Based on Pascal, Ada was originally intended to be a standard language for weapons systems. However, it has been used successfully in commercial applications.

    The advantage of Ada is that it is a structured language, with a modular design; thus, pieces of a large program can be written and tested separately. Moreover, because it has features that permit the compiler to check it for errors before the program is run, programmers are more apt to write error-free programs.

In writing a program, you have to follow the correct *syntax,* the rules of the programming language. Programming languages have their own grammar just as human languages do. But computers are probably a lot less forgiving if you use these rules incorrectly. Even a typographical error can constitute faulty syntax.

Programming languages are also called *high-level languages.* For the computer to be able to understand them, they must be translated into the low-level language called machine language. **Machine language is the basic language of the computer, representing data as 1s and 0s.** Machine-language programs vary from computer to computer; that is, they are *machine-dependent.*

Machine language

```
11110010 01110011 1101 001000010000 0111 000000101011
11110010 01110011 1101 001000011000 0111 000000101111
11111100 01010010 1101 001000010010 1101 001000011101
11110000 01000101 1101 001000010011 0000 000000111110
11110011 01000011 0111 000001010000 1101 001000010100
10010110 11110000 0111 000001010100
```

COBOL

```
MULTIPLY HOURS-WORKED BY PAY-RATE GIVING GROSS-PAY ROUNDED
```

A high-level language allows users to write in a familiar notation, rather than numbers or abbreviations. *(See ● Panel A.4.)* Most high-level languages are not machine-dependent—that is, they can be used on more than one kind of computer. Examples are FORTRAN, COBOL, BASIC, Pascal, and C. For a high-level language to work on a computer, it needs a *language translator* to translate it into machine language. Depending on the language, either of two types of translators may be used—a *compiler* or an *interpreter*.

- Compiler—execute later: **A _compiler_ is a language translator that converts the entire program of a high-level language into machine language BEFORE the computer executes the program.** The programming instructions of a high-level language are called the *source code.* The compiler translates it into machine language, called *object code.* The object code can then be saved and executed later (as many times as desired), rather than run right away. *(See ● Panel A.5.)*

  Examples of high-level languages using compilers are COBOL, FORTRAN, Pascal, and C.

- Interpreter—execute immediately: **An _interpreter_ is a language translator that converts each high-level language statement into machine language and executes it IMMEDIATELY, statement by statement.** In contrast to the compiler, no object code is saved. Therefore, interpreted code generally runs more slowly than compiled code. However, code can be tested line by line.

  An example of a high-language language using an interpreter is BASIC.

A.4 Object-Oriented Programming & Visual Programming

KEY QUESTION
How do OOP and visual programming work?

Two developments have made programming a bit easier—*object-oriented programming* and *visual programming.*

Object-Oriented Programming

Imagine you're writing a program in BASIC or a similar language, creating your coded instructions one at a time. As you work on some segment of the program (such as how to compute overtime pay), you may think, "I'll bet some other programmer has already written something like this. Wish I had it. It would save a lot of time."

Fortunately, a kind of recycling technique now exists. This is object-oriented programming, as used in C++, for example. Let us explain this approach in four steps:

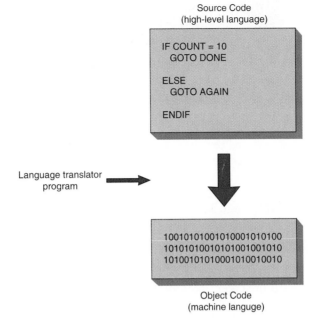

Source Code
(high-level language)

```
IF COUNT = 10
    GOTO DONE

ELSE
    GOTO AGAIN

ENDIF
```

Language translator program

```
10010101001010001010100
10101010010101001001010
10100101010001010010010
```

Object Code
(machine languge)

Conventional Programs

Object-Oriented Programs

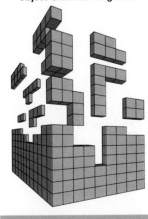

● PANEL A.6
Conventional versus object-oriented programs

1. **What OOP is:** In *__object-oriented programming__* (OOP, pronounced "oop"), **data and instructions for processing that data are combined into a self-sufficient "object" that can be used in other programs.** *(See ● Panel A.6.)* Objects provide powerful tools for programmers.

2. **What an "object" is:** An *object* is a block of preassembled programming code in a self-contained module. The module contains both (1) a chunk of data and (2) processing instructions that may be performed on that data.

3. **When an object's data is processed—sending the "message":** Once the object becomes part of a program, the processing instructions are activated when a "message" is sent. A *message* is an alert sent to the object when the program needs to perform an operation involving that object.

4. **How the object's data is processed—the "methods":** The message need only identify the operation. The processing instructions that are part of the object contain information on how the operation is actually to be performed. These instructions are called the *methods*.

Once you've written a block of program code (that computes overtime pay, for example), it can be reused in any number of programs. Thus, with OOP, in contrast to traditional programming, you don't have to reinvent the wheel each time.

Object-oriented programming takes longer to learn than traditional programming because it involves internalizing a whole new way of thinking. The beauty of OOP, however, is that it speeds up development time and lowers costs, because an object can be used repeatedly in different applications and by different programmers.

Visual Programming

__Visual programming__ **is a method of creating programs in which the programmer makes connections between objects by drawing, pointing, and clicking on diagrams and icons and by interacting with flowcharts.** Thus, the programmer can create programs by clicking on icons that represent common programming routines. Visual programming enables users to think more about the problem solving than about handling the programming language. There is no need to learn syntax or write code.

The most popular visual programming language is Visual Basic. Developed by Microsoft Corp. in the early 1990s, *Visual Basic* offers a visual environment for program construction, allowing you to build various components using buttons, scroll bars, and menus.

A.5 Internet Programming: HTML, XML, VRML, Java, & ActiveX

KEY QUESTION

What are the features of HTML, XML, VRML, Java, and ActiveX?

Many of the thousands of Internet data and information sites around the world are text-based only. However, the World Wide Web permits virtually unlimited use of graphics, animation, video, and sound.

One way to build such multimedia sites on the Web is to use some fairly recently developed programming languages and standards: HTML, XML, VRML, and Java.

HTML—for Creating 2-D Web Documents & Links

HTML (Hypertext Markup Language, discussed in Chapter 2) lets people create onscreen documents for the Internet that can easily be linked by words and pictures to other documents. Not considered a "real" programming language, HTML is a type of code that embeds simple commands within standard ASCII text documents to provide an integrated, two-dimensional display of text and graphics. In other words, a document created in any word processor and stored in ASCII format can become a Web page with the addition of a few HTML commands.

One of the main features of HTML is the ability to insert hypertext links into a document. Hypertext links enable you to display another Web document simply by clicking on a link area—usually underlined or highlighted—on your current screen. One document may contain links to many other related documents. The related documents may be on the same server as the first document, or they may be on a computer halfway around the world. A link may be a word, a group of words, or a picture.

XML—for Making the Web Work Better

The chief characteristics of HTML are its simplicity and its ease in combining plain text and pictures. But, in the words of journalist Michael Krantz, "HTML simply lacks the software muscle to handle the business world's endless and complex transactions."[2]

Enter XML. Whereas HTML makes it easy for humans to read Web sites, *XML (extensible markup language)* makes it easy for machines to read Web sites. At present, when you use your browser to find a Web site, search engines generate too many options, so that it's difficult to turn up the specific site you want—say, one with a recipe for a low-calorie chicken dish for 12. Says Krantz, "XML makes Web sites smart enough to tell other machines whether they're looking at a recipe, an airline ticket, or a pair of easy-fit blue jeans with a 34-inch waist." To do so, XML lets Web site developers put "tags" on their Web pages that describe information in, for example, a food recipe as "ingredients," "calories," "cooking time," and "number of portions." Thus, your browser can read those tags and search much more effectively for that low-calorie poultry recipe for 12.

VRML—for Creating 3-D Web Pages

VRML rhymes with "thermal." *VRML (Virtual Reality Modeling Language)* is a type of programming language used to create three-dimensional Web pages. Even though VRML's designers wanted to let nonprogrammers create their own virtual spaces quickly and painlessly, it's not as simple to describe a three-dimensional scene as it is to describe a page in HTML. However, many existing modeling and CAD tools now offer VRML support, and new VRML-centered software tools are arriving.

Java—for Creating Interactive Web Pages

Available from Sun Microsystems and derived from C++, Java is a major departure from the HTML coding that makes up most Web pages. Sitting atop markup languages such as HTML and XML, *Java* is an object-oriented programming language that allows programmers to build applications that can run on any operating system. With Java, big applications programs can be broken into mini-applications, or "applets," that can be downloaded off the Internet and run on any computer. Moreover, Java enables a Web page to deliver applets that, when downloaded, can make Web pages interactive.

ActiveX—also for Creating Interactive Web Pages

ActiveX was developed by Microsoft as an alternative to Java for creating interactivity on Web pages. Indeed, Java and ActiveX are the two major contenders in the Web-applet war for transforming the Web into a complete interactive environment.

ActiveX is a set of controls, or reusable components, that enables programs or content of almost any type to be embedded within a Web page. Whereas Java requires you to download an applet each time you visit a Web site, with ActiveX the component is downloaded only once, then stored on your hard drive for later and repeated use.

Thus, the chief characteristic of ActiveX is that it features *reusable* components—small modules of software code that perform specific tasks (such as a spelling checker), which may be plugged seamlessly into other applications. With ActiveX you can obtain from your hard disk any file that is suitable for the Web—such as a Java applet, animation, or pop-up menu—and insert it directly into an HTML document.

Programmers can create ActiveX controls or components in a variety of programming languages, including C, C++, Visual Basic, and Java. Thousands of ready-made ActiveX components are now commercially available from numerous software development companies.

PRACTICAL ACTION BOX
How You Can Easily Create Your Own Web Site

Now that you're on the Internet, wouldn't it be cool to promote your business or yourself by creating your own Web site?

To do so, you once needed to learn the basics of HTML. No more. Although knowledge of HTML's page-layout codes still helps, now you can point-and-click to create Web pages with easy-to-use Web-authoring tools. Moreover, many locations on the Internet will post your Web pages for free or at low cost.

Web-Authoring Tools

All the following let you create Web pages, using icons and menus:[a]

- **Internet service providers, portals, and other free sources:** Internet service providers such as America Online, Earthlink, AT&T WorldNet Service, and Juno Web offer simple page-creation programs that let you type in text, specify background colors, and submit photographs. Often this service is free when you sign up with the ISP.

  Portals such as Yahoo!, Lycos, Go Network, and About.com also allow you to create your Web site at no charge, if you're willing to register (giving personal information) and allow the portal to run advertising on your site.

  With these programs from ISPs and portals, you can create a Web page in a couple of hours. However, the appearance of your page will be limited in the number of colors, borders, fonts, and page templates available. Another alternative is Homestead.com, a free Web publishing service that offers many specific templates.

- **Browsers:** Both Netscape Communicator and Microsoft Explorer offer Web-authoring tools. For example, Netscape Communicator provides Composer, which allows you to design Web pages by choosing items from menus; the program automatically inserts HTML tags for each component you select.

- **Commercial applications software:** Most commercial applications programs allow you to save documents in HTML format. With the word processing program Microsoft Word, you can use the Help feature to find out how to translate your document into a Web page, although not all elements will look the same.

- **Web-authoring software:** Several Web-authoring software packages available for $50—$150 can serve not only professional Webmasters but also beginners. In this category are Microsoft FrontPage 2000, Claris Home Page, Adobe Page Mill, Corel Web Designer, IBM's TopPage, and Ixla's WebEasy.

Creating a Good Web Page

The design templates and options offered allow you to do a lot of things. But your objective is to get people to *use* your Web site. Some suggestions:

- **Decide what you want to say:** The first thing to do is decide on the purpose of your Web site. Entertain your friends? Post news for your extended family? Attract customers to your small business? You can get ideas by looking at other people's sites, but ultimately you need to have a clear idea of the goal of your site and know what you want to say.[b]

- **Be concise:** "For the next several years, the vast majority of users will access the Internet through slow modems," writes Jakob Nielsen, author of *Designing Web Usability: The Practice of Simplicity.* "All [Web] pages must download quickly, or users may not only become reluctant to follow the links, but they may also have trouble navigating the site and finding areas they've previously visited. People get lost more often on slow sites than on fast ones and are more likely to leave them and never come back."[c] The implications: People want to get their information quickly, so you have to write concise text that is easy to scan, using highlighted keywords, subheads, and bulleted lists. Give facts, not fluff. Be aware that most users will spend only a few seconds at the site.

- **Build links:** The software allows you to simply highlight a block of text, then type in the address of any link you want. Links, of course, are what will help people find your site, using search engines or connections from other Web pages.

- **Make your site follow good design principles:** Avoid going hog wild with crazy type fonts. Don't put in sound files that play automatically. Use black type for extended sections of text. Don't have bloated graphics, such as files exceeding 30 kilobytes, which take too long to load. (Incidentally, if you don't have a scanner, you can scan photos onto a floppy disk at a copy shop such as Kinko's, then upload them to the site later.)

Publishing Your Web Page

Once you've created your Web site, you'll need to "publish" it—upload it to a Web server in order for it to be viewed on the Internet. You can get upload instructions from your ISP. Some ISPs will give you free space on their servers. Submit-Shack.com is a free service that lets you submit your site to dozens of search engines.[d] Angel.com will also post your site for free.

Note: Don't include any information on your site that you wouldn't put on a sign on your front door. That includes the names or location of family members shown in any photographs, especially those of children.

Index

Index

Notes

Chapter 1

1. Frederick Allen, "Technology at the End of the Century," *Invention & Technology*, Winter 2000, pp. 11–16.
2. Bruce Sterling, "Learning to Love Obsolescence," *Newsweek*, January 1, 2000, p. 68.
3. Fred Abatemarco, "Hand Weapons of the Modern Age," *Popular Science*, September 1999, pp. 9–10.
4. Vivian S. Toy, "Teen-Agers and Cell Phones: It's All Talk, and All the Time," *New York Times*, August 2, 1999, pp. A1, A17.
5. Michael Specter, "Your Mail Has Vanished," *New Yorker*, December 6, 1999, pp. 96–103.
6. Data from International Data Corp., in "Like It or Not, You've Got Mail," *Business Week*, October 4, 1999, pp. 178–184.
7. "Like It or Not, You've Got Mail," 1999.
8. Robert Rossney, "E-Mail's Best Asset–Time to Think," *San Francisco Chronicle*, October 5, 1995, p. E7.
9. Adam Gopnik, "The Return of the Word," *New Yorker*, December 6, 1999, pp. 49–50.
10. Peter H. Lewis, "The Good, the Bad, and the Truly Ugly Faces of Electronic Mail," *New York Times*, September 6, 1994, p. B7.
11. Gopnik, 1999.
12. David A. Whittle, *Cyberspace: The Human Dimension* (New York: W. H. Freeman, 1997).
13. "Living Online," *The Futurist*, July–August 1997, p. 54.
14. Edward Iwata, "The Net at 30," *USA Today*, December 14, 1999, pp. 1A, 2A.
15. Kevin Maney, "The Net Effect: Evolution or Revolution?" *USA Today*, August 9, 1999, pp. 1B, 2B.
16. Center for Communication Policy (www.ccp.ucla.edu), reported in David Plotnikoff, "Study Asks How Wired World Has Changed the Way We Live," *San Jose Mercury News*, June 13, 1999, pp. 1F, 7F.
17. December 10–13, 1999, survey by The Strategis Group, reported in Dru Sefton, "The Big Online Picture: Daily Web Surfing Now the Norm," *USA Today*, March 22, 2000, p. 3D.
18. Teenage Research Unlimited, reported in Eryn Brown, "The Future of Net Shopping? Your Teens," *Fortune*, April 12, 1999, p. 152.
19. Brown, 1999.
20. Kevin Murphy, Gartner Group, quoted in Timothy J. Mullaney, "Death to E-Words Everywhere," *Business Week*, November 15, 1999, p. 10.
21. Adapted from Dan Gillmor, "Electronic Appliances We Have—We Need Embedded Values, Too," *San Jose Mercury News*, October 3, 1999, pp. 1E, 3E.
22. Andy Reinhardt, Steven V. Brull, Peter Burrows, and Catherine Yang, "The Soul of a New Refrigerator," *Business Week*, January 17, 2000, p. 42; Cox News Service, "Internet Can Be Found in Tiny Places," *San Francisco Chronicle*, October 11, 1999; and Steven Butler, "Smart Toilets and Wired Refrigerators," *Newsweek*, June 7, 1999, p. 48.
23. David Einstein, "Custom Computers," *San Francisco Chronicle*, April 15, 1999, pp. B1, B3.
24. Einstein, 1999.
25. Tammy Joyner, "Turning Old Jobs into Hot Ones," *San Francisco Chronicle*, January 17, 2000, p. B3.
26. Laurence Hooper, "No Compromises," *Wall Street Journal*, November 16, 1992, p. R8.
27. Tom Forester and Perry Morrison, *Computer Ethics: Cautionary Tales and Ethical Dilemmas in Computing* (Cambridge, MA: The MIT Press, 1990), pp. 1–2.
28. Norman Solomon, "The Media's Role in the Commercialization of Cyberspace," *San Francisco Chronicle*, January 27, 2000, p. A25.
29. Kevin Maney, "What's on a Snack Web Site? A Whole Lot of Nothing," *USA Today*, February 2, 2000, p. 3B.
30. Joseph N. Pelton, "The Fast-Growing Global Brain," *The Futurist*, August–September, 1999, pp. 24–27.

Bookmark It! Box

a. Study by Impulse Research for *Iconoclast* newsletter, reported in D. Plotnikoff, "E-Mail: A Critical Medium Has Reached Critical Mass," *San Jose Mercury News*, April 4, 1999, pp. 1F, 2F.
b. K. Clark, "At Least the Coffee and Pens Are Still Free," *U.S. News & World Report*, June 7, 1999, p. 70.
c. S. Shostak, "You Call This Progress?" *Newsweek*, January 18, 1999, p. 16.
d. A. Markels, "Don't Manage by E-Mail," *San Francisco Examiner*, August 11, 1996, p. B-5, reprinted from *The Wall Street Journal*; "Don't Overuse Your E-Mail," *CPA Client Bulletin*, March 1999, p. 3; B. Fryer, "E-Mail: Backbone of the Info Age or Smoking Gun?" *Your Company*, July/August 1999, pp. 73–76; S. Armour, "Boss: It's in the E-Mail," *USA Today*, August 10, 1999, p. 3B; L. Guernsey, "Attachments #@%&#@ Are Full #+@&*¢# of Surprises," *New York Times*, July 22, 1999, p. D11.

Chapter 2

1. Greg Miller, "Ethernet Is Changing Dorm Life," *Los Angeles Times*, January 14, 2000, pp. A1, A10.
2. Miller, 2000.
3. Virginia Brooks, quoted in Andy Reinhardt, Peter Elstrom, and Paul Judged, "Zooming Down the I-Way," *Business Week*, April 7, 1997, pp. 76–87.
4. Some of this discussion was adapted from Kate Murphy, "Cruising the Net—in Hyperdrive," *Business Week*, January 24, 2000, pp. 170–172.
5. Joel Dreyfuss, "Set Yourself Free: You Have Nothing to Lose but Your Wires," *Fortune*, January 24, 2000, pp. 139–140.
6. William J. Holstein and Fred Vogelstein, "You've Got a Deal!" *U.S. News & World Report*, January 24, 2000, pp. 34–40.
7. Paul Davidson, "Companies Rush to Bring You Net for Free," *USA Today*, December 16, 1999, p. 1B; and Larry Armstrong, "Something for Nothing," *Business Week*, April 26, 1999, p. 94.
8. Paul Davidson, "Start-Up Pitches Free High-Speed Net Access," *USA Today*, January 4, 2000, p. 1B.
9. Stephen H. Wildstrom, "What Keeps AOL on Top," *Business Week*, January 24, 2000, p. 25.
10. Lawrence Magid, "AOL Not Only Way America Can Get Online," *San Francisco Examiner*, January 23, 2000, pp. B-5, B-6.
11. Study by Impulse Research for Iconoclast newsletter, reported in D. Plotnikoff, "E-Mail: A Critical Medium Has Reached Critical Mass," *San Jose Mercury News*, April 4, 1999, pp. 1F, 2F.
12. Kelley, 1999.
13. Elizabeth Weise, "Instant Battles Keep Raging Over Messaging," *USA Today*, August 4, 1999, p. 5D.
14. Deborah Kong, "Cross E-Mail with a Phone, That's Instant Messaging," *San Jose Mercury News*, August 1, 1999, pp. 1E, 3E.
15. Tom Stein, "Instant Message Nirvana," *San Francisco Chronicle*, August 2, 1999, pp. B1, B6.
16. Michelle Slatalla, "The Office Meeting that Never Ends," *New York Times*, September 23, 1999, pp. D1, D8.
17. Jeanne Hinds, quoted in Slatalla , 1999.
18. Pitney Bowes study, cited in Slatalla, 1999.
19. Editors of *PC Computing*, "End E-Mail Insanity Forever," *PC Computing*, November 1999, pp. 170–198;

Jennifer Powell, "E-Mail: Common Sense Plays a Big Role," *Smart Computing*, March 2000, p. 47; and Kenneth M. Morris, *User's Guide to the Information Age* (New York: Lightbulb Press, 1999), pp. 111, 153.

20. David Lazarus, "Fan Spam Is Hard to Shake," *San Francisco Chronicle*, February 7, 2000, pp. C1, C2.

21. Julian Haight, quoted in Lazarus, 2000, p. C1.

22. Calculations by NEC Research and Inktomi, reported in Elizabeth Weise, "Web Changes Direction to People Skills," *USA Today*, January 24, 2000, pp. 1D, 2D.

23. Jennifer Powell, "Lessons from the Browser War (1995–2000)," *PC Computing*, March 2000, p. 64.

24. Weise, "Web Changes Direction to People Skills," 2000; and Saul Hansell, "Obsessively Independent, Yahoo Is the Web's Switzerland," *New York Times*, August 23, 1999, pp. C1, C10.

25. Janet Kornblum, "Portals Suffer as Internet Surfers Get More Savvy," *USA Today*, July 2, 1999, p. 1B; and Dan Gillmor, "Small Portals Prove that Size Matters," *San Jose Mercury News*, December 6, 1998, pp. 1E, 3E.

26. Study by Steve Lawrence and C. Lee Giles in journal *Nature*, reported in Peter Svensson, "Search Engines Can't Keep Up," *San Francisco Chronicle*, July 8, 1999, p. B3; and in Mike Snider, "Web Growth Outpacing Search Power," *USA Today*, July 8, 1999, p. 1A;

27. Weise, "Web Changes Direction to People Skills," 2000.

28. Elizabeth Weise, "Successful Net Search Starts with a Need," *USA Today*, January 24, 2000, p. 3D; Timothy Hanrahan, "The Best Way to . . . Search Online," *Wall Street Journal*, December 6, 1999, p. R25; and Weise, "Web Changes Direction to People Skills," 2000.

29. Timothy Hanrahan, "The Best Way to Search Online," *Wall Street Journal*, December 6, 1999, p. R25; Reva Basch, "Cutting Through the Clutter," *Smart Computing*, July 1999, pp. 88–91; Matt Lake and Dylan Tweney, "Find It on the Web," *PC World*, June 1999, pp. 168–182; and Matt Lake, "Desperately Seeking Susan OR Suzie NOT Sushi," *New York Times*, September 3, 1998, p. D1

30. Lake, 1998.

31. Lake and Tweney, 1999.

32. Jon Van, "Internet Firms Take Aim at Phone Giants," *San Francisco Examiner*, February 6, 2000, p. B-3; reprinted from *Chicago Tribune*.

33. Laurie Bryan, "Wired for Shopping," *San Francisco Chronicle*, February 9, 2000, zone 7, pp. 1, 4.

34. Morris, 1999, p. 100.

35. Megan Doscher, "The Best Way to Find Love," *Wall Street Journal*, December 6, 1999, p. R34.

36. Sharon Cleary, "The Best Way to Find an Old Friend," *Wall Street Journal*, December 6, 1999, pp. R26, R45.

37. Elizabeth Weise, "Cultivating a Sense of Community On Line," *USA Today*, June 30, 1999, p. 6D.

38. Wendy M. Grossman, "Language Is a Virus," *PC Computing*, March 2000, p. 62.

39. William M. Bulkeley, "The Best Way to Go to School," *Wall Street Journal*, December 6, 1999, pp. R18, R22.

40. Bulkeley, 1999; and Faith Bremner, "On-line College Classes Get High Marks Among Students," *USA Today*, November 16, 1998, p. 16E.

41. Jane Manners, "Web Health Checkup," *Brill's Content*, February 2000, pp. 114–116; and Lisa Bransten, "The Best Way to Stay Healthy," *Wall Street Journal*, December 6, 1999, pp. R36, R45.

42. Study by Cyber Dialogue, reported in Evan Ramstad, "The Best Way to Amuse Yourself," *Wall Street Journal*, December 6, 1999, p. R43.

43. Lorrie Grant, "Internet Has Become Integral Part of Everyday Life," *USA Today*, April 21, 1999, p. 6B.

44. Haya El Nasser, "Main Street Enters Mainstream," *USA Today*, November 16, 1999, p. 3A.

45. Denise Caruso, "On-Line Day Traders Are Starting to Have an Impact on a Few Big-Cap Internet Stocks in What Some Call a 'Feeding Frenzy,'" *New York Times*, December 14, 1998, C3.

46. Gary McWilliams, "The Best Way to Find a Job," *Wall Street Journal*, December 6, 1999, pp. R16, R22.

47. Del Jones, "E-Purchasing Saves Businesses Billions," *USA Today*, February 7, 2000, pp. 1B, 2B.

48. John Rubino, "Measures," *Individual Investor*, March 2000, pp. 37–42.

49. Rubino, 2000.

Bookmark It! Box

a. Karen Jacobs, "The Best Way to Pick a Provider," *Wall Street Journal*, December 6, 1999, pp. R8, R10; Tracy Baker, "Free Access to the Internet," *Smart Computing*, June 1999, pp. 44–46; Tina Kelley, "Choosing an ISP: Convenience, Cost, and Service," *San Jose Mercury News*, May 30, 1999, pp. 1F, 4F; reprinted from the *New York Times*; and "Going Online: Best Ways to Get Started," *Consumer Reports*, July 1998, p. 64.

Chapter 3

1. David Einstein, "Software Coming Online," *San Francisco Chronicle*, November 18, 1999, p. B3.

2. Anita Hamilton, "Scheduling Snafu," *Time*, May 10, 1999, p. 96.

3. Paul Davidson, "Online Software Catches On," *USA Today*, February 4, 2000, pp. 1B, 2B.

4. Davidson, 2000.

5. Luc Hatlestad, "Small Beginnings," *Red Herring*, March 2000, pp. 208–212.

6. J. William Gurley, "The New Market for 'Rentalware,'" *Fortune*, May 10, 1999, p. 142.

7. Gary Bloom, quoted in Lawrence M. Fisher, "Software Evolving into a Service Rented Off the Net," *New York Times*, December 20, 1999, p. C37.

8. Joshua Quittner, "Aqua: The Movie," *Time*, January 31, 2000, p. 82.

9. Brent Schlender, "Steve Jobs' Apple Gets Way Cooler," *Fortune*, January 24, 2000, pp. 66–71.

10. Rebecca Buckman, "Microsoft Ready to Demonstrate a 'Me' Generation of Windows," *Wall Street Journal*, April 7, 2000, p. B3; and Edward C. Baig, "'Millennium': Cleaner Windows?" *USA Today*, January 10, 2000, p. 3D.

11. Steve Hamm, Peter Burrows, and Andy Reinhardt, "Is Windows Ready to Run E-Business?" *Business Week*, January 24, 2000, pp. 154–160.

12. Stephen H. Wildstrom, "Windows 2000: Worth the Wait," *Business Week*, February 21, 2000, p. 20; Hamm, Burrows, and Reinhardt, 2000; John Markoff, "Microsoft Facing a Skeptical Market with Windows 2000," *New York Times*, February 14, 2000, pp. C1, C6; Ellis Booker, "Users Look Forward to Deploying Windows 2000," *InternetWeek*, February 18, 2000 (online); and David Bank, "Microsoft and Market Expect Slower Start for Windows 2000 Sales," *Wall Street Journal*, February 14, 2000, p. B8.13 Hamm, Burrows, and Reinhardt, 2000; and Lawrence M. Fisher, "Sun Plans to Start Shipping Operating System Next Month," *New York Times*, January 27, 2000, p. C12.

14. Irving Wladawsky-Berger, quoted in Deborah Solomon, "Could Linux Outdo Windows?" *USA Today*, March 9, 2000, pp. 1B, 2B.

15. Stanley Holmes, "Companies Form Group to Promote Linux," *Los Angeles Times*, March 10, 2000, pp. C2, C4; Sam Jaffe, "Will Linux Investors Be Left Out in the Cold?" *Business Week*, March 2000, pp. 178–180; David Kirkpatrick, "Dell to Wintel: Your Hegemony Is Over," *Fortune*, February 21, 2000, pp. 50, 52; Lawrence M. Fisher, "Looking for a New Life in Linux," *New York Times*, February 14, 2000, p. C4; Lawrence M. Fisher, "A View that Needs No Windows," *New York Times*, February 6, 2000, sec. 3, p. 2; and Bloomberg News, "Linux-Related Shares Soar on News of Revised Version," *Los Angeles Times*, February 3, 2000, p. C7.

16. David Rynecki, "Is Palm's IPO Really the One to Catch?" *Fortune*, February 7, 2000, pp. 213–214; Roy Furchgott, "Using the Net to Soup Up Your Palm Top Computer," *Business Week*, November 8, 1999, pp. 165–166; Walter S. Mossberg, "A Pilot Rival Organizes Your Life, Then Morphs into Something Else," *Wall Street Journal*, September 16, 1999, p. B1; and Kevin Maney, "Palm Has the Whole World in Its Hand," *USA Today*, September 14, 1999, pp. 1B, 2B.

17. Edward C. Baig, "4 PDAs View to Be Your Assistant," *USA Today*, December 15, 1999, p. 3D.

18. Henry Norr, "Microsoft's Pocket PC Hits Store Shelves Next Month," *San Francisco Chronicle*, March 6, 2000, pp. B1, B3, B7.

19. Autumn de Leon, "Portable Computing at Hand," *Time*, November 29, 1999, p. 96.
20. Michel Marriott, "Newcomer in the OS Ranks," *New York Times*, February 10, 2000, p. D8.
21. Seth Schiesel, "AT&T to Invest $250 Million in Application Services Sector," *New York Times*, January 27, 2000, p. C11.
22. Margaret Trejo, quoted in Richard Atcheson, "A Woman for *Lear's*," *Lear's*, November 1993, p. 87.
23. John Ennis, quoted in Peter Plagens and Ray Sawhill, "Throw Out the Brushes," *Newsweek*, September 1, 1997, pp. 76–77.
24. Bernie Ward, "Computer Chic," *Sky*, April 1993, pp. 84–90.

Bookmark It! Box
a. Associated Press, quoted in Ellen Goodman, "Computercide on My Mind," *San Francisco Chronicle*, July 31, 1997, p. A21; reprinted from *Boston Globe*.

Chapter 4
1. Stephen H. Wildstrom, "The High Cost of 'Free' PCs," *Business Week*, September 13, 1999, p. 20.
2. Matt Richtel, "Consumers Warm to 'Free' PC's," *New York Times*, April 15, 1999, pp. D1, D8; Peter Burrows, "Why Pay More? Well . . .," *Business Week*, November 15, 1999, pp. 214–220; Karl Taro Greenfeld, "Giving Away the E-Store," *Time*, November 22, 1999, pp. 58–60; and Jube Shriver, Jr., "Net Firms Offering Freebies Are Paying Dearly: Consumers Expect Even More," *Los Angeles Times*, February 7, 2000, pp. C1, C9.
3. "Computers, Then and Now," *Consumer Reports*, May 2000, p. 10.
4. Data from PC Data Inc., cited by Gary McWilliams, "Reversing Course, Home-PC Prices Head Higher," *Wall Street Journal*, January 13, 2000, pp. B1, B4; and John Simons, "Cheap Computers Bridge Digital Divide," *Wall Street Journal*, January 27, 2000, p. A22.
5. Stephen C. Miller, "Another Way to Get on Line without Buying a Computer," *New York Times*, October 14, 1999, p. D3; Ashley Dunn, "Pumped-Up Appliance or No-Frills Computer? i-Opener a Shortcut to Net," *Los Angeles Times*, March 9, 2000, pp. C1, C7; and Amy Harmon, "Courtesy of Amateurs, a $99 Personal Computer," *New York Times*, March 18, 2000, pp. B1, B2.
6. Associated Press, "Playing Games and Doing Banking on PlayStation Console," *San Francisco Chronicle*, March 31, 2000, p. C5.
7. Henry Norr, "The NC—with a Twist," *San Francisco Chronicle*, February 28, 2000, pp. B1, B2.
8. Henry Norr, "Why Thin Computing Is In," *San Francisco Chronicle*, February 28, 2000, p. B2.
9. Henry Norr, "Do You Need the Fastest Processor?" *San Francisco Chronicle*, April 17, 2000, pp. E1, E2, E3; Peter H. Lewis, "With 2 New Chips, the Gigahertz Decade Begins," *New York Times*, March 9, 2000,

pp. D1, D3; Henry Norr, "Intel Ships Fast Chip, Catches Up with AMD," *San Francisco Chronicle*, March 9, 2000, pp. B1, B4; and Edward C. Baig, "Gunning the Engine on 1-Gig PCs," *USA Today*, March 22, 2000, p. 3D.
10. John Markoff, "IBM to Show a Breakthrough in Chip Making," *New York Times*, April 3, 2000, pp. C1, C2; and Edward Iwata, "IBM Boasts of Better, Faster Chip," *USA Today*, April 3, 2000, p. 1B.
11. Baig, 2000; and Walter S. Mossberg, "Mossberg's Mailbox," *Wall Street Journal*, April 13, 2000, p. B9.
12. Ken Hawk, "Peripheral Issues: The USB," *Reno Gazette-Journal*, March 13, 2000, p. 3E.
13. Mossberg, 2000.
14. Becky Waring, "Taking the Desktop with You," *San Francisco Chronicle*, April 13, 2000, pp. D1, D3; Walter Mossberg, "Mossberg's Mailbox," *Wall Street Journal*, April 6, 2000, B10.
15. Edward Baig, "Be Happy, Film Freaks," *Business Week*, May 26, 1997, pp. 172–173.
16. Edmund Sanders, "Makers of Smart Cards Are Betting Big on U.S.," *Los Angeles Times*, February 28, 2000, pp. C1, C6.
17. Edward C. Baig, "Tapping Out the Quick Keys to a Comfortable Keyboard," *USA Today*, March 29, 2000, p. 3D.
18. Bruce Headlam, "Spills Are Not a Problem for a Flexible Keyboard," *New York Times*, April 6, 2000, p. D3.
19. Becky Waring, "Microsoft Builds a Magical Mouse," *San Francisco Chronicle*, March 30, 2000, p. B3.
20. Claudia H. Deutsch, "There's Gold in Those Old Photos in the Attic," *New York Times*, June 30, 1997, p. C6.
21. Anita Hamilton, "Broadcasting from New Ork," *Time*, April 17, 2000, p. 80.
22. Philip Robinson, "Flat Panels: Are They Worth It?" *Reno Gazette-Journal*, March 13, 2000, p. 3E.

Bookmark It! Box
a. Walter S. Mossberg, "How to Buy a Laptop: Some Basic Guidelines in a Dizzying Market," *Wall Street Journal*, October 21, 1999, p. B1.
b. Stephen H. Wildstrom, "How to Shop for a Laptop," *Business Week*, April 3, 2000, p. 25.
c. Wildstrom, 2000; Mossberg, 1999; "The New Laptops," *Consumer Reports*, May 2000, pp. 12–16; Walter S. Mossberg, "Buying Your Next PC? Get the Most Memory, Not the Fastest Chip," *Wall Street Journal*, April 6, 2000, p. B1; and Bill Howard, "Notebook PCs," *PC Magazine*, August 1999, pp. 154–155.

Chapter 5
5. Michel Marriott, "WebTV Puts It All Together," *New York Times*, December 30, 1999, pp. D2, D5.
2. Bruce Haring, "Step Right Up for the Next Push in Remote Control," *USA Today*, September 15, 1999, p. 7D.
3. Mike Snider, "Waves of Interference Slow Digital TV," *USA Today*, October 27, 1999, p. 7D; Neil Gross,

Richard Siklos, and Heidi Dawley, "HDTV: You're Not Going to Like this Picture," *Business Week*, October 25, 1999, p. 50; and Andrew J. Glass, "HDTV Still Stationed at the Starting Gate," *San Francisco Chronicle*, September 27, 1999, p. E5.
4. "What Does 'Digital' Mean in Regard to Electronics?" *Popular Science*, August 1997, pp. 91–94.
5. Simon Romero, "Weavers Go Dot-Com, and Elders Move In," *New York Times*, March 28, 2000, pp. A1, A4.
6. Survey by Pew Research Center for the People and the Press, June 2000, reported in Will Lester, "Poll: Young, Educated Americans Turn to Internet for News," *Reno Gazette-Journal*, June 12, 2000, p. 6A.
7. Fred Vogelstein, "Is It Sharing or Stealing?" *U.S. News & World Report*, June 12, 2000, pp. 38–40.
8. Marcia Vickers, "Don't Touch that Dial: Why Should I Hire You?" *New York Times*, April 13, 1997, sec. 3, p. 11.
9. International Data Corp., cited in "Do Dirty Dishes, Dust Bunnies Violate OSHA Rules?" [editorial], *USA Today*, January 7, 2000, p. 14A.
10. Study by John J. Heldrich Center for Workforce Development, Rutgers University, cited in Kirstin Downey Grimsley, "How to Get Permission to Telecommute," *San Francisco Chronicle*, February 16, 2000, p. B3.
11. David Kline, quoted in W. James Au, "The Lonely Long-Distance Worker," *PC Computing*, February 2000, pp. 42–43.
12. David Leonhardt, "Telecommuting to Pick Up as Workers Iron Out Kinks," *New York Times*, December 20, 1999, p. C6.
13. Maryanne Murray Buechner, "Superconnected," *Time*, March 22, 1999.
14. Mike Romano, "Brave New Home," *U.S. News & World Report*, April 5, 1999, pp. 60–62.
15. Yahoo!, cited in Del Jones, "Cyberporn Poses Workplace Threat," *USA Today*, November 27, 1995, p. B1.
16. Lawrence J. Magid, "Be Wary, Stay Safe in the On-line World," *San Jose Mercury News*, May 15, 1994, p. 1F.
17. Peter H. Lewis, "Limiting a Medium without Boundaries," *New York Times*, January 15, 1996, pp. C1, C4.
18. John M. Broder, "Making America Safe for Electronic Commerce," *New York Times*, June 22, 1997, sec. 4, p. 4; Margaret Mannix and Susan Gregory Thomas, "Exposed Online," *U.S. News & World Report*, June 23, 1997, pp. 59–61; and Noah Matthews, "Shareware," *San Jose Mercury News*, October 12, 1997, p. 4F.
19. Survey by Equifax and Louis Harris & Associates, cited in Bruce Horovitz, "80% Fear Loss of Privacy to Computers," *USA Today*, October 31, 1995, p. 1A.

Bookmark It! Box
a. Michel Marriott, "For Extra Cheese, Ctrl + Pizza," *New York Times*, February 10, 2000, pp. D1, D8; Walter S. Mossberg, "A Simple New Gadget Lets You Go Online without Using a

PC," *Wall Street Journal*, January 27, 2000, p. B1.
b. Howie Hunger, quoted in John Yaukey, "Small, Simple PCs Take the Day," *Reno Gazette Journal*, May 8, 2000, p. 1E.
c. Roy Furchgott, "Web to Go—Sort of," *Business Week*, February 14, 2000, pp. 144–145.
d. Dori Jones Yang, "A Boob Tube with Brains," *U.S. News & World Report*, March 13, 2000, pp. 42–43; David Lieberman, "Malone: Cable Firms Must Tune in to Interactive TV," *USA Today*, April 17, 2000, p. 10B.
e. Kenneth Terrell, "PlayStation 2: Hold Your Fire—and Your Dollars," *U.S. News & World Report*, May 8, 2000, pp. 66–67; Kenneth Terrell, "Sega Takes Its Play Online," *U.S. News & World Report*, May 8, 2000, pp. 66–67; Jay Greene, Irene Kunii, and Janet Rae-Dupree, "Get Ready to Rumble," *Business Week*, March 20, 2000, p. 48; Dean Takahashi, "Sega Will Give Away Dreamcast Players to Lure Subscribers to the Web," *Wall Street Journal*, April 4, 2000, pp. B1, B4; Steven Levy, "Here Comes PlayStation 2," *Newsweek*, March 6, 2000, pp. 54–59; N'Gai Croal, "The Art of the Game," *Newsweek*, March 6, 2000, pp. 60–62; and Kelly Zito, "Sega's Got Game on the Web," *San Francisco Chronicle*, April 4, 2000, pp. C1, C5.
f. Kevin Maney, "Without Neat Plug-in Modules, Visor Suffers," *USA Today*, April 24, 2000, p. 7B; N'gai Croal, "A PC in Every Pocket," *Newsweek*, May 1, 2000, pp. 521–53; Henry Norr, "Technicolor Palm," *San Francisco Chronicle*, February 24, 2000, pp. C1, C4; Stephen H. Wildstrom, "Loosening Palm's Grip," *Business Week*, May 1, 2000, p. 28; and Joshua Quittner, "PCs? Forget 'Em!" *Time*, May 8, 2000, p. 105.
g. Chris O'Malley, "Two-Way Sharing," *Popular Science*, November 1999, p. 58.
h. "E-Mail When You're Far from Home," *Business Week Frontier*, March 27, 2000, p. F.6.
i. Guy Richardson, "Feeling Wired Because You're Not Wired?" *Reno Gazette-Journal*, March 13, 2000, p. 3E.
j. Michel Marriott, "Out of the Mouths of Babes, Wirelessly," *New York Times*, March 23, 2000, pp. D1, D6.

Chapter 6
1. Carole A. Lane, quoted in Leslie Miller, "You Are a Database and Access Abounds," *USA Today*, June 9, 1997, p. 6D.
2. Robert D. Atkinson and Randolph H. Court, *The New Economy Index: Understanding America's Economic Transformation* (Washington, D.C.: Progressive Policy Institute, 1999).
3. See J. Quittner, "Tim Berners-Lee," *Time*, March 29, 1999, pp. 193–194.
4. Forrester Research Inc., reported in "E-Commerce: It's Clicking" [editorial], *Business Week*, January 11, 1999, p. 154.
5. Odyssey, reported in S. Lohr, "Survey Suggests Consumers Are Taking to

E-Commerce," *New York Times*, March 22, 1999, p. C4.
6. Sarah E. Hutchinson and Stacey C. Sawyer, *Computers, Communications, and Information: A User's Introduction*, rev. ed. (Burr Ridge, IL: Irwin/McGraw-Hill, 1998), pp. E1.1–E1.3.
7. Jeff Bezos, quoted in K. Southwick, interview, October 1996, *www.upside.com*.
8. D. Levy, "On-line Gamble Pays Off with Rocketing Success," *USA Today*, December 24, 1998, pp. 1B, 2B.
9. Kara Swisher, "Why Is Jeff Bezos Still Smiling?" *Wall Street Journal*, April 24, 2000, pp. B1, B10.
10. Jonathan Berry, John Verity, Kathleen Kerwin, and Gail DeGeorge, "Database Marketing," *Business Week*, September 5, 1994, pp. 56–62.
11. Hedberg, 1995.
12. Cheryl D. Krivda, "Data-Mining Dynamite," *Byte*, October 1995, pp. 97–103.
13. Edmund X. DeJesus, "Data Mining," *Byte*, October 1995, p. 81.
14. Forester Research, cited in Denise Caruso, "Taking Stock of the Differences Between the Consumer Internet Market and Its Business-to-Business Cousin," *New York Times*, February 28, 2000, p. C5; and Gartner Group, cited in William J. Holstein, "Rewiring the 'Old Economy,'" *U.S. News & World Report*, April 10, 2000, pp. 38–40.
15. Don Tapscot, "Virtual Webs Will Revolutionize Business," *Wall Street Journal*, April 24, 2000, p. A38.
16. Carolyn Said, "Online Middlemen," *San Francisco Chronicle*, April 10, 2000, pp. C1, C3; Holstein, 2000; Claudia H. Deutsch, "Another Economy on the Supply Side," *New York Times*, April 8, 2000, pp. B1, B4; and Kelly Zito, "Online Exchange for Shops," *San Francisco Chronicle*, March 9, 2000, pp. B1, B4.
17. Clay Shirky, "Haggling Goes High-Tech," *Wall Street Journal*, April 10, 2000, p. A46.
18. Tapscot, 2000.
19. John Chambers, quoted in Del Jones and Beth Belton, "Cisco Chief: Virtual Close to Hit Big," *USA Today*, October 12, 1999, p. 3B.
20. "Connecting the Dot-Coms," *Newsweek*, April 17, 2000, p. 68G.
21. William Safire, "Art vs. Artifice," *New York Times*, January 3, 1994, p. A11.
22. Cover, *Newsweek*, June 27, 1994; and cover, *Time*, June 27, 1994.
23. Jonathan Alter, "When Photographs Lie," *Newsweek*, July 30, 1990, pp. 44–45.
24. Fred Ritchin, quoted in Alter, 1990.
25. Robert Zemeckis, cited in Laurence Hooper, "Digital Hollywood: How Computers Are Remaking Movie Making," *Rolling Stone*, August 11, 1994, pp. 55–58, 75.
26. Woody Hochswender, "When Seeing Cannot Be Believing," *New York Times*, June 23, 1992, pp. B1, B3.
27. Bruce Horowitz, "Believe Your Eyes? Ads Bend Reality," *USA Today*, April 24, 2000, pp. 1B, 2B.

28. Penny Williams, "Database Dangers," *Quill*, July/August 1994, pp. 37–38.
29. Lynn Davis, quoted in Williams, 1994.

Chapter 7
1. Katie Hafner, "We're Not All Connected, Yet," *New York Times*, January 27, 2000, pp. D1, D9.
2. Michael Dertouzos, "The Net Revolution Spawns a 'Fast Caste,'" *Los Angeles Times*, January 20, 2000, p. A15.
3. Survey by Inktomi Corp. and NEC Research Institute, reported in Ashley Dunn, "It's a Very Wide Web: 1 Billion Pages' Worth," *Los Angeles Times*, January 20, 2000, p. C7.
4. Alan Murray, "Trying to Make World Safe for E-Commerce," *Wall Street Journal*, November 29, 1999, p. A1.
5. Jube Shiver, Jr., "Web Offers Few Riches for Poor," *Los Angeles Times*, March 16, 2000, pp. C1, C7; Katie Hafner, "A Credibility Gap in the Digital Divide," *New York Times*, March 5, 2000, sec. 4, p. 4; Janet Kornblum, "Poor Aren't Profiting, Non-profit Group Says," *USA Today*, March 14, 2000, p. 3D; and Jube Shiver, Jr., "Big Bandage for a Narrowing Internet Gap," *Los Angeles Times*, January 29, 2000, pp. A1, A12.
6. Scott Carlson, "High-Speed Network Will Serve Universities for 3 More Years," *Chronicle of Higher Education*, April 21, 2000, p. A49.
7. Laurence H. Tribe, "The FCC vs. the Constitution," *Wall Street Journal*, September 5, 1997, p. A8.
8. John M. Broder, "Let It Be," *New York Times*, June 30, 1997, pp. C1, C9.
9. *A Framework for Global Electronic Commerce*, quoted in Steven Levy, "Bill and Al Get It Right," *Newsweek*, July 7, 1997, p. 80.
10. Jim Hornthal, quoted in Jon Schwartz, "Clinton Advocates Net Self-Rule," *San Francisco Chronicle*, July 2, 1997, pp. B1, B2.
11. Eli M. Noam, "An Unfettered Internet? Keep Dreaming," *New York Times*, July 11, 1997, p. A2.
12. Mike France, "What's in a Name.com? Plenty," *Business Week*, September 6, 1999, pp. 86–90.
13. We are grateful to Prof. John Durham for contributing these ideas.
14. John Allen Paulos, "Smart Machines, Foolish People," *Wall Street Journal*, October 5, 1999, p. A26.
15. Arthur M. Louis, "Nasdaq's Computer Crashes," *San Francisco Chronicle*, July 16, 1994, pp. D1, D3.
16. Stephen H. Wildstrom, "Oh, for a Bug-free Browser," *Business Week*, November 1, 1999, p. 24.
17. Henry K. Lee, "UC Student's Dissertation Stolen with Computer," *San Francisco Chronicle*, January 27, 1994, p. A15.
18. Thomas J. DeLoughry, "2 Students Are Arrested for Software Piracy," *Chronicle of Higher Education*, April 20, 1994, p. A32.
19. John T. McQuiston, "4 College Students Charged with Theft Via Computer," *New York Times*, March 18, 1995, p. 38.

20. Janet Rae-Dupree and Richard J. Newman, "A Twisted Kind of Love," *U.S. News & World Report*, May 15, 2000, p. 24; Brad Stone, Mark Hosenball, and Stefan Theil, "Bitten by Love," *Newsweek*, May 15, 2000, pp. 42–43; and Lev Grossman et al., "Attack of the Love Bug," *Time*, May 15, 2000, pp. 49–56.

21. Edward Iwata, "Mutating Computer Virus Hits," *USA Today*, May 19, 2000, p. 1A.

22. David Carter, quoted in Associated Press, "Computer Crime Usually Inside Job," *USA Today*, October 25, 1995, p. 1B.

23. Bryan Burrough, "Invisible Enemies," *Vanity Fair*, June 2000, pp. 172–177, 208–214.

24. Steven Bellovin, cited in Jane Bird, "More Than a Nuisance," *The Times* (London), April 22, 1994, p. 31.

25. Eugene Carlson, "Some Forms of Identification Can't Be Handily Faked," *Wall Street Journal*, September 14, 1993, p. B2.

26. Justin Matlkick, "Security of Online Markets Could Well Be at Stake," *San Francisco Chronicle*, September 16, 1997, p. A21.

27. John Holusha, "The Painful Lessons of Disruption," *New York Times*, March 17, 1993, pp. C1, C5.

28. The Enterprise Technology Center, cited in "Disaster Avoidance and Recovery Is Growing Business Priority," special advertising supplement in *LAN Magazine*, November 1992, p. SS3.

Bookmark It! Box

a. Lee De Cesare, "Virtual Term Papers" [letter], *New York Times*, June 10, 1997, p. A20.

b. Bruce Leland, quoted in Peter Applebome, "On the Internet, Term Papers Are Hot Items," *New York Times*, June 8, 1997, sec. 1, pp. 1, 20.

c. Eugene Dwyer, "Virtual Term Papers" [letter], *New York Times*, June 10, 1997, p. A20.

d. Mitchell Zimmerman, "How to Track Down Collegiate Cyber-Cheaters" [letter], *New York Times*, June 15, 1997, sec. 4, p. 14.

e. Tanya Schevitz, "Point, Click, Plagiarize," *San Francisco Chronicle*, November 5, 1999, pp. A1, A19.

f. William L. Rukeyser, "How to Track Down Collegiate Cyber-Cheaters" [letter], *New York Times*, June 14, 1997, sec. 4, p. 14.

g. David Rothenberg, "How the Web Destroys the Quality of Students' Research Papers," *Chronicle of Higher Education*, August 15, 1997, p. A44.

Chapter 8

1. Stephen Cohen, quoted in Carolyn Lochhead, "The Engines of a New Economy," *San Francisco Examiner & Chronicle*, "Sunday" section, February 27, 2000, p. 2.

2. Stephen L. Talbott, quoted in Lisa Guernsey, "Editor Explores Unintended, and Negative Side, of Technology," *New York Times*, November 25, 1999, p. D7.

3. Guy Richardson, "Time to Curl Up and Take a Napster," *Reno Gazette-Journal*, May 29, 2000, p. 4E.

4. Survey by Webnoize Inc., Cambridge, Mass., reported in Martin Peers, "Survey Studies Napster's Spread on Campuses," *Wall Street Journal*, May 15, 2000, p. B8.

5. Randall E. Stross, "Napster Nonsense," *U.S. News & World Report*, May 29, 2000, p. 49.

6. Judith Anne Gunther, "An Encounter with AI," *Popular Science*, June 1994, pp. 90–93.

7. William A. Wallace, *Ethics in Modeling* (New York: Elsevier Science, 1994).

8. Christine Borgman, quoted in Gary Chapman, "What the Online World Really Needs Is an Old-Fashioned Librarian," *San Jose Mercury News*, August 21, 1995, p. 3D; reprinted from *Los Angeles Times*.

9. Campus Computing Project 1997 survey, reported in Lisa Guernsey, "E-Mail Is Now Used in a Third of College Courses, Survey Finds," *Chronicle of Higher Education*, October 17, 1997, p. A30; and Edward C. Baig, "A Little High Tech Goes a Long Way," *Business Week*, November 10, 1997, p. E10.

10. Student Monitor LLC, cited in Danielle Sessa, "For College Students, Web Offers a Lesson in Discounts," *Wall Street Journal*, January 21, 1999, p. B7.

11. U.S. Department of Education, cited in Steve Rhodes, "Classrooms with Class—and Possibly Espresso Machines," *Newsweek*, December 14, 1998, p. 20.

12. *Does It Compute?*, study by Harold Wenglinsky, reported in Ethan Bronner, "Computers Help Math Learning, Study Finds," *New York Times*, September 30, 1998, p. A16.

13. Dulcie Leimbach, "Encouraging Creativity, without the Mess," *New York Times*, November 12, 1998, p. D12.

14. Kelly McCollum, "Web Site Lets Thousands of Schoolchildren Follow Research on Albatrosses," *Chronicle of Higher Education*, February 13, 1998, p. A32.

15. Kelly McCollum, "High School Students Use Web Intelligently for Research, Study Finds," *The Chronicle of Higher Education*, December 4, 1998, p. A25.

16. Kelly McCollum, "How a Computer Program Learns to Grade Essays," *Chronicle of Higher Education*, September 4, 1998, pp. A37–A38; and William H. Honan, "High Tech Comes to the Classroom," *New York Times*, January 27, 1999, p. A22.

17. Nanette Asimov, "Home-Schoolers Plug into the Internet for Resources," *San Francisco Chronicle*, January 29, 1999, pp. A1, A15.

18. Mary Beth Marklein, "Distance Learning Takes a Gigantic Leap Forward," *USA Today*, June 4, 1998, pp. 1D, 2D; and Godie Blumenstyk, "Leading Community Colleges Go National with New Distance-Learning Network," *Chronicle of Higher Education*, July 10, 1998, pp. A16–A17.

19. Rebecca Quick, "Software Seeks to Breathe Life into Corporate Training Classes," *Wall Street Journal*, August 6, 1998, p. B8.

20. April Lynch, "Bleeding Sailor Performs Self-Surgery Via E-Mail," *San Francisco Chronicle*, November 19, 1998, pp. A1, A10.

21. Kate Murphy, "Telemedicine Getting a Test in Efforts to Cut Costs of Treating Prisoners," *New York Times*, June 8, 1998, p. C3.

22. Associated Press, "Dad Uses Internet to Find Wonder Drug for His Son," *San Francisco Chronicle*, May 29, 1998, p. A22.

23. Jim Hudak, quoted in Heather Green and Linda Himelstein, "A Cyber Revolt in Health Care," *Business Week*, October 19, 1998, pp. 154–56.

24. Tanya Schevitz, "Pregnant in Cyberspace," *San Francisco Chronicle*, July 29, 1998, pp. A13, A17.

25. Rick Satava, quoted in Gary Taubes, "Surgery in Cyberspace," *Discover*, December 1994, pp. 85–94.

26. Denise Grady, "Software to Compute Women's Cancer Risk," *New York Times*, January 26, 1999, p. D4.

27. Rita Beamish, "Computers Now Helping to Screen for Troubled Teen-Agers," *New York Times*, December 17, 1998, p. D9.

28. Jamie Beckett, "Sorting Out Cybershrinks," *San Francisco Chronicle*, August 11, 1998, p. C3.

29. Vincent Kiernan, "Using the Web, Epidemiologist Aims to Improve Public Health in Developing Nations," *Chronicle of Higher Education*, January 30, 1998, pp. A21–A22.

30. John O'Neil, "Implanted Chip Offers Hope of Simplifying Drug Regimens," *New York Times*, February 2, 1999, p. D6.

31. Nancy Ann Jeffrey, "Hydraulics and Computers Help Artificial Limbs Get 'Smarter,'" *Wall Street Journal*, August 14, 1998, p. B1.

32. Associated Press, "Implant Transmits Brain Signals Directly to Computer," *New York Times*, October 22, 1998, p. G9.

33. Myron Magnet, "Who's Winning the Information Revolution," *Fortune*, November 30, 1992, pp. 110–117.

34. Tony Rutkowski, quoted in Patricia Schnaidt, "The Electronic Superhighway," *LAN Magazine*, October 1993, pp. 6–8.

35. Hal Lancaster, "Technology Raises Bar for Sales Job; Know Your Dress Code," *Wall Street Journal*, January 21, 1997, p. B1.

36. Ingred Wickelgren, "Treasure Maps for the Masses," *Business Week/Enterprise*, 1996, pp. ENT22–ENT24.

37. Frank Barnako, "Online Sales Seen Worth $2 Trillion in 2003," CBS MarketWatch, May 1, 2000.

38. Tim McCollum, "New Horizons in Communications," *Nation's Business*, August 1996, pp. 38–39; and Laura Casteneda and Jon Swartz, "Kicking Virtual Tires," *San Francisco Chronicle*, October 15, 1996, p. C4.

39. Robert D. Hof, "Will Shoppers Take to Cyber Groceries?" *Business Week*, February 23, 1998, p. 110R.

40. 1998 Andersen Consulting study, cited in "Market for On-Line Supermarkets," *USA Today*, January 6, 1999, p. 1B.

41. David Leonhardt, "The Meat and Potatoes of Online Shopping?" *Business Week*, December 7, 1998, p. 46; George Anders, "Co-Founder of Borders to Launch Online Megagrocer," *Wall Street Journal*, Feburary 22, 1999, pp. B1, B4; Lorrie Grant, "Soon, On-Line Grocers Will Deliver the Goods to Your Door," *USA Today*, March 8, 1999, p. 1B; and George Anders, "Amazon.com Buys 35% Stake of Seattle Online Grocery Firm," *Wall Street Journal*, May 18, 1999, p. B8.

42. Brian Bremner, Joan Warner, and Jonathan Ford, "Hold It Right There, Citibank," *Business Week*, March 25, 1996, p. 176; and Adam Zagorin, "Cashless, Not Bankless," *Time*, September 23, 1996, p. 52.

43. Paul Saffo, quoted in Jared Sandberg, "CyberCash Lowers Barriers to Small Transactions at Internet Storefronts," *Wall Street Journal*, September 30, 1996, p. B6.

44. Stewart Alsop, "The First Powerhouse Bank of the Virtual World," *Fortune*, September 7, 1998, pp. 159–160.

45. Christine Dugas, "Direct Deposit Wins for Safety, Speed," *USA Today*, January 25, 1999, p. 5B.

46. Arthur M. Louis, "The Check's on the Net," *San Francisco Chronicle*, June 22, 1999, pp. C1, C3; Christine Dugas, "Bank Deal Puts Stamp on Bill Delivery on Line," *USA Today*, June 24, 1999, p. 1B; and Martha Brannigan, "Bill Payments Via the Internet Get a Big Boost," *Wall Street Journal*, January 28, 1999, pp. B1, B14.

47. Denise Caruso, "On-Line Day Traders Are Starting to Have an Impact on a Few Big-Cap Internet Stocks in What Some Call a 'Feeding Frenzy,'" *New York Times*, December 14, 1998, p. C3.

48. Rebecca Buckman and Aaron Lucchetti, "Electronic Networks Threaten Trading Desks on Street," *Wall Street Journal*, December 23, 1998, pp. C1, C15.

49. Fred Vogelstein, "A Virtual Stock Market," *U.S. News & World Report*, April 26, 1999, pp. 47–48.

50. Eileen Glanton, "Will Extended Hours Benefit Small Investors?" *San Francisco Examiner*, June 6, 1999, p. B-3.

51. "Net Catches Merrill Lynch, but On-Line Waters Remain Rough" editorial, *USA Today*, June 2, 1999, p. 14A.

52. Allan Sloan, "Long Live the Middleman," *Newsweek*, June 14, 1999, p. 46.

53. Peter H. Lewis, "Play It Again, RAM," *New York Times*, November 12, 1998, pp. D1, D3.

54. Benny Evangelista, "AOL Trying to Catch Internet Music Wave," *San Francisco Chronicle*, June 2, 1999, pp. B1, B12.

55. Charles Bermant, "Musicians Tap Rich Lode of Sheet Music, Sold and Shared," *New York Times*, April 23, 1998, p. D5.

56. Lee Gomes, "Free Tunes for Everyone!" *Wall Street Journal*, June 15, 1999, pp. B1, B4.

57. Deidre Pike, "Reno Musicians: It's Business, Even Artists Need to Pay Bills," *Reno Gazette-Journal*, May 29, 2000, pp. 1E, 3E.

58. Jon Pareles, "Musicians Want a Revolution Waged on the Internet," *New York Times*, March 8, 1999, pp. B1, B8.

59. Stross, 2000.

60. Peter Stack, "An Animated Future," *San Francisco Chronicle*, December 17, 1996, pp. E1, E3.

61. Peter Stack, "The Digital Divide," *San Francisco Chronicle*, May 19, 1999, p. E1.

62. David Ansen and Ray Sawhill, "The New Jump Cut," *Newsweek*, September 2, 1996, pp. 64–66.

63. Bruce Haring, "Digitally Created Actors: Death Becomes Them," *USA Today*, June 24, 1998, p. 8D.

64. Jonathan Marshall, "Surfing the Internet Can Land You a Job," *San Francisco Chronicle*, July 17, 1995, pp. D1, D3.

65. Marshall, 1995.

66. Belle Wise, "Magna Cum Market—Popular Jobs for College Grads," *USA Today*, April 12, 2000, p. 7A.

67. Deborah Mathis, "Learn or Be Left Behind," *Reno Gazette-Journal*, January 1, 2000, p. 41G.

Bookmark It! Box

a. Shelly Freierman, "For Lifelong Learning, Click Here," *New York Times*, August 25, 1997, p. C6.

b. Pam Dixon, *Virtual College* (Princeton, NJ: Peterson's, 1996).

c. Glenn R. Jones, *Cyberschools: An Education Renaissance* (Jones Digital Century, 1997).

d. Joseph B. Walter, quoted in William H. Hanon, "Northwestern University Takes a Lead in Using the Internet to Add Sound and Sight to Courses," *New York Times*, May 28, 1997, p. A17.

e. Robert L. Johnson, "Extending the Reach of 'Virtual' Classrooms," *Chronicle of Higher Education*, July 6, 1994, pp. A19–A23.

f. Bruce Weber, "Notes from Cyberclass," *New York Times*, January 3, 1999, sec. 4A, pp. 15, 46.

Appendix

1. Gary Webb, "Potholes, Not 'Smooth Transition,' Mark Project," *San Jose Mercury News*, July 3, 1994, p. 18A.

2. Michael Krantz, "Keeping Tabs Online," *Time*, November 10, 1997, pp. 81–82.

Bookmark It! Box

a. Catherine Greenman, "You Want to Be in Pictures? Will You Settle for a Web Page?" *New York Times*, March 16, 2000, p. D9; Chris O'Malley, "Prefab Home Pages," *Popular Science*, September 1999, p. 56; and Deborah Kong, "Building a Web Site," *San Jose Mercury News*, April 1999, pp. 1F, 2F.

b. Thomas E. Weber, "There's No Place Like Home: A Reporter Builds a Web Page," *Wall Street Journal*, February 16, 1998, p. B22.

c. Jakob Niesen, "Make Your Site User-Friendly," *FSB*, Spring 2000, p. e8.

d. Deborah Kong, "Building a Web Site," *San Jose Mercury News*, April 11, 1999, pp. 1F, 2F.

Photo Credits

Page 2, *middle* courtesy of Intel; 2, *top* courtesy of Motorola Corporation; 3, courtesy of Sun Microsystems; 5, courtesy of Motorola Corporation; 6, Inge Yspeert/Corbis; 9, *bottom* Tom Tracy/Photophile; 9, *top* Tony O'Brien; 10, *bottom* courtesy of Intel; 10, *bottom left* courtesy of IBM; 10, *bottom right* courtesy of Sharp Electronics; 10, *middle left* courtesy of Toshiba; 10, *middle right* courtesy of Palm Inc.; 10, *top* courtesy of Sun Microsystems; 10, *top right* R. Ian Lloyd; 14, courtesy of Intel; 19, courtesy of Microsoft; 20, courtesy of Unisys Archives; 12, *bottom* Mark Richards/PhotoEdit; 21, *top* Don Mason/The Stock Market; 22, *bottom* Fujifotos/The Image Works; 22, *top* Eric Draper/AP/Wide World; 25, *bottom left* courtesy of Sony Corporation; 25, *bottom right* courtesy of Motorola Corporation; 25, *top right* Don Mason/The Stock Market; 27, *middle left* Tom Tracy/Photophile; 27, *middle right* courtesy of Intel; 27, *top right* courtesy of Toshiba; 28, *bottom* Tony O'Brien; 28, *top* courtesy of Palm Inc.; 29, *bottom* courtesy of Sun Microsystems; 33, *bottom* Richard T. Nowitz/Corbis; 43, courtesy of CIDCO; 73, Richard T. Nowitz/Corbis; 101, courtesy of Apple Computer; 102, courtesy of Apple Computer; 105, courtesy of Handspring; 135, courtesy of Handspring; 143, John S. Reid; 144, Coco McCoy/Rainbow; 145, courtesy of Sun Microsystems; 146, courtesy of Netpliance; 150, courtesy of Intel; 151, courtesy of Intel; 155, *bottom* Tom Pantages; 155, *top* Will & Deni McIntyre/Photo Researchers Inc.; 161, John S.Reid; 163, *bottom* John S. Reid; 164, Courtesy of IBM Corp.; 165, Coco McCoy/Rainbow; 171, *bottom* courtesy of LandWare Inc.; 171, *top* Coco McCoy/Rainbow; 172, *bottom left* courtesy of IBM; 172, *top* Marble FX by Logitech; 173, *bottom left* courtesy of Compaq Computer; 173, *bottom right* courtesy of Aqcess Technologies, Inc., Irvine, CA; 173, *top* courtesy of AT&T Global Info/Solution; 174, *bottom* courtesy of Cal-Comp Ultraslate; 174, *top* courtesy of FTG Data Systems; 175, *top* Charles Gupton/Stock Boston; 175, *(image scanner)* Melissa Farlow; 177, courtesy of Intel and Konica; 178, *bottom* Tom Burdete/USGS; 178, *top* courtesy of IBM Corporation; 181, *left* courtesy of Sun Microsystems; 181, *right* courtesy of ViewSonic Corp.; 184, Michael Newman/PhotoEdit; 178, David Hanover/Stone; 189, courtesy of Intel; 190, courtesy of CalComp Ultraslate; 191, *top* Melissa Farlow; 192, courtesy of IBM Corp.; 193, *bottom* Coco McCoy/Rainbow; 193, *top* Michael Newman/PhotoEdit; 195, John S. Reid; 196, *top* courtesy of Aqcess Technologies, Inc., Irvine, CA; 197, *bottom* Tom Burdete/USGS; 197, *top* Coco McCoy/Rainbow; 199, *bottom* Marble FX by Logitech; 199, *top* courtesy of AT&T Global Info/Solution; 200, *bottom* John S. Reid; 200, *top* courtesy of IBM Corporation; 204, AFP/Corbis; 212, *bottom* AFP/Corbis; 212, *top* Richard Hamilton Smith/Corbis; 213, *bottom* Bruce Ayres/Stone; 213, *top* Hulton Getty/Stone; 218, *bottom* courtesy of AT&T; 218, *middle* courtesy of AT&T; 218, *top* courtesy of AT&T; 227, Ian Shaw/Stone; 227, *bottom* PhotoEdit; 230, *bottom* AFP/Corbis; 230, *top* UPI/Bettmann/Corbis; 232, Bruce Ayres/Stone; 239, Michael St. Maur Sheil/Corbis; 256, *bottom* courtesy of Princeton Video Inc.; 256, *top* Paul Higdon/New York Times; 264, *bottom* Bettmann Corbis; 264, *top* PhotoEdit; 266, PhotoEdit; 267, James Balog; 269, Bettmann Corbis; 270, *top* AFP/Corbis; 281, Kaku Karita; 282, *top* J. Kyle Keener; 283, Kaku Karita; 285, *left* Rick Allen & Cindy Burnham/Nautilus Productions; 285, *medium* Lori Grinker/Contact Press Images; 285, *right* Lori Grinker/Contact Press Images; 286, *bottom* Kaku Karita; 286, *top* Rudi Meisel/Visum; 287, *left* Denise Rocco/U.C. Berkeley; 287, *right* Mark Richards/PhotoEdit; 291, J. Kyle Keener; 292, Fritz Hoffmann; 299, *bottom* Rudi Meisel/Visum; 299, *middle* Mark Richards/PhotoEdit; 299, *top* Kaku Karita; 300, Kaku Karita; 301, PhotoDisc; 302, PhotoDisc; 303, PhotoDisc; 304, PhotoDisc; 305, PhotoDisc; 306, PhotoDisc; 307, PhotoDisc; 315, PhotoDisc; 316, PhotoDisc; 317, PhotoDisc.

Thanks to Michael Flores and Dereck McDonald, Wired Solutions, Incline Village, NV, for their help in photographing equipment.

Panel 2.8, p. 64, adapted from Elizabeth Weise, "Successful Net Search Starts with Need," *USA Today*, January 24, 2000, p. 3D. Panel 6.1, p. 241, adapted from Carolyn Said, "Online Middlemen," *San Francisco Chronicle*, April 10, 2000, pp. C1, C3. **Photos on pp. 108, 149, 158, 159, 163 (Panel 4.12), 167, 169 (Panel 4.19, right), p. 172 (touchpad), 183, 219, 249, 270 (laptop) by Brian K. Williams.**